Selenium WebDriver 3.0 自动化测试框架实战教程

吴晓华　王晨昕　编著

清華大學出版社
北京

内容简介

本书主要讲解开源的 Web 自动化测试工具 WebDriver 的原理、API 接口实例、pytest、五大框架实战(行为驱动、分布式测试框架、数据驱动测试框架、关键词驱动测试框架和混合模式测试框架)、常见自动化测试的问题及处理方法。

本书既可让初学者从入门到精通,循序渐进;也可帮助中高级测试工程师夯实技能,从形象到抽象,提供测试思想中更多的可能性。

本书封面贴有清华大学出版社防伪标签,无标签者不得销售。
版权所有,侵权必究。举报:010-62782989,beiqinquan@tup.tsinghua.edu.cn。

图书在版编目(CIP)数据

Selenium WebDriver 3.0 自动化测试框架实战教程/吴晓华,王晨昕编著.—北京:清华大学出版社,2022.7

ISBN 978-7-302-61244-5

Ⅰ.①S… Ⅱ.①吴… ②王… Ⅲ.①软件工具-自动检测-教材 Ⅳ.①TP311.56

中国版本图书馆 CIP 数据核字(2022)第 110266 号

责任编辑:贾 斌
封面设计:何凤霞
责任校对:胡伟民
责任印制:杨 艳

出版发行:清华大学出版社
网 址:http://www.tup.com.cn,http://www.wqbook.com
地 址:北京清华大学学研大厦 A 座 **邮 编**:100084
社 总 机:010-83470000 **邮 购**:010-62786544
投稿与读者服务:010-62776969,c-service@tup.tsinghua.edu.cn
质量反馈:010-62772015,zhiliang@tup.tsinghua.edu.cn
课件下载:http://www.tup.com.cn,010-83470236
印 装 者:大厂回族自治县彩虹印刷有限公司
经 销:全国新华书店
开 本:185mm×260mm **印 张**:26.25 **字 数**:660 千字
版 次:2022 年 8 月第 1 版 **印 次**:2022 年 8 月第1 次印刷
印 数:1~2500
定 价:99.00 元

产品编号:090290-01

前言

随着互联网的高速发展，中国互联网达到了一个空前的繁荣水平，数亿量级用户的产品登上了中国的互联网发展舞台，阿里巴巴、腾讯、百度等多个互联网巨头也开始在世界的互联网舞台崭露头角，互联网行业的从业人员也达到了上百万人的规模，中国互联网产品已经深入到网民生活的各个方面。

随着互联网行业在中国的迅猛发展，对于中国的软件开发和测试行业也提出了更高的技术要求与质量要求，软件测试从业者的技术水平也被提升到空前的高要求阶段。以往我们看到测试人员的招聘重点都是仅限于对测试用例设计和业务的理解，现今所看到的更多测试职位对测试人员提出了更高的技术能力要求。例如，精通一门编程语言，熟悉 MySQL 或者 Oracle 数据库，精通自动化测试和性能测试，能独立开发测试工具等。为了能更好地适应互联网社会的发展潮流，软件测试从业者必须在技术能力上不断提升自己，才能真正站在职业发展的巅峰。

自动化测试技术对测试人员来说，是一个必要的高级技能要求，越来越多的测试从业者并不甘于手工测试，都非常希望通过自动化的方式来减少枯燥无味且不断重复的手工测试劳动。尽管主流的 Web 自动化测试开源工具 Selenium WebDriver，已经成为众多软件测试从业者学习的热点，但是市面上针对 Selenium 自动化测试方面的书籍很少，基于实践方式来讲解 Selenium 应用技术的书籍更是凤毛麟角。我有幸受吴老邀请，将我工作中实践内容与吴老教学内容融合，一起编写了这本基于 Python 语言实践操作的 Selenium 3 教学书籍，来解决软件测试人员学习自动化测试的需求。

Selenium 是一个开源的测试工具，代表了未来测试工具的趋势；而 Python 则是全世界都在用的一门简洁、高效、易用、优雅的编程语言，初学者只需要花少量的时间就能上手。本书着重点在讲解 Selenium 的使用技巧上，对学习 Python 语言有需要的朋友，请参考相关资料或者利用丰富的网络资源。

本书采用图文并茂的方式分步骤讲解 Selenium 的各种实用技巧，并且提供被测试对象的实现代码或者被测试对象的访问网址，方便读者在本地搭建自己的测试环境或者访问互联网上的被测试网址，从而能顺利地进行自动化测试技术的实践。经过我们数月的不懈努力，此书终于跟大家见面了，希望能够让读者通过本书深入熟悉 Selenium 3 的使用技巧，帮助大家在自动化测试方向上能大显身手。我们相信，通过我们不断的努力，一定可以改变中国测试行业技术含量低的现状。

2015 年，在一个机缘巧合下认识了吴老，他丰富的测试经验和过硬的技术都让我折服，我就像是一粒罅隙中的种子遇到了阳光，拼命地将自己置身于这片透着温暖的金银色的阳光浴中。好不容易遇见这么一位低调奢华有内涵的大神，岂能放过，于是我就死皮赖脸地揪着吴老，套他的各种本领。当时的每一天除了吃饭、工作及少量的睡觉时间，其余时间不是吃斋（看书），就是念佛（敲代码），那日子甚是枯燥，甚是无味，期盼着咸鱼能有翻身的一天。我待 Code 如初恋，Code 虐我千百遍，半年时间过去了，被虐得千疮百孔的我也算是摸清 Code 的脾气了，闲来无事也能写上几百行了，算是有点欣慰吧。

有了一定的 Code 功底后，开始正式进入自动化学习。自动化测试的学习是一个不断实践，不断总结，不断积累的过程。很多人会有一个错误的认识，认为自动化测试不就是一个工具的使用，一个调用别人写好的 API 的过程么，需要那么深厚的 Code 功底有何用？殊不知你此时仅是一个 ToolBoy 或者 ToolGirl，要想随心所欲地完成各种自动化测试，无论是 Web 自动化、移动端自动化，或是性能自动化，都必须在拥有一定的 Code 功底后，方能理解这些工具底层实现的原理，并且能在当工具本身不能满足测试需求时，还能随心所欲地扩充或更改。理解了这些工具的实现思想，也就为你搭建属于自己的测试框架打下了坚实的基础。经过半年的蹂躏与被蹂躏，加之工作中的不断实践，也能搭建那么几个自己还算满意的测试框架，做得还甚是开心，而且还能被吴老盯上，叫来给大家编写这本书，也着实有点小开心。

这是我第一次写专著技术类的书籍，深深地感觉到把知识点用通俗易通的语言描述清楚是一件多么不容易的事，为此我投入了大量的时间与精力来组织本书的语言，如果还是存在不那么浅显易懂的语句，请先尝试从代码层面进行理解，如果在实践本书中代码时发生了错误，请不要怀疑是我们代码的问题，请先检查你的环境是否有问题，浏览器版本与驱动版本是否匹配，所使用的 Python 包版本之间是否存在版本兼容的问题，Python 代码是否存在缩进问题等。如果仍然解决不了你的问题，欢迎把错误丢到笔者的脸上，我们定会马不停蹄地帮你解决。

冰冻三尺，非一日之寒，希望大家能在自动化测试学习的道路上做到博观而约取，厚积薄发。最后，祝大家工作顺利，万事如意。

各章内容介绍：

第一篇 “基础篇”：第 1～8 章。

第 1 章介绍了 Selenium 的发展历史及组成 Selenium 的工具套件，列举了 Selenium 1 和 Selenium 2 支持的浏览器和平台，讲解了 Selenium RC 和 WebDriver 的实现原理，同时也介绍了 Selenium 1、Selenium 2 和 Selenium 3 各自的特点及区别。

第 2 章介绍了在日常测试工作中常见的自动化测试目标，讲解了如何获得公司管理层对于开展自动化测试的支持，介绍了如何衡量自动化测试工作的投入产出比及在敏捷开发中的应用，讲解了自动化测试工作的分工及测试工具的选择与推广，分享了在实际项目中最佳实践经验，说明了学习 Selenium 工具的能力要求。

第 3 章介绍了使用 Selenium 工具时所需要的相关辅助插件 FireBug 和 FirePath 的安装及使用方法。

第 4 章介绍了 Selenium IDE 的安装、界面选项的含义、IDE 的使用方法、录制脚本及导出脚本等。

第 5 章介绍了 Python 开发环境和 PyCharm 集成开发环境的安装、配置及使用。

第 6 章介绍了 WebDriver 的安装和配置方法。

第 7 章主要介绍了单元测试的基本知识，如何在自动化测试中使用以及生产测试报告。

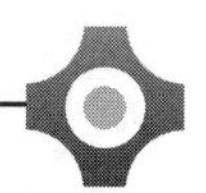

第 8 章主要讲解了自动化测试过程中使用的页面元素定位方法，包括 ID 定位、Name 定位、链接文字定位、Class 定位、XPath 定位及 CSS 定位，推荐使用 XPath 作为页面元素定位的主要方法。

第二篇 "实战应用篇"：第 9～11 章。

第 9 章讲解了如何使用 WebDriver 工具分别驱动 IE 浏览器、Chrome 浏览器以及 Firefox 浏览器，进行自动化测试。

第 10 章通过实例全面讲解 WebDriver 基础 API。

第 11 章通过实例全面讲解 WebDriver 高级 API，并提供了一些解决实际问题的方法。

第三篇 "自动化测试框架搭建篇"：第 12～15 章。

第 12 章讲解了数据驱动的概念，并基于 Excel、XML、MySQL 及单元测框架结合 ddt 进行数据驱动测试。

第 13 章讲解了 lettuce 行为驱动框架在自动化测试中的使用，分别基于英文和中文进行了实例讲解。

第 14 章通过实例全面讲解如何基于 Selenium Grid 进行分布式自动化测试。

第 15 章深入讲解了如何从零开始搭建一个数据驱动测试框架、关键字驱动测试框架以及数据驱动与关键字驱动混合的测试框架，并提供完整的框架代码。此章节为本书最综合、最重要的章节，建议读者在阅读完前面所有章节后再阅读此章节。

第四篇 "常见问题和解决方法"：第 16 章。

第 16 章讲解了 WebDriver 使用过程中常见的疑难问题及解决办法，方便读者在使用 WebDriver 过程中遇到问题时进行查阅。

特别致谢：

感谢我们的好朋友陈良军、李江和王浩花费大量的时间与精力帮我们校对书稿，发现了不少书写的错误、晦涩难懂的语句以及代码的错误，在此我们真诚地感谢他们对本书做出的贡献，帮助我们完成这件非常有意义的事情。

王晨昕

2022 年 5 月

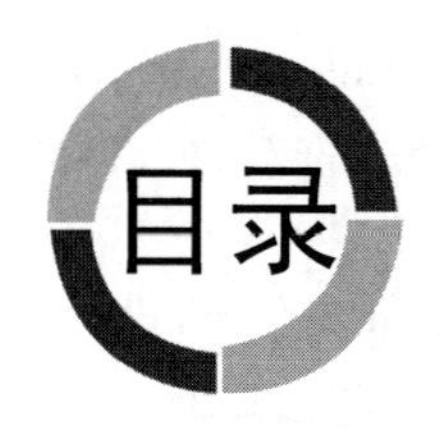

目录

第一篇 基础篇

第二篇 实战应用篇

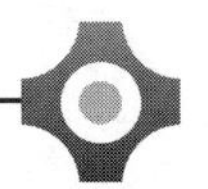

第四篇　常见问题和解决方法

第一篇 基础篇

第1章 Selenium简介

Selenium 工具诞生的时间已经超过 10 年，目前在软件开发公司中已得到大规模的应用，但是很少有人能够清楚描述此工具的发展历史和特点，本章介绍让读者和 Selenium 工具来一次亲密的接触，以便了解它的前世今生及其特点。

1.1 Selenium 的前世今生

2004 年在 ThoughtWorks 公司，一个名为 Jason Huggins 的软件工程师为了减少手工测试的工作量，自己实现了一套基于 JavaScript 的代码库，使用这套库可以进行页面的交互操作，并且可以重复地在不同浏览器上进行各种测试操作。通过不断地改进和优化，这个代码库逐步成为了 Selenium Core。Selenium Core 为 Selenium Remote Control（RC）和 Selenium IDE 提供了坚实的核心基础能力。

当时的自动化测试工具比较稀少，现有的工具也无法灵活地支持各种复杂的测试操作，大部分测试人员只能使用手工的方式完成 Web 产品的测试工作。开发人员不断地开发代码，测试人员不断地发现 bug，开发人员不断地修改 bug，测试人员不断地回归测试来确认 bug 已经被正确修正，并且确认程序没有引入新的 bug。这样的产品开发模式，导致测试人员必须经常手工回归测试系统的大部分功能，由此产生了大量的重复性手工操作。Jason Huggins 想改变这样的现状，所以他开发了基于 JavaScript 的代码库，希望帮助测试人员从日常的重复性工作中解脱出来，经过不断的努力，Selenium 1 版本诞生了。

Web 自动化测试工具 Selenium 是跨时代的，因为它允许测试工程师使用多种开发语言来控制不同类型的浏览器，从而实现不同的测试目标。Selenium 是开源工具软件，用户无需付费就可以使用它，甚至可以根据自己的使用需求来进行深入的定制化，改写其原有的一些功能。基于以上这些优点，越来越多的测试人员开始使用此工具来进行 Web 系统的自动化测试工作。在短短几年时间内，全世界范围内都出现了 Selenium 工具的忠实拥护者，目前中国的几大互联网公司均使用 Selenium 作为 Web 自动化测试实施的主要工具。

但是随着互联网技术的不断发展以及浏览器对于 JavaScript 语言的安全限制，Selenium 的发展也遇到很多难以解决的困难。由于其自身实现的机制，Selenium 无法突破浏览器沙盒的限制，导致很多测试场景的测试需求难以被实现。

2006 年 Google 的工程师 Simon Stewart 开启了一个叫作 WebDriver 的项目，此项目可以直接让测试工具调用浏览器和操作系统本身提供的内置方法，以此来绕过 JavaScript

环境的沙盒限制，WebDriver 项目的目标就是为了解决 Selenium 的痛处，并且也做到了。2008 年，Selenium 和 WebDriver 这两个项目进行了合并，至此 Selenium 2 出现了，也就是我们现在常常看到的 Selenium WebDriver(简称：WebDriver)。

```
Selenium 2 = Selenium 1 + WebDriver
```

Selenium 的官网地址是 www.seleniumhq.org，网站提供了 Selenium WebDriver 的安装文件和使用教程。Selenium 2 是 Selenium 1 的升级版本，它本身向下兼容 Selenium 1 的所有功能，同时又提供了更多新 API 来完成自动化测试的各种复杂需求。现阶段，Selenium 1 已经退出历史舞台，大部分 Web 自动化测试人员已经转向使用 Selenium 2 (WebDriver)来搭建自己的自动化测试框架。

2016 年 10 月，Selenium 3 诞生。开发者在 Selenium 2 的基础上做了很多了不起的工作，这个版本有很多新特性，主要实现了把核心 API 跟客户端 Driver 进行分离，同时去掉越来越较少使用的 Selenium RC 功能。为了迎合历史的发展潮流，本书全部的案例均基于 Windows 7 操作系统上的 Selenium 3 的 WebDriver 的 API 进行讲解。

1.2 Selenium 工具套件介绍

- Selenium 2(Selenium Webdriver)：提供了极佳的特性，例如，面向对象 API，同时提供 Selenium 1 的接口用于向下兼容。
- Selenium 1(Selenium RC 或 Remote Control)：支持更多的浏览器，支持更多的编程语言(Java、JavaScript、Ruby、PHP、Python、Perl 和 C#)。
- SeleniumIDE(集成开发环境)：Firefox 插件，提供图形界面来录制和回放脚本。但此插件只是用来做原型的工具，此插件需要使用第三方的 JavaScript 代码库才能支持循环和条件判断，并不希望测试工程师使用此工具来运行大批量的测试脚本。
- Selenium-Grid 可以在多个测试环境以并发的方式执行测试脚本，实现测试脚本的并发执行，缩短大量测试脚本的执行时间。

1.3 Selenium 支持的浏览器和平台

Selenium 的一大特点就是能够在多种操作系统上支持多种浏览器的自动化测试，下面我们将介绍此工具能够支持的操作系统和浏览器类型。

1.3.1 Selenium IDE、Selenium 1 和 Selenium RC 支持的浏览器和平台

表 1-1 列出了 Selenium IDE、Selenium 1 和 Selenium RC 支持的浏览器和操作系统。

表 1-1

浏 览 器	Selenium IDE	Selenium 1(即 SeleniumRC)	操 作 系 统
Firefox 3.x	录制脚本和回放脚本	启动浏览器 运行测试脚本	Windows、Linux、Mac

续表

浏　览　器	Selenium IDE	Selenium 1 （即 SeleniumRC）	操 作 系 统
Firefox 3	录制脚本和回放脚本	启动浏览器 运行测试脚本	Windows、Linux、Mac
Firefox 2	录制脚本和回放脚本	启动浏览器 运行测试脚本	Windows、Linux、Mac
IE 8	仅能通过 Selenium 1(RC)来运行测试脚本	启动浏览器 运行测试脚本	Windows
IE 7	仅能通过 Selenium 1(RC)来运行测试脚本	启动浏览器 运行测试脚本	Windows
IE 6	仅能通过 Selenium 1(RC)来运行测试脚本	启动浏览器 运行测试脚本	Windows
Safari 4	仅能通过 Selenium 1(RC)来运行测试脚本	启动浏览器 运行测试脚本	Windows、Mac
Safari 3	仅能通过 Selenium 1(RC)来运行测试脚本	启动浏览器 运行测试脚本	Windows、Mac
Safari 2	仅能通过 Selenium 1(RC)来运行测试脚本	启动浏览器 运行测试脚本	Windows、Mac
Opera 10	仅能通过 Selenium 1(RC)来运行测试脚本	启动浏览器 运行测试脚本	Windows、Linux、Mac
Opera 9	仅能通过 Selenium 1(RC)来运行测试脚本	启动浏览器 运行测试脚本	Windows，Linux，Mac
Opera 8	仅能通过 Selenium 1(RC)来运行测试脚本	启动浏览器 运行测试脚本	Windows、Linux、Mac
Google Chrome	仅能通过 Selenium 1(RC)来运行测试脚本	启动浏览器 运行测试脚本	Windows、Linux、Mac

1.3.2　Selenium 2(WebDriver)支持的浏览器

官网中并没有明确列出 WebDriver 支持浏览器的所有版本号，仅仅列出浏览器的名称。下面结合笔者个人的实际使用情况列出 WebDriver 支持的浏览器版本，请读者在测试实践中进行再次确认。

- Google Chrome。
- IE 6、7、8、9、10 和 11。
- Mac 操作系统的 Safari 默认版本均支持。
- Firefox 的大部分版本。
- Opera。
- HtmlUnit。
- Android 手机操作系统的默认浏览器。
- iOS 手机操作系统的默认浏览器。

1.4 Selenium RC 和 WebDriver 的实现原理

当执行 Selenium 自动化测试脚本时，测试人员可以看到浏览器中发生神奇一幕，页面上会自动进行各种操作，例如，打开新窗口、在输入框中输入文字、寻找下拉列表框等。测试人员为此不禁要多问一句："它到底是怎么实现的？"为了解 Selenium 工具的神奇之处，读者必须深入了解它的实现原理和机制。

1.4.1 Selenium RC 的实现原理

Selenium RC 实现原理如图 1-1 所示。

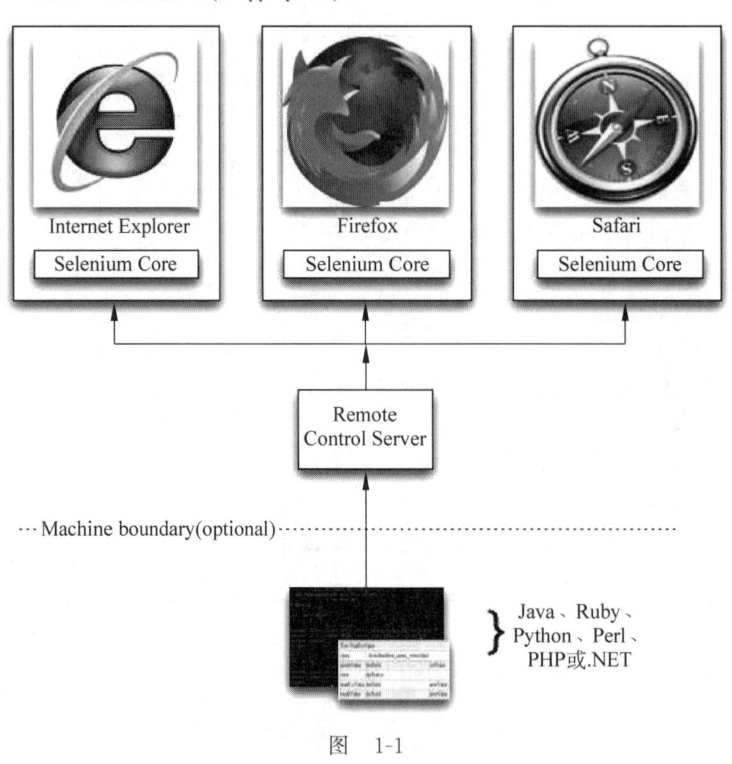

图 1-1

Selenium 1 的自动化测试执行步骤如下：

第一步：测试人员基于 Selenium 支持的编程语言编写好测试脚本程序。

第二步：测试人员执行测试程序。

第三步：测试脚本程序发送访问网站的 HTTP 请求给 Remote Control Sever(RC)。

第四步：Remote Control Sever(RC)收到请求后，访问被测试网站并获取网页数据内容，并在网页中插入 Selenium Core 的 JavaScript 代码库，然后返回给测试人员执行测试的浏览器。

第五步：测试脚本在浏览器内部再调用 Selenium Core 来执行测试代码逻辑，最后记录测试的结果，完成测试。

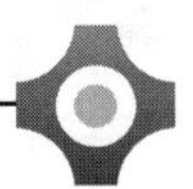

参阅以上几个步骤，有人会疑惑为什么要执行第四步？我们需要先学习一下浏览器的JavaScript 安全机制——同源策略。浏览器访问了某个域名的网站后，浏览器会打开此网站的网页，获取到此网站的网页内容。网页内容中包含了要在网页里面执行的 JavaScript 语句或外部引用的 JavaScript 文件，浏览器会执行属于此域名下的 JavaScript 语句和文件。如果外部引用的 JavaScript 文件 URL 和当前网页的域名不一致，那么浏览器会拒绝执行此 JavaScript 文件中的代码。通过此方式，浏览器就可以防止一些恶意的 JavaScript 文件被加载到用户的浏览器中，起到一定的安全防护作用。

Selenium 1 工具的核心部分是基于 JavaScript 代码库实现的，这个库肯定默认地和被测试网站分离，也就是说，这个 JavaScript 库的 URL 和被测试网站的域名肯定是不一致的。参阅上面提到的浏览器同源策略，Selenium 1 的 JavaScript 库肯定是被禁止执行的，这样就无法实现对网站的自动化测试了。为了绕过浏览器的安全机制，Selenium 1 的作者使用了代理方法来解决此问题，图 1-2 详细说明了 Selenium 1 代理模式的实现机制。

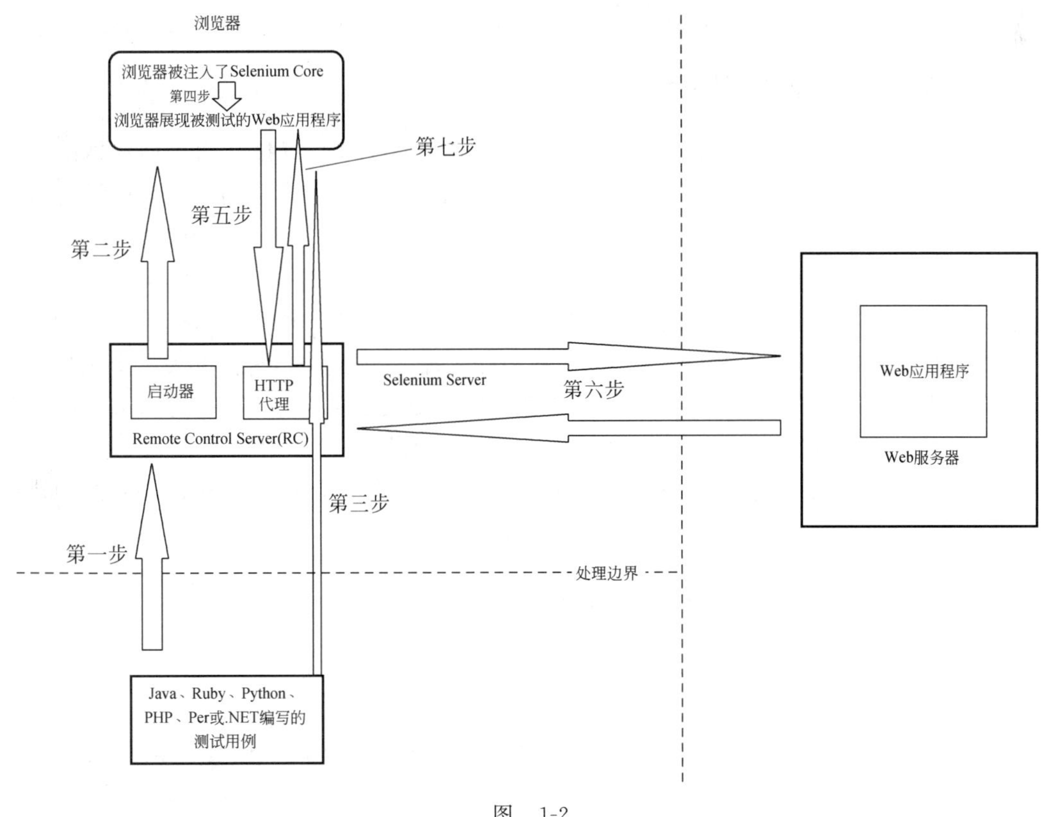

图 1-2

Selenium 1 代理模式的实现机制具体如下。

第一步：执行测试脚本，脚本向 Selenium Server 发起请求，要求和 Selenium Server 建立连接；

第二步：Selenium Server 的 Launcher 启动浏览器，向浏览器中插入 Selenium Core 的 JavaScript 代码库，并把浏览器的代理设置为 Selenium Server 的 HTTP Proxy；

第三步：测试脚本向 Selenium Server 发送 HTTP 请求，Selenium Server 对请求进行解析，

然后通过 HTTP Proxy 发送 JavaScript 命令通知 Selenium Core 执行操作浏览器的动作；

第四步：Selenium Core 接收到指令后，执行测试脚本指定的网页操作命令；

第五步：浏览器收到新的页面请求信息（在第四步中，Selenium Core 的操作可能引发新的页面请求），于是发送 HTTP 请求给 Selenium Server 的 HTTP Proxy，请求新的 Web 页面；

第六步：由于 Selenium Server 在启动浏览器时将浏览器的代理访问地址设置为 Selenium Server 的 HTTP Proxy，所以 Selenium Server 会接收到所有由它启动的浏览器发送的请求。Selenium Server 接收到浏览器发送的 HTTP 请求后，重组 HTTP 请求，获取对应的 Web 页面；

第七步：Selenium Server 的 HTTP Proxy 把接收到的 Web 页面返回给浏览器。

通过以上步骤，达到了将 Selenium Core 的 JavaScript 代码库插入到被测试网页的目的，然后就可以基于此代码库在被测试网页中进行各种自动化测试操作了。此种方式是一种非常巧妙的"欺骗"，必须由衷地赞扬一下 Selenium 1 作者的聪明智慧。

1.4.2 WebDriver 的实现原理

WebDriver 与之前 Selenium 1 的 JavaScript 注入实现不同，直接利用了浏览器的内部接口来操作浏览器。对于不同平台中的不同浏览器，必须依赖浏览器内部的原生组件（Native Component）来实现把对 WebDriver API 的调用转化为对浏览器内部接口的调用。

Selenium 1 采用 JavaScript 的合成事件来处理网页元素的操作，比如要单击某个页面元素，要先使用 JavaScript 定位到这个元素，然后触发单击事件。而 WebDriver 使用的是系统的内部接口或函数，首先是找到这个元素的坐标位置，并在这个坐标点触发一个鼠标左键的单击操作。由此，可以看出 WebDriver 能更好地模拟真实的环境，但仅能测试那些可见的页面元素。也正因为这个区别，有些隐藏的页面元素是可以使用 Selenium 1 进行操作的，而尝试使用 WebDriver 单击的某个隐藏的页面元素，将会引发 cannot clickable 的错误提示信息。

1.4.3 Selenium 1 和 WebDriver 的特点

1. Selenium 1 的缺点

（1）无法调用和触发本机的键盘和鼠标事件。

（2）由于浏览器的同源策略，只能使用插入 JavaScript 方式来进行模拟网页操作的测试。

（3）无法处理基本身份认证、自签名的证书以及文件上传/下载的框体。

2. Selenium 2（WebDriver）的优点

（1）Selenium 必须操作真实浏览器，但 WebDriver 却可以使用 HTMLUnit 进行测试，在不打开浏览器的情况下进行快速测试。

（2）WebDriver 基于浏览器的内部接口来实现自动化测试，更接近用户使用的真实情况。

（3）WebDriver 提供了更简洁的面向对象 API，提高了测试脚本的编写效率。

（4）WebDriver 在使用过程中无须单独启动 Selenium Server。

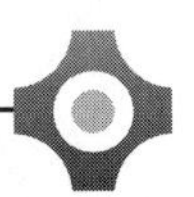

1.5 Selenium 3 的新特性

Selenium 3 将以开发一款聚焦于 Web 端和移动端的自动化测试工具为目标，WebDriver API 是 Selenium 2 的主要插件，Selenium 3 依然沿用，并且现在是基于 W3C 标准。Selenium 会不断扩充 WebDriver API，提供移动端的测试套件，以提高不同项目间的互操作性。Selenium 3 同时也更关注系统的稳定性，移除原始的 Selenium 核心实现，丢弃 Selenium RC API。相比 Selenium 2，Selenium 3 新特性如下：

(1) Selenium 3 去掉了 Selenium RC，这是 Selenium 3 最大的变化。

(2) Selenium 3 只支持 Java 8 版本以上。

(3) Selenium 3 不再提供默认浏览器支持，所有支持的浏览器均由浏览器官方提供支持，也就是由官方提供相应的 Driver 进行支持，由此提高了自动化测试的稳定性。Selenium 3.0 以前版本能直接启动 Firefox 浏览器，而 Selenium 3.0 版本开始需要下载 Firefox 官方提供的 geckodriver 驱动才能启动 Firefox 浏览器，并且 Firefox 浏览器必须是 48 版本以上。

(4) Selenium 3 通过 Apple 自己的 safaridriver 支持 Mac OS 上的 Safari 浏览器。Safari 浏览器的驱动直接被集成到 Selenium Server 上，也就是说想在 Safari 浏览器上执行自动化测试脚本，必须使用 Selenium Server。

(5) Selenium 3 通过 Microsoft 官方提供的 MicrosoftWebDriver 支持 Edge 浏览器。由此在 Windows 10 系统中就可以实现 Edge 浏览器自动化测试，只需要在 https://developer.microsoft.com/en-us/microsoft-edge/tools/webdriver/网址下载相应版本的驱动程序即可实现。

(6) Selenium 3 只提供支持 IE 9 及以上版本，早期版本也许还能工作，但不再提供支持。

Selenium 3 让 Web 自动化测试运行更稳定，性能更高，支持的浏览器更多、更新。

第 2 章 关于自动化测试

虽然很多测试工程师都了解一些"自动化测试"的知识，但是鲜有人能够准确地回答如下问题：

- 在自动化实施的过程中应该如何设定自动化测试的目标？
- 如何衡量自动化测试的投入产出比？
- 需要什么样的人员分工？
- 自动化测试的最佳实践是什么？

要想知道上面问题的答案，请仔细阅读本章内容，它其实比你搭建一个自动化测试框架的意义还要大。

2.1 自动化测试目标

我们做任何事情都应该有个目的，有了目的就会产生一个对应的目标，然后再基于这个目标进行相关活动的实施，以此来达到目的。类似地，我们在进行自动化实施的时候，首先要明确自动化测试的目标，即实现了自动化测试到底能为我们带来什么好处？解决了什么问题？我们不能为了自动化而自动化，必须在实施自动化测试之前明确自动化测试的目标。

笔者基于多年的自动化测试实践，列出了一些相对通用的自动化测试目标。

1. 提高测试人员的工作成就感和幸福感，减少手工测试中的重复性工作

目前，在中国的大部分中小企业中，手工测试占日常测试工作的大部分比例，测试人员必须跟随开发团队一起不断地进行迭代式开发和测试，一个功能模块可能在整个开发周期中被重复测试超过 10 次以上，测试人员在执行了如此多的重复工作之后，常常会对"IT 民工"这个词有着更加深刻的理解。

如何改变这个现状呢？使用自动化测试肯定是个很好的选择，脚本写好以后，可以不断地重复运行，测试人员只需要点一下按钮就可以开始测试工作了，然后去喝喝茶看看报纸，一会儿回来看一下测试结果，就完成了以往需要手工测试花费很长时间的工作。测试工作的成就感和幸福感油然而生，测试人员也会有精力和意愿去主动地推进自动化测试在不同项目中的深入实施。

如何验证达到了此目的呢？可以通过测试人员的满意度调查来判定是否提高了测试人员的成就感和满意度。

2. 提高测试用例的执行效率，实现快速的自动化回归测试，快速给开发团队质量反馈

使用手工方式来执行测试用例，速度必然是很慢的。人是高级动物，不是机器，工作时间长了必然会觉得劳累，测试执行的速度自然地就慢下来了，在测试用例非常多的情况下，完整地测试一遍所有测试用例的时间成本就会相当高了。

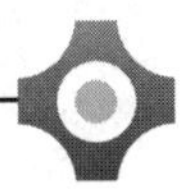

使用自动化测试取代手工测试，那么测试用例的执行者就变成了机器执行，机器可以24 小时不停地执行，它可以毫无怨言地不知疲倦地快速地完成测试脚本指派给它的测试任务。此种方式势必可以大大提高测试执行的效率，减少测试用例的执行时间，提高测试执行的准确性。

目前，敏捷开发模式也在各类软件企业中开始普及和应用。敏捷开发对于被开发产品的质量反馈有着很高的要求，需要每星期甚至每天开发出一个 build 版本，并且部署在测试环境上，希望测试人员能够给出快速的质量反馈。目前，只有通过自动化测试的方式才能真正实现对于大型敏捷开发项目快速的质量反馈需求，缺少自动化测试的敏捷开发项目会大大增加项目失败的风险。

如何验证达到了此目的呢？可以和以前手工测试的执行时间进行比对，看看是否明显缩短了测试用例的执行时间，询问开发人员项目的质量反馈速度是否为快速地发布产品带来很大帮助。

3. 减少测试人员的数量，提高开发和测试的比例，节省企业的人力成本

在大部分 IT 企业的运营成本中，差不多 50%～70%的成本是人工成本，如何能够有效地控制人工成本，对于企业的发展有着重要作用。使用自动化测试方式，势必会减少手工测试的工作量，从而达到减少测试人员的目的，进而降低企业的人工成本，增强企业的盈利能力。

如何验证达到了此目的呢？在相同级别测试工作量的情况下，企业可以测算在使用自动化测试后，项目中是否减少了测试人员投入数量和工作时长。

4. 在线产品的运行状态监控

在完成产品开发和测试工作后，产品会被发布到生产环境，正式地为用户提供服务。但是产品在生产环境的运营过程中，总是会由于各类原因造成这样或者那样的运行问题或故障。如何快速发现这样的问题呢？有人说“出了问题一定会有用户给客服打电话进行投诉的，那么我们就可以发现生产环境中的问题了。”如果采用这样的处理方式，势必会降低用户对于产品使用的满意度。另外，如果没有热心的用户进行投诉，那么生产环境问题被发现的时间会被大大推迟，所以依靠客户投诉的方式是不可取的。

为了保证快速及时地发现生产环境的不定期问题，建议采用拨测的方式来监控产品的运行状态，可以编写自动化测试脚本测试产品的主要功能逻辑，定时去运行测试脚本检查产品系统是否依旧可以正常工作，如果运行测试脚本后没有发现任何的问题，则休眠等待一段时间后再运行测试脚本检测产品系统的运行状态。如果测试脚本发现了产品系统的运行问题，在重试几次之后确认产品系统的问题依旧存在，则测试脚本是否会自动发出报警邮件和短信给系统运维的值班人员进行系统报警，相关人员收到报警后可以人工去处理系统出现的运行故障，这样就达到实时监控产品系统的目的，实现在第一时间发现和处理系统的故障。

如何验证达到了此目的呢？在生产环境运行的产品系统出现问题，则系统可以在几分钟内实现自动报警给相关人员。

5. 插入大量测试数据

在系统级别的测试过程中，经常要插入大量的测试数据来验证系统的处理能力，比如测

试人员想要插入 100 个注册用户，并且每个用户都有特定的 10 条用户数据，那么需要插入的数据量足有 1000 条之多，使用手工的方式来插入这些数据势必会花费很长的时间和精力。测试人员可以通过三种自动化的方式来实现上述测试数据插入要求。

第一种方式：测试人员编写数据库的存储过程脚本，在数据库的不同数据表中插入测试数据，使用这样的方式可以实现海量数据的快速插入。当然此方式也有缺点，如果搞不清楚数据库中各个表的逻辑关系和数据格式的插入要求，很可能插入错误数据，导致无法被前台的程序所正确展示和使用。

第二种方式：按照系统接口的调用规范要求，在测试系统的接口层编写测试脚本调用插入数据的系统接口，实现测试数据的快速插入，速度虽然不一定有第一种方式快，但是能够基本保证插入数据的正确性。如果被测试系统没有接口层，那么此方式就无法实施了。

第三种方式：使用前台的自动化测试工具，在系统的前台界面模拟用户的真实操作行为来输入各类测试数据，然后再提交到测试系统中。此方式的优点是可以真正模拟用户插入数据的行为，保证数据插入的准确性和完整性，包含前台界面的系统均可使用此方式。此方式的缺点是插入数据的速度要比前两种方式慢很多。

针对被测试系统的实际情况，测试人员可以使用三种方式之一实现测试数据的插入需求。

6. 常见的错误目标：使用自动化完全替代手工测试，使用自动化测试发现更多的新 bug

很多测试人员都有一个错误的想法，就是想用自动化测试完全替代手工测试，如果设定此目标则会给自动化测试的实施带来极大的困难。测试工作本身就是一种艺术，需要使用测试人员的智慧去探索系统中可能出现的问题，并且需要在测试过程中使用不同的测试方法、测试数据和测试策略来发现更多问题。而自动化测试的实施方式则是使用固定的方法和固定数据去实施测试，无法像人一样根据测试系统的响应情况做出及时的测试策略调整，势必会造成测试逻辑的低覆盖率。另外，测试用例中有很多异常操作很难使用程序来进行模拟，若要完全实现自动化测试来模拟则会带来极大的技术难度挑战。所以，只要设定自动化测试能够替代一定比例的手工测试工作为目标即可，千万不可对自动化测试的覆盖度设定过高的比例要求。

还有的测试人员期望使用自动化测试来发现更多的新 bug，这也是一个常见误区。虽然在编写自动化测试用例的过程中会发现大部分的 bug，但是自动化测试本身的作用不是用来发现新 bug，而是用来验证以前能够正常工作的功能是否依旧可以正常工作。举一个例子，一个被测试系统有 100 个功能点，由 5 万行代码来实现，这 100 个功能在上一个版本中均通过测试，在下一个迭代的版本开发中，程序员根据产品的 5 个新需求修改了 5 个复杂的功能点，并且新增和修改了 500 行代码，那么测试人员针对这样的场景如何来测试这个版本的产品呢？因为测试人员不知道被修改的 500 行代码到底会怎样影响整体的 100 个功能点，所以只能把 100 个功能点都测试一遍才能放心地让这个版本进行发布和上线。100 个功能点的测试工作量就这样产生了，如果采用手工测试的方式，则测试用例的执行周期肯定会是个很长的周期，并且测试人员发现了新 bug 后，程序员又修改了 100 行代码，那么是不是又要重新测试这 100 个功能点呢？如果再次测试，那么测试人员就陷入了周而复始的重复劳动中，如果不测试全部 100 个功能点，那么被修改代码产生的不确定性又难以得到评估。如果测试人员拥有了这 100 个功能点的自动化测试脚本，就不会出现进退两难的境地

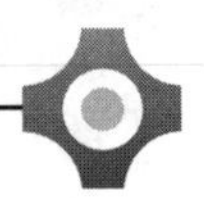

了，测试人员可以使用自动化测试脚本快速验证原有的 95 个功能点是否正常工作。自动化测试可以大大降低手工测试的重复性，测试人员只要手工测试 5 个被修改的功能即可。测试人员充分测试这 5 个功能点并确认没有 bug 产生后，可以新增编写这 5 个功能点的自动化测试用例，用于下一个版本的自动化测试即可。从上例可以看出，自动化测试更适合用于回归测试，而不是用来发现新 bug。

基于以上 6 个常见的自动化测试目标，测试人员应根据测试项目的具体要求正确地设定自动化测试目标。

2.2 管理层的支持

在一个企业中推广自动化测试是一件非常困难的任务，因为打破旧有的手工测试习惯、工作模式和工作流程，必然会让整个开发和测试团队有不适应的地方，难免会遇到各种的抵制、不理解和不合作。若能够借助高层的力量，那么自动化测试的推广工作势必会事半功倍。如果你想在企业中推广自动化测试，首先要寻求高层的支持，让高层管理人员在开发和测试团队中宣传和贯彻自动化测试的意义和实施目标，并要求公司和团队给予必要的资源和时间支持。缺乏高层的支持，自动化测试的推广基本上会无疾而终。

在自动化测试的实施过程中，要先选择合适的项目进行试点实施，建议选择开发进度不太紧张，且产品需求相对稳定的项目进行实施。在实施过程中，要合理地设定自动化测试的实施目标，并争取在实施结束后实现目标。将试点项目的自动化测试成果汇报给相关管理层，让他们进一步理解自动化测试的意义、成果和作用。基于试点项目的杰出成果，让管理层再进行其他项目的宣传和推广，逐步在全公司开展自动化测试工作。

2.3 投入产出比

大部分软件企业或互联网企业的经营都是为了谋取尽可能多的利润，它们都希望投入尽可能少的成本来获取尽可能多的利润，所以要从这个角度来谈一下自动化测试的投入产出比问题。

在尝试自动化测试实践时，测试团队需要分析投入哪些资源，比如：技术人员的工时投入、购买相关软件版权的费用、机柜、带宽和服务器的投入等，需要列出具体的资源需求列表。结合项目的实际情况，测试团队评估自动化测试的短期目标和长期目标，并描述出可能获取到的收益，再提交给研发团队的管理层进行投入产出比评估。若管理层认为投入产出比比较高，那么就可以开始实施工作了，如果觉得不高，则很可能无法进行实施。

建议测试团队要从以下几个方面考虑自动化测试的成本投入：

(1) 项目本身是否适合实施自动化测试，测试脚本的编写和维护成本是否较高？

(2) 现有测试团队成员是否具备自动化测试的实施能力？如果不具备，是否可以采用培训的方式来提升，还是进行外部招聘有能力的自动化测试实施人员？

(3) 使用何种自动化测试软件，是否需要购买版权？

(4) 现有的测试环境硬件要求是否符合自动化测试的实施要求？

(5) 研发团队管理层对于自动化测试的潜在期望和要求？

建议测试团队要从以下几个方面中重点考虑自动化测试的产出：

（1）从短期和长期来分析能够节省多少测试人力资源的投入？

（2）是否能够开发出比较成熟的自动化测试框架，解决测试脚本编写和维护成本高的问题？

（3）自动化测试脚本是否可以快速地被执行，并确认具体量化指标？

（4）自动化测试的引入是否会提高开发人员的开发效率和质量，并确认具体量化指标？

2.4 敏捷开发中的自动化测试应用

目前，敏捷开发模式已经在国内众多的开发团队中盛行，开发团队已经逐步享受到敏捷开发带来的高效和价值，其中敏捷团队全员的质量负责方式和大规模的自动化测试引入成为了现在的热度话题。敏捷开发的本质到底是什么呢？为什么大家开始高度认可它的价值呢？本小节我们做一个简单的解释说明。

首先，我们要讲一下传统开发模式遇到的问题，以往传统的开发模式大部分都是按照长周期和里程碑的方式进行管理，有明确的需求、设计、开发、测试和上线的几个阶段，产品的发布周期也比较长，一般2到3个月，长的甚至有1到2年的时间。虽然每个阶段都有明确的目标和工作范围，但是令人困扰的是需求总是在不断地产生变化，不断影响项目的设计、开发、测试等多个阶段，导致项目设计人员在初期就要想办法做出各种冗余的系统设计来防止未来变更的需求带来的负面影响。然而，计划总是赶不上变化，需求的变化和不确定性依然会带来各种问题，导致项目被不断地延迟，团队成员也越来越抵制需求的变更，项目质量也会不断下降，对于大型项目来说总是危机重重。

敏捷开发和传统开发模式完全不同，它只会实现明确的需求，拥抱变化，使用自动化测试和重构的方式来响应不断变化的需求，实现每月、每周甚至每天发布新版本，解决传统开发模式的很多问题。

敏捷开发的核心理念是小步快跑，它具有如下6个特点：

（1）鼓励团队成员的面对面沟通，敏捷开发模式认为人和人的相互交流胜于任何流程和工具。

（2）客户协作胜过合同谈判。

（3）把工作重点放在可执行的程序上，而不是写大量的文档。

（4）团队协作和团队激励，团队对产品的发布承担责任，明确团队的统一目标。

（5）响应变化胜过遵循计划。

（6）使用持续集成和自动化测试方式快速反馈项目质量，及时适应新的需求，保证产品的正确性。

其中，自动化测试是敏捷开发中很重要的一个环节，因为敏捷开发模式一般会在每天提交开发的代码到代码版本控制系统，为了保证所有提交的代码都是正确的，开发团队通常都会使用自动化测试手段来进行回归测试，验证所有代码修改没有影响到以前版本的功能。通过自动化测试手段，开发团队可以实现每日代码集成的开发任务，并保证每天的代码开发质量。自动化测试是敏捷开发模式的基础，如果缺少自动化测试，那么敏捷测试通常会失败，因为项目本身无法控制持续集成过程中出现的代码修改风险，也无法对项目的不断重构

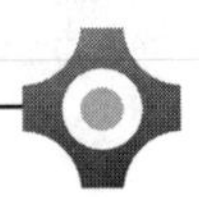

提供快速测试的支持，势必会引发项目延期、质量下降等一系列问题，无法真正实现小步快跑的目标。

敏捷开发中通常使用测试驱动开发的方法（Test-Driven Development，TDD），这种方法不同于传统软件开发流程的开发方法，要求在编写某个功能的代码之前先编写测试代码，开发人员只编写使测试通过的功能代码，通过测试来推动整个开发的进行，此方式可以确保开发人员集中精力在明确的需求上，防止过度设计，尽可能保持代码的简洁性，提高开发效率。

还有一种敏捷开发中常用的技术就是行为驱动开发（Behavior-Driven Development，BDD）。BDD是测试驱动开发的进化，可以有效改善设计，并在系统的演化过程中为团队指明前进方向。BDD使用客户和开发者通用的语言来定义系统的行为，从而做出符合客户需求的设计，避免其他开发模式中常见的客户和开发双方对于需求理解的不一致性。

敏捷开发中的测试可以从如图2-1所示的三个层级进行。

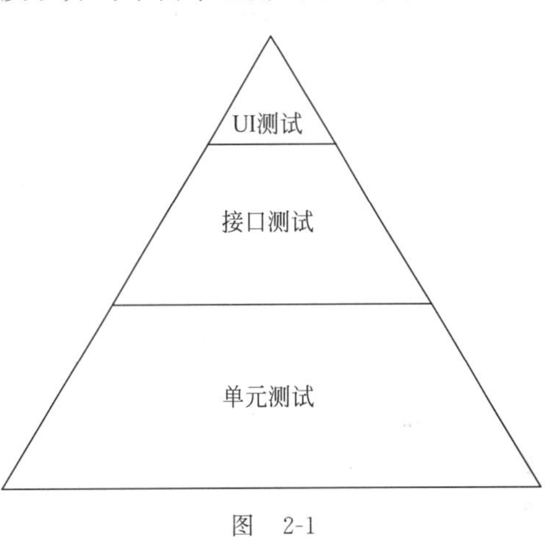

图 2-1

图2-1是一个三角形的示意图，三层中的每一层区域大小代表着每一个层级测试的收益大小，我们可以看出单元测试的收益是最大的，接口测试其次，UI测试的收益最小。单元测试的颗粒度是最小的，测试范围集中在类和方法，测试用例编写相对简单，并且出现bug后，定位问题相对快速，可以在开发初期发现大部分问题，并且单元测试执行的速度最快，通常在毫秒级别运行就可以得到测试结果。接口测试的颗粒度更粗了一些，测试范围集中在模块、子模块间的数据交互，定位问题相对复杂，涉及分析的代码量很大，测试执行速度也比单元测试慢许多。UI测试的收益最小，测试通常在系统测试和验收测试阶段中进行，基于全部的系统代码进行测试，测试出现问题后定位和分析困难。UI测试通常在用户使用的界面进行，测试执行相对于单元测试和接口测试慢很多，并且因为UI界面经常发生变化和调整，自动化执行和维护成本也很高。

敏捷开发中的自动化测试可以基于这三种方式进行，基于上述的收益说明，敏捷测试更鼓励在单元测试和接口测试上投入更多资源，以此来实现快速编写、快速执行、快速定位问题的测试目的，能够快速地给予项目质量反馈。UI测试虽然相对来说收益最低，但是UI层对于用户来说是最直观的感知，所以也要在这个层级实现一定程度的自动化测试，尽可能模拟用户的各种真实操作，确保用户的最佳产品体验。

2.5 自动化测试人员分工

自动化测试通常涉及3种分工角色：

（1）测试框架开发人员。

（2）基于测试框架编写测试脚本的人员。

(3) 编写需要自动化测试用例以及测试框架需求的人员。

三种角色可以根据测试团队人员的实际水平进行角色合并,通常情况下测试开发人员承担第一种和第二种角色,非测试开发人员承担第三种角色。测试框架搭建人员的技术能力要求最高,通常在人才市场上处于非常抢手的情况,优秀的测试开发人员年薪一般在20~30万元,并且他们的职业发展空间比传统的手工测试更大,更容易上升到测试团队的管理层。

2.6 自动化测试工具的选择和推广使用

2.6.1 自动化测试工具的选择

高效实施自动化的前提是要选择一个适合测试团队使用的自动化测试工具,优秀的测试工具会让自动化测试工作的实施事半功倍,反之则可能给自动化测试的实施工作带来灭顶之灾。选择自动化测试工具要持谨慎态度,需要结合工具特点和测试团队的实际情况进行综合分析,再最终决定选择哪个工具作为测试团队使用的测试工具。

表 2-1 列出选择自动化测试工具时需要考虑的关键点。

表 2-1

工具特点	收费/开源	测试工具支持的编程语言	测试工具的兼容性	工具学习成本	是否支持持续集成工具	工具运行的稳定性
团队现状	团队是否有预算购买?团队是否有优化测试工具的能力?	团队成员是否具备相关编程语言的基础?	是否满足被测试对象的兼容性要求?	根据团队成员能力评估工具学习的时间成本和人工成本	评估是否易于和持续集成工具进行集成?	是否可以在无人值守状态下稳定不间断运行?

2.6.2 Selenium WebDriver 和 QTP 工具的特点比较

目前,主流的 Web 自动化测试工具是 Selenium WebDriver 和 QTP,下面详细比较这两种工具的特点,如表 2-2 所示。

表 2-2

比较项	说明
用户仿真	Selenium:在浏览器后台执行,执行时可以最小化,可以在一台机器上同时执行多个测试
	QTP:完全模拟终端用户,独占屏幕,只能开启一个独占的实例
UI 元素组件的支持	Selenium:支持主要的组件,但是某些事件、方法和对象属性的支持不够
	QTP:良好的支持,提供对.NET 的组件支持
UI 对象的管理和支持	Selenium:需要自写代码实现,相对复杂
	QTP:做了很好的支持,支持录制添加

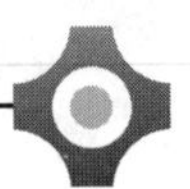

续表

比　较　项	说　　明
对话框的支持	Selenium：只支持一部分浏览器的弹出框，需要调用其他三方工具来进行操作
	QTP：基本都支持
浏览器的支持	Selenium：支持多种主流浏览器 IE、Firefox、Chrome、Opera 和 Safari
	QTP：只支持 IE 和 Firefox
面向对象语言和扩展性支持	Selenium：支持多种编程语言和外部库 Java、Python、C# 等
	QTP：只能使用 VBScript 编写脚本，不支持其他语言和外部库
支持的操作系统/平台	Selenium ：支持跨平台
	QTP：只支持 Windows
脚本创建难易	Selenium ：创建脚本相对困难
	QTP：创建脚本相对简单
版权费用	Selenium ：免费
	QTP：按照安装的机器台数计费，版权费用昂贵
持续集成工具	Selenium ：支持主流的持续集成工具
	QTP：不支持

综上因素比较，具备一定编程能力的测试团队更适合选择 Selenium WebDriver 作为团队的主要 Web 自动化测试工具，对于预算充裕且团队成员编程能力一般的测试团队更适合选择 QTP 工具作为团队的主要 Web 自动化测试工具。

2.7　在项目中实施自动化的最佳实践

自动化测试在大部分企业的推行过程中都会遇到各种困难，在不合适的项目和不适当的项目阶段实施自动化，导致自动化测试实施效果不佳，自动化测试团队会被质疑其存在的价值。自动化测试的实施是一个复杂的过程，须结合企业文化、研发流程、团队技术能力、项目情况以及实施成本等多种因素来逐步实施。以下列出了一些企业在自动化实施过程中的十个最佳实践，供广大自动化测试的爱好者参考。

(1) 在自动化测试实施前，建立可衡量和易达到的自动化测试实施目标，不要在初期制定过高的目标和期望。

俗话说"好的开始是成功的一半"，为了后续更好地推广和实施自动化测试，须在初期就让研发团队和相关参与者看到自动化测试带来的好处，增强大家成功实施自动化测试的信心。使用可衡量的目标，有助于参与各方有效地评估自动化测试的效果；易达到的目标会进一步鼓励自动化测试实施者按部就班地开展实施工作，避免采用急功近利和好高骛远的实施方法。

(2) 选择适合公司普遍使用的测试工具，可以是一个工具或者一组工具，做出选择后需要针对选定工具进行深入研究。

每种测试工具都具有其独特的优点和缺点，每种工具都具有某些独特适用的使用场景，建议对工具充分了解后再进行团队内部的使用技巧培训，夯实自动化测试实施的技术基础。另外，建议中小公司尽可能选择使用开源的测试工具，降低购买商业测试工具产生的相关成本。

（3）分析测试项目的特点，编写适合项目特点的自动化测试框架，减少编写测试脚本的重复性和复杂性，降低其他测试人员编写自动化测试脚本的门槛。

每个测试项目都是独特的，总是会有一些很独特的测试需求，需要仔细分析其特点后，由测试开发团队实现适合当前项目使用的自动化测试框架。一个优秀的定制化测试框架可以有效地推动自动化测试在项目中的实施。由于大部分测试人员的编程能力存在一定局限性，须依靠优秀的测试框架来降低编写自动化脚本的难度，从而让更多的测试人员从自动化测试中受益，更好地调动团队积极性去支持自动化测试的进一步实施。

（4）聘用具备丰富开发经验的工程师承担测试框架的开发工作，并根据测试框架的推广程度进行不断优化。

测试框架的意义已无须赘述，为了更好地服务于测试人员，团队应该聘用最优秀的技术开发人员来承担测试框架的开发工作，良好设计的测试框架会极大地增加自动化测试实施的成功率。优秀的测试框架也不会短周期内被迅速开发出来，必须经过一个长期的优化过程，才能打磨出一套适应公司大多数项目的自动化测试框架，因此建议长期投入优化测试框架的人力。

（5）在自动化测试工作开始大规模推广实施前，必须在中小类型项目进行充分试点实施，充分评估实施自动化测试的风险和产出，总结试点实施中的问题和收益，并在后期推广过程中尽可能扬长避短。

为了降低自动化测试实施过程中的风险，测试团队应该提前进行风险分析，做好针对性的风险应对计划，并在中小类型项目中进行试点实施，证明测试团队已经做好了实施自动化测试的充分准备。在试点过程中尽可能多地发现问题，并通过不断地解决问题来完善自动化测试实施的方法和流程，为后续的大规模推广做好充分准备。

（6）获得开发团队的协作支持，提高开发代码的可测试性，降低自动化测试实施的难度。

由于测试工具本身的局限性、测试人员的编程能力以及被测试对象的复杂度，很可能需要开发团队做出一定程度的配合才能实现较为复杂的自动化测试脚本。在自动化测试实施前，测试团队应该和开发团队对代码的可测试性要求达成共识，建议制定代码开发的可测试性标准或规范，并在自动化测试实施过程中不断完善。

（7）在需求相对稳定的阶段，开始 UI 层级大规模自动化测试脚本的编写。

项目启动阶段，项目需求一般都是不太稳定的，UI 层的需求变化很大，如果在项目启动阶段就开始编写大量 UI 级别自动化测试脚本，一旦需求发生了大范围的变更，那么自动化测试脚本的维护工作量也会随之产生，这不但会降低自动化测试人员的实施积极性，也会增加自动化测试投入的人工成本。自动化测试工程师会质疑自己为什么每天都要维护以前可以正常运行的自动化测试脚本。为了降低自动化脚本维护的成本，须在项目需求稳定阶段，且大部分严重 bug 已经修改完毕的情况下，再进行大规模自动化测试脚本的编写，尽量降低维护测试脚本的工作量，使自动化测试脚本可以使用更长的周期。

（8）在测试过程中，使用局部自动化测试的实施策略。

有的时候大规模实施自动化可能会带来各种实施的难度和困难，维护大量的自动化测试脚本可能也没有太多人力和时间去完成，这样可能会导致测试团队抵制使用自动化测试

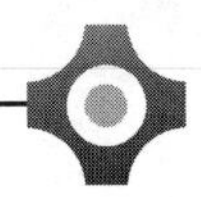

技术。测试人员可以尝试使用局部自动化测试的实施策略，找到相对重复的手工劳动过程，然后编写自动化测试脚本来替代重复性的手工劳动。少量的测试脚本编写和调试会比较容易，耗时更少，并且更易于传递给其他测试人员使用。如果能够减少一些测试人员的手工测试工作量，测试团队何尝不多做一些这样的尝试呢？小脚本积累多了也是一笔很可观的财富，总会有爆发的那一天。

（9）全面提高自动化测试实施人员的技术素质。

实施自动化测试的技术要求很高，为了能够保证自动化测试的创新和普遍使用，测试团队负责人须尽可能提高自动化测试实施人员的技术素质。每个人的技术基础打好了，后续才能发挥每个人的主观能动性，结合项目应用场景，因地制宜地编写出优秀的自动化测试框架和高质量的测试脚本。我们要认识到企业中的“人”才是最重要的资产，须让这些重要的资产不断增值，而不是让他们贬值。

（10）定期做好自动化测试最佳实践的总结。

自动化测试的实施过程不能一蹴而就，也不可能一帆风顺，总是会遇到各种困难和问题，作为自动化测试的实施团队应该定期总结一段时间内自动化测试的得失，不断形成团队最佳实践的自动化测试知识库，这样才能让自动化测试技术在企业的实施更加深入和全面，确保企业在人员流失的时候不至于丢掉宝贵的最佳实践经验。建议最佳实践经验的资料都放到团队内部的培训文档中，让更多的后来者能够站在前人的肩膀上不断成长，为企业降低更多的自动化测试实施成本，提高人员的劳动生产率。

2.8　学习 Selenium 工具的能力要求

自动化测试相对于手工测试人员来说需要更多的知识和编程技能，基于 Python 编程语言，列出一些在使用 Selenium WebDriver 工具时常遇到的一些知识领域：HTML、XML、CSS、JavaScript、Ajax、MySQL 数据库、unittest、Jenkins/Hudson、Lettuce 测试框架。

建议 Selenium WebDriver 的工具使用者都尽可能地深入学习以上知识领域，尤其要增加学习编程技能的时间，编程能力的高低直接决定你是否可以写出优秀的自动化测试框架。真正的自动化测试高手，从技术能力上来说比中等开发人员的水平还要高，所以想成为一个能够独当一面的自动化测试工程师，须不断地进行学习各类开发知识。不是每个测试工程师都可以成为自动化测试工程师，要想改变常年手工测试的命运，必须坚持不懈地学习和实践，才能让我们离自动化测试的巅峰越来越近，最终有一天我们会站在顶峰摇旗呐喊。

第 3 章 自动化测试辅助工具

Selenium 工具本身虽然很强大,但是它也需要一些辅助工具来解决一些特定的问题。本章主要介绍和 Selenium 工具配合使用的辅助工具。

3.1 安装 Firefox 浏览器

Firefox 浏览器的安装步骤如下。

(1) 浏览器访问网址 http://www.Firefox.com.cn/。

(2) 单击浏览器页面中的"立即下载"链接,下载 Firefox 浏览器安装文件,下载页面如图 3-1 所示。

图 3-1

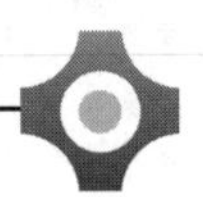

(3) 下载完成后，在下载文件保存目录会生成一个文件名为 Firefox-latest. exe 文件。

(4) 鼠标左键双击 Firefox-latest. exe 安装文件，按照安装向导一步一步地进行安装，如无特殊安装路径要求，则不断单击“下一步”按钮即可完成 Firefox 浏览器安装。

(5) 安装完毕后，桌面会显示 Firefox 浏览器的快捷方式图标。

更多说明：

- 本书实例所使用的 Firefox 浏览器版本是 71.0 (64 位)。
- 更多 Firefox 历史版本选择访问：http://ftp. mozilla. org/pub/firefox/releases/。

3.2　安装 Firebug 插件

3.2.1　打开工具箱

(1) 方法 1：打开火狐浏览器后，使用快捷方式打开，按下组合键：Ctrl+Shift+I 来打开或者关闭工具箱。

(2) 方法 2：单击浏览器地址栏区域最右侧的“打开菜单”按钮，如图 3-2 和图 3-3 所示。再单击菜单中的“Web 开发者”，然后单击“切换工具箱”。

显示出 Firefox 的工具箱(即：开发者工具)，如图 3-4 所示。

图 3-2

图 3-3

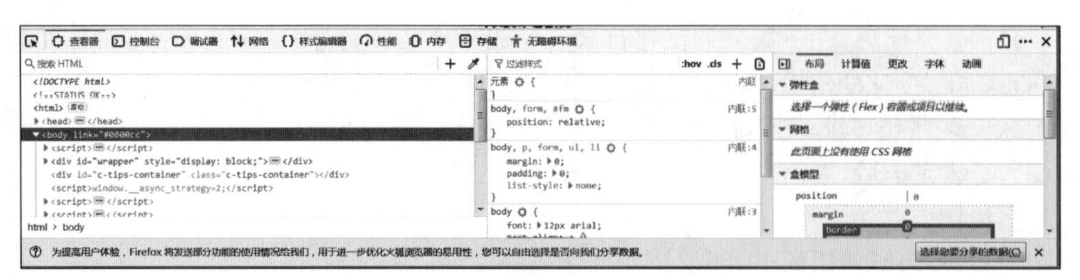

图　3-4

3.2.2　定位页面元素的 HTML 代码

（1）打开 Firefox 浏览器，访问 https://www.baidu.com。

（2）打开 Firefox 浏览器的工具箱，单击指针图标 ，使指针处于选中状态，鼠标单击百度页面中的搜索框，这样可以在工具箱中找到被高亮显示的搜索框对应的 HTML 源码。根据对应的 HTML 源码可以编写定位元素的定位表达式，如图 3-5 所示。

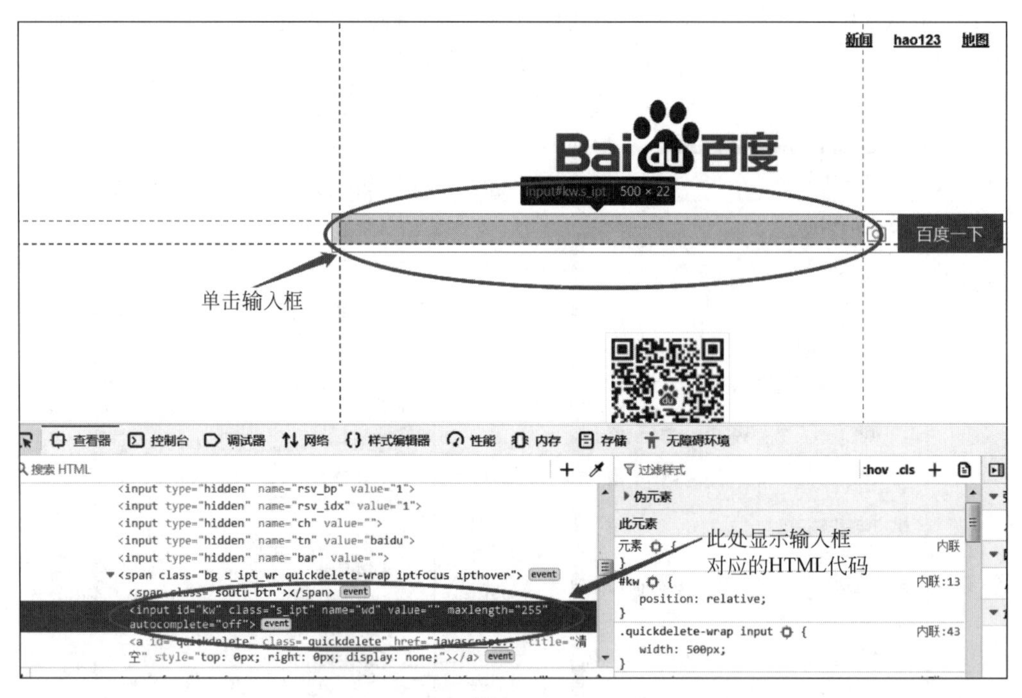

图　3-5

3.3　使用定位页面元素的 Firefox 浏览器插件

本节主要介绍 Firefox 浏览器页面元素定位插件的安装和使用方法。

3.3.1　安装 Firebug 元素定位插件

（1）通过菜单项“附加组件”打开插件搜索界面，如图 3-6 所示，或者使用组合键 Ctrl+Shift+A 打开。

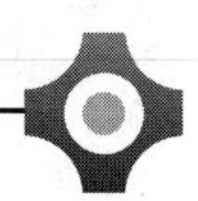

图　3-6

（2）在插件搜索界面中的搜索框中输入 xpath 后按回车键，如图 3-7 所示。

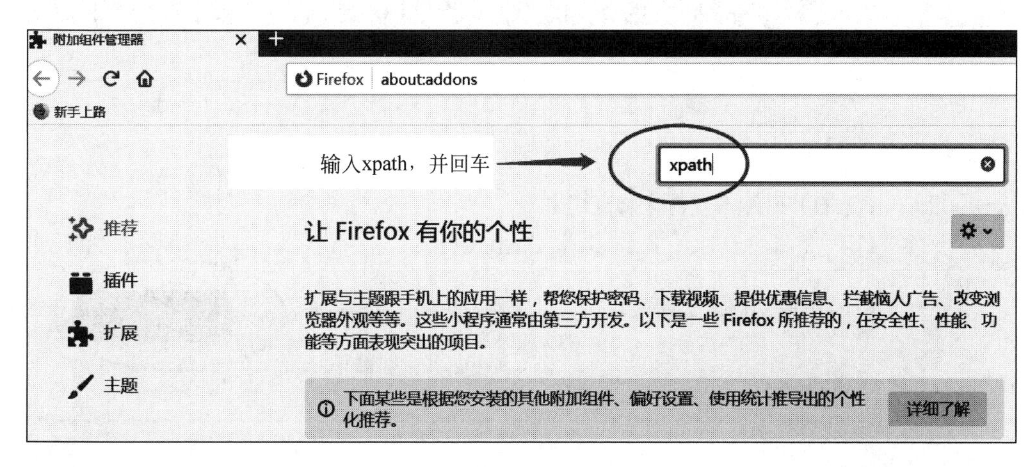

图　3-7

（3）回车后，可以看到在列表中显示很多定位元素的插件。

(4) 选择五星插件 Ruto-XpathFinder 进行安装,单击这个标题即可进入安装界面,如图 3-8 所示。

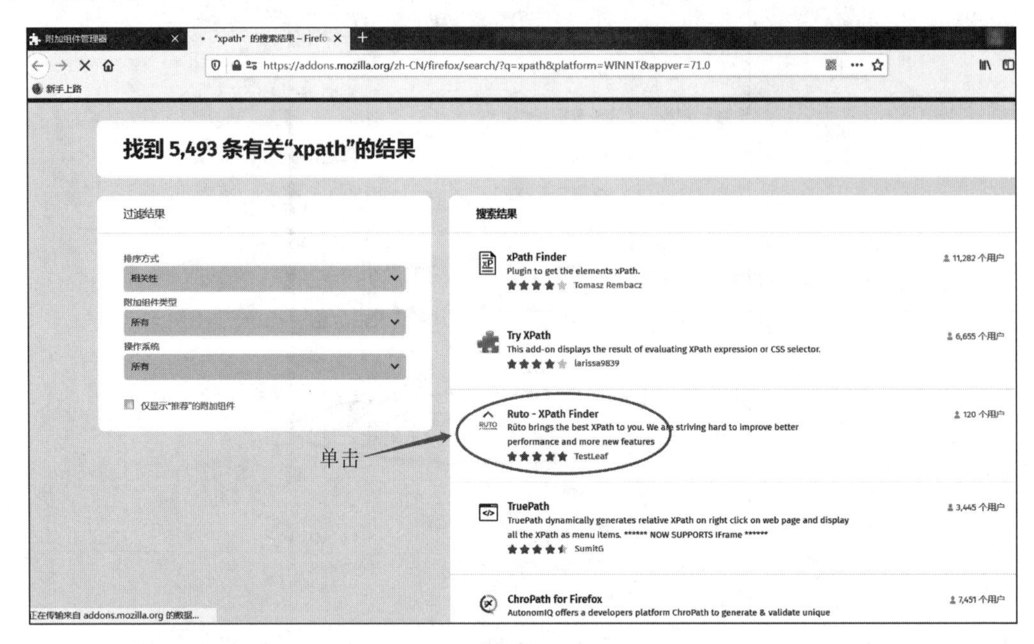

图 3-8

(5) 进入安装界面后,单击"添加到 Firefox"按钮即可,如图 3-9 所示。

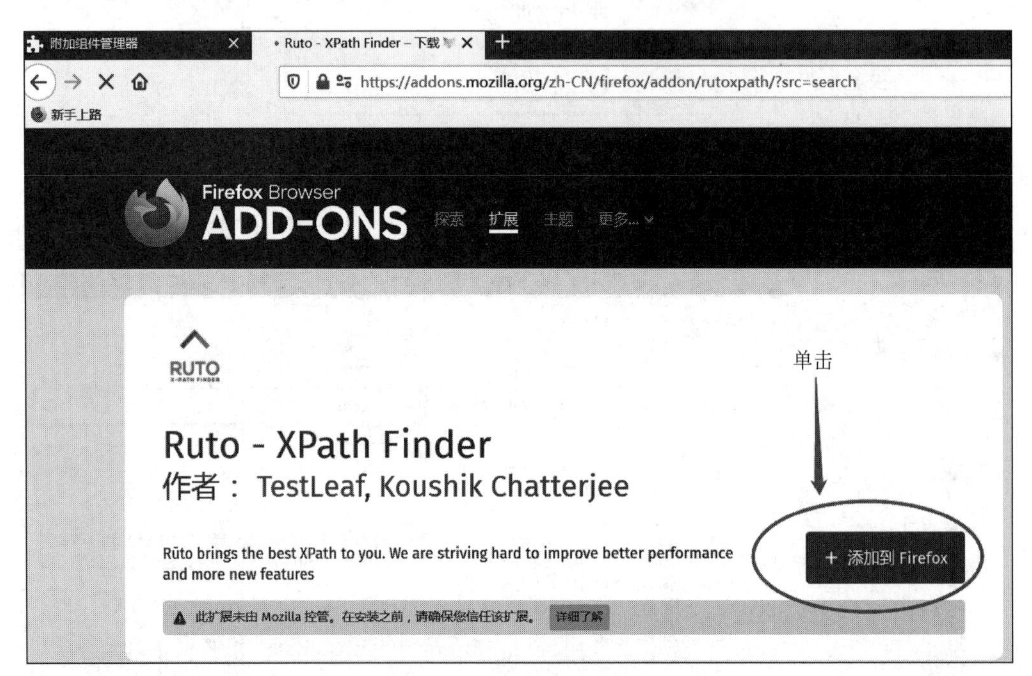

图 3-9

(6) 单击后界面显示权限确认框,单击"添加"按钮即可,如图 3-10 所示。

(7) 安装完成后,浏览器的工具栏中显示插件按钮,后续可以进行使用,如图 3-11 所示。

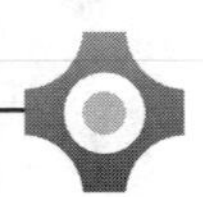

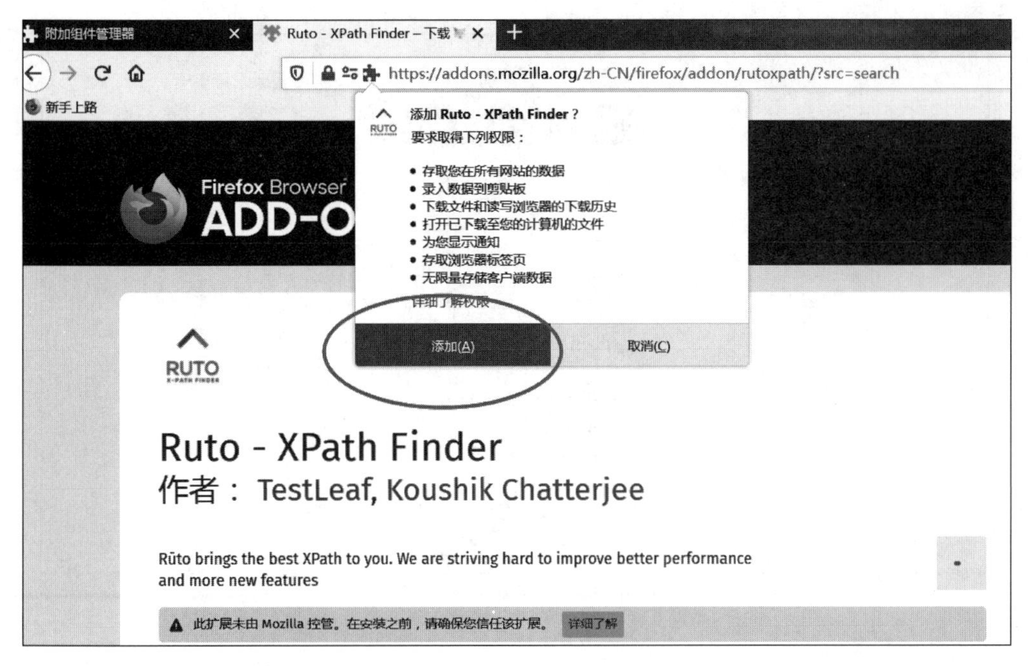

图　3-10

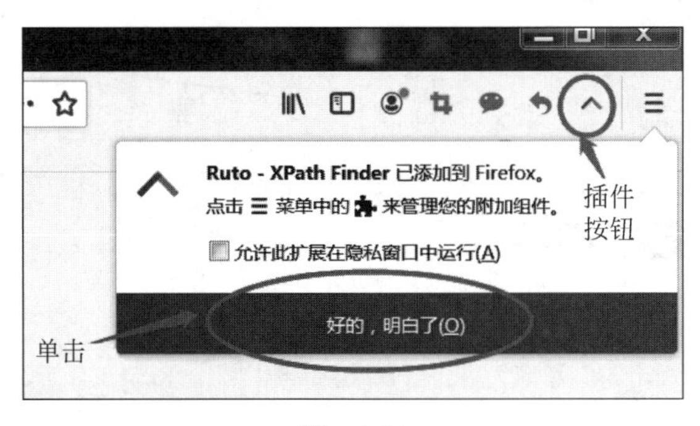

图　3-11

(8) 可以根据自己偏好，从插件列表中选择不同的插件进行安装和使用。

3.3.2　使用 Ruto-XPath Finder 进行页面元素定位

以定位百度首页的搜索输入框为例，请参阅下面步骤：

(1) 打开 Firefox 浏览器，输入网址 https://www.baidu.com。

(2) 在输入框上方，右击打开快捷菜单，如图 3-12 所示。

(3) 单击浏览器工具栏中的 Ruto-XPath Finder 插件按钮，即可看到定位后的各种定位表达式，可以根据自己的需要选择使用，如图 3-13 所示。

3.3.3　使用 XPath Finder 插件进行页面元素定位

使用 3.3.1 小节介绍的插件安装方法，安装好 XPath Finder 插件，在 Firefox 浏览器工具栏会显示图标，使用此插件方法如下：

图 3-12

图 3-13

(1) 打开 Firefox 浏览器，输入网址 https://www.baidu.com。

(2) 单击 Firefox 浏览器工具栏中的 XPath Finder 插件图标，使插件处于选中状态，鼠标光标会变为十字形状。

(3) 在百度页面上单击搜索输入框后，在网页的左下角可以看到 XPath 表达式，如图 3-14 所示。

(4) 鼠标选中表达式后，按组合键 Ctrl+C 就可以复制定位表达式了。

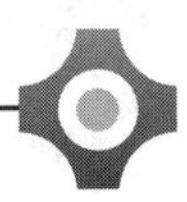

图　3-14

3.4　IE 浏览器自带的辅助开发工具

在 IE 8 以上版本中，均自带辅助开发工具，功能类似于 Firefox 浏览器中的工具箱，可用于查看页面元素。但是 IE 的辅助开发工具不支持 XPath 表达式定位，所以无法使用它来获取页面元素的 XPath 定位表达式。

启动 IE 浏览器后，按 F12 键即可打开 IE 浏览器的辅助开发工具，如图 3-15 所示。

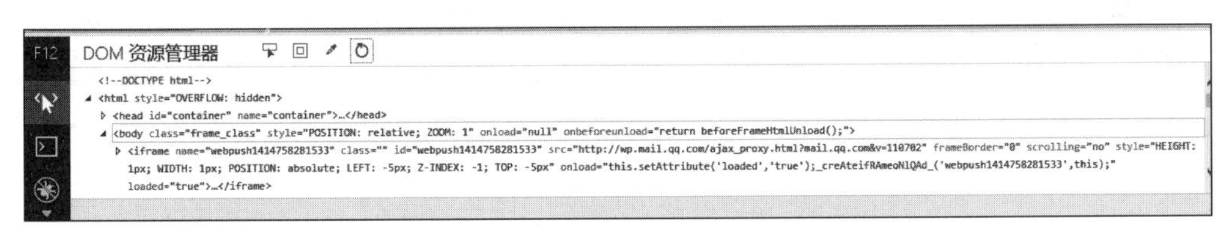

图　3-15

在自动化测试脚本开发过程中，此辅助开发工具主要用于查看页面元素的 HTML 代码，在 Firefox 浏览器不能正常显示页面元素时，可以结合此工具来查看页面元素的 HTML 代码，以便后续编写页面元素的 XPath 或者 CSS 定位表达式。

3.5　Chrome 浏览器自带的辅助开发工具

作为最流行的浏览器之一，Chrome 浏览器已经被广大网民广泛使用，它自带的开发者工具也非常出色，除了定位页面元素对应的 HTML 源码外，主要用来验证手写的 XPath 元素定位表达式是否正确，具体使用步骤如下：

（1）打开 Chrome 浏览器。

（2）输入网址 https://www.baidu.com。

（3）从键盘按组合键 Shift+Ctrl+I，打开开发者工具。

（4）从键盘再按组合键 Ctrl+F，开发者工具里面会显示一个表达式输入的框，在里面输入 xpath 定位表达式//input[@id='kw']，然后回车，即可看到百度首页中的搜索输入框被高亮显示，并且 HTML 中黄色高亮显示搜索框对应的 HTML 代码部分，如图 3-16 所示。

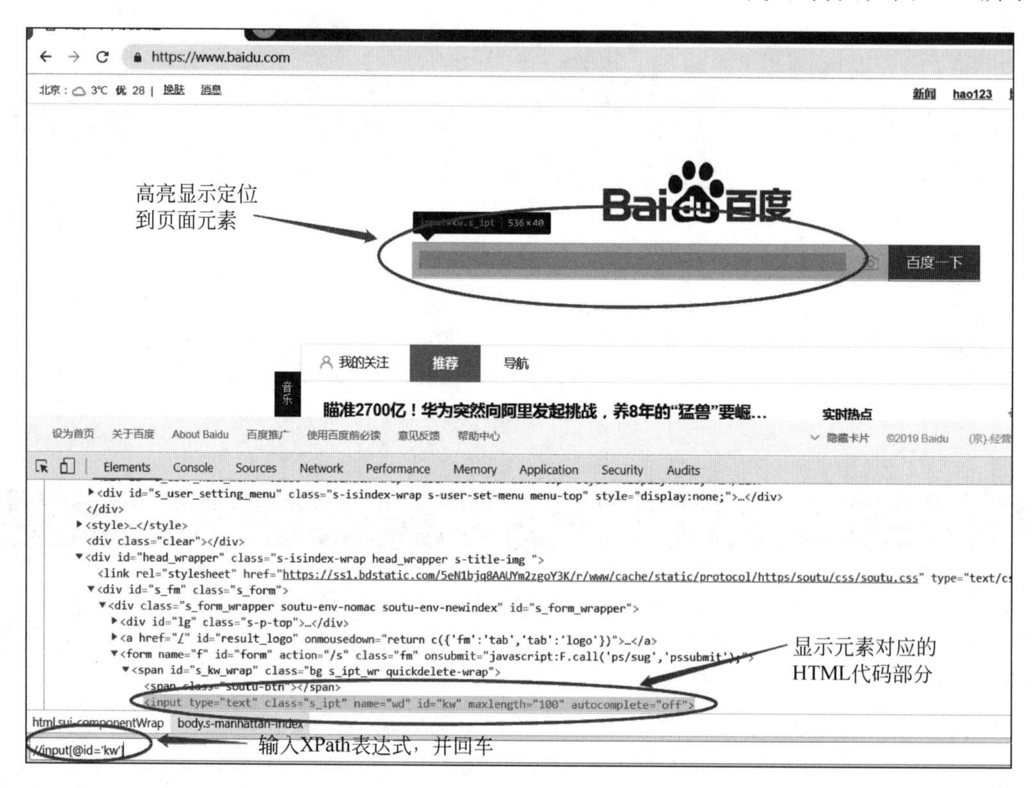

图 3-16

第4章 搭建Python 3环境和PyCharm集成开发环境

Python 语言编写的 Selenium 自动化测试脚本，笔者选择在 PyCharm 集成开发环境中运行，读者也可以选择直接在文本编辑器中编写，然后通过 CMD 执行。如果做大型项目开发，推荐大家使用 PyCharm 工具，其功能很强大，能在编写程序时给我们带来很多方便，特别是代码补全，可调用函数查看等。

本书所有的环境搭建以及 Selenium 脚本编写默认都是基于 Windows 7 64 位系统，读者如果使用的是 Windows 10 或者其他类型的系统，本书中部分内容可能不匹配，请读者根据具体情况做相应修改。

4.1 安装和配置 Python 3 环境

在使用一种编程语言编写程序之前，首先需要搭建支持这个编程语言的开发环境，本节先讲解 Python 环境的搭建和配置。

4.1.1 下载并安装 Python 3 解释器

具体操作步骤如下。

（1）访问 https://www.python.org/。

（2）单击菜单栏中的 Downloads，出现如图 4-1 所示的界面。

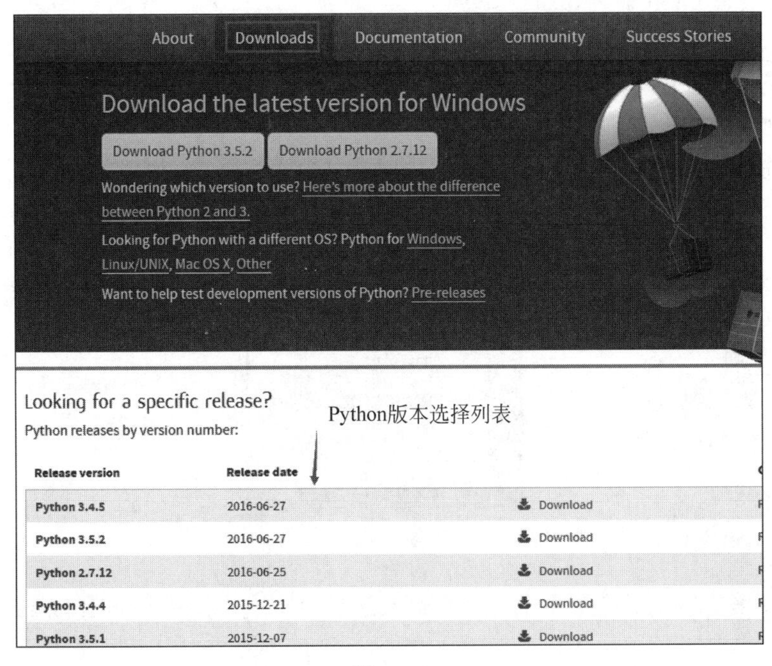

图 4-1

(3) 本书选择 Python3.5.4 版本,单击后面的下载图标 Download 。

(4) 选好 Python 版本并进入其下载界面后,选择适合你的平台及版本位数(笔者的系统为 Windows7 x64),直接单击链接就可以进行下载,如图 4-2 所示。

Version	Operating System	Description
Gzipped source tarball	Source release	
XZ compressed source tarball	Source release	
Mac OS X 32-bit i386/PPC installer	Mac OS X	for Mac OS X 10.5 and later
Mac OS X 64-bit/32-bit installer	Mac OS X	for Mac OS X 10.6 and later
Windows help file	Windows	
Windows x86-64 embeddable zip file	Windows	for AMD64/EM64T/x64
Windows x86-64 executable installer	Windows	for AMD64/EM64T/x64
Windows x86-64 web-based installer	Windows	for AMD64/EM64T/x64
Windows x86 embeddable zip file	Windows	
Windows x86 executable installer	Windows	
Windows x86 web-based installer	Windows	

图 4-2

(5) 下载完后会得到一个扩展名为 exe 的安装程序,双击该安装程序,然后一路单击"下一步"按钮进行安装,安装完成后在操作系统的 C 盘会出现一个 Python35 目录。

至此,完成 Python 开发环境的安装。

4.1.2 配置 Python 3 环境

具体步骤如下。

(1) 右击"计算机"桌面图标,在弹出的菜单中,单击"属性",如图 4-3 所示。

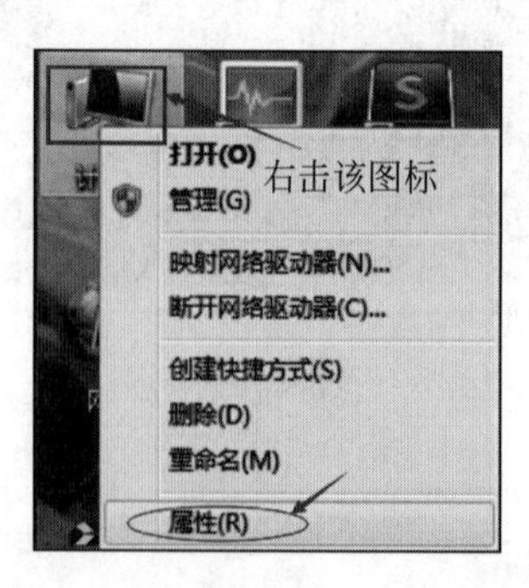

图 4-3

(2) 在弹出的系统对话框中,单击"高级系统设置"项 高级系统设置 ,调出"系统属性"配置对话框,如图 4-4 所示。

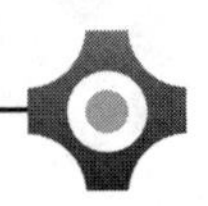

(3) 进入“系统属性”窗口的“高级”菜单界面，单击对话框右下角的“环境变量”，调出系统环境变量配置界面，如图 4-4 所示。

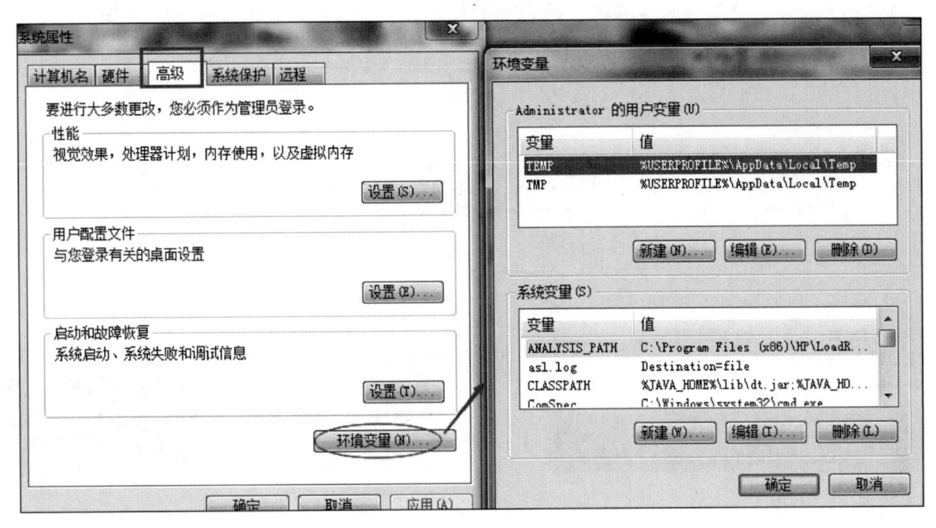

图　4-4

(4) 系统环境变量配置对话框中，在系统变量列表中找到变量名为 Path 的行，双击该行，调出“编辑系统变量”对话框，如图 4-5 所示。

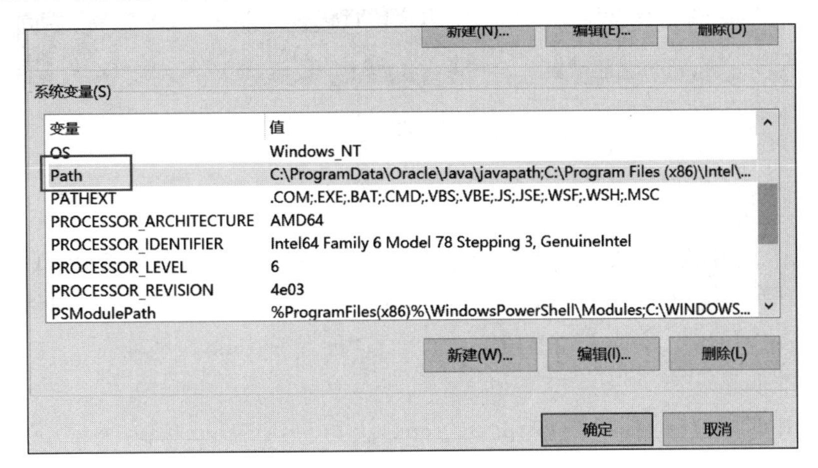

图　4-5

(5) 在“编辑系统变量”对话框中的“变量值(V)：”行右边的输入框中，在最前面添加 Python 的安装路径“C:\Python35”，然后单击“确定”按钮后完成 Python 环境配置步骤，如图 4-6 所示。

Python 安装路径后面有一个分号(;)，表示路径间的分隔符。

(6) 在“运行”输入框中输入“cmd”然后直接按回车键，如图 4-7 所示，弹出 CMD 对话框，如图 4-8 所示。

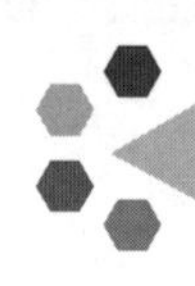

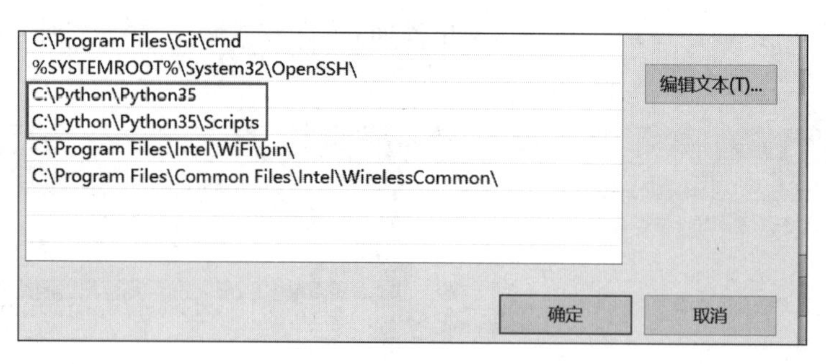

图 4-6

图 4-7

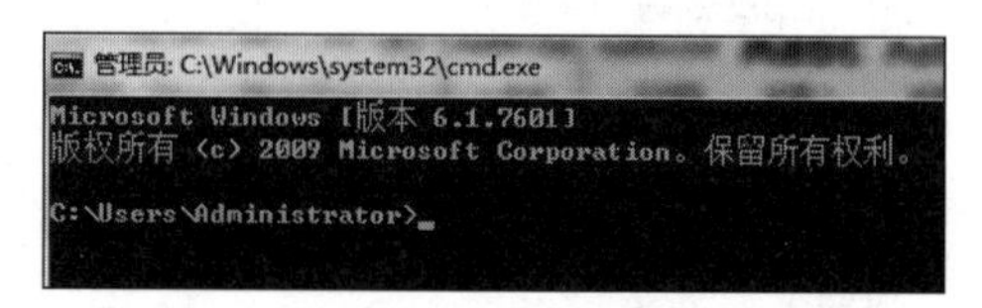

图 4-8

（7）在弹出的 CMD 对话框中输入 python，按回车键后会显示如图 4-9 所示的信息，表示 Python 环境配置成功。

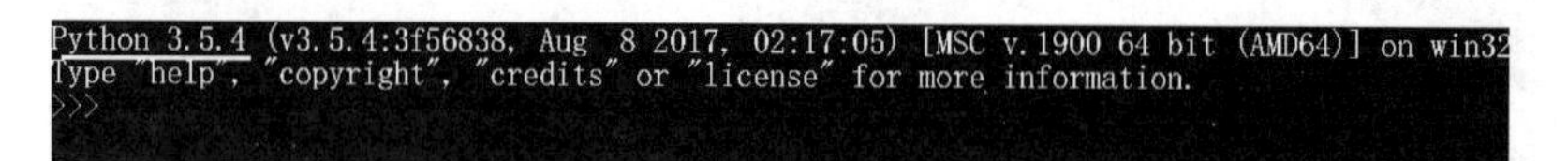

图 4-9

4.1.3 安装 pip

一般 Python 默认会自带一个低版本的 pip 工具，如果读者选择使用默认的 pip 工具，可以忽略本章节，只需要将“C:\Python35\Scripts”目录添加到系统的 Path 环境变量中，即可在 CMD 下直接使用，添加方法参照第 4.1.2 节“配置 Python 3 环境”。

具体操作步骤如下。

（1）在浏览器中访问网址：https://pypi.python.org/pypi/pip，下载 pip 源码包，如图 4-10 所示。

图 4-10

（2）单击图 4-10 所示的 pip 源码包，下载 pip 包，等待下载完成后，会得到一个扩展名为 gz 的压缩文件，比如 pip-8.1.2.tar.gz。

（3）解压 pip-8.1.2.tar.gz 文件，保存到 D 盘 pip-8.1.2 目录下，CMD 命令窗口下，使用 cd 命令将当前的工作目录切换到 D:\pip-8.1.2，然后输入 python setup.py install 并回车，安装 pip 工具，如图 4-11 所示。

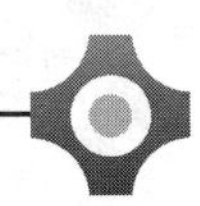

(4) 待 pip 安装成功后的界面如图 4-12 所示。

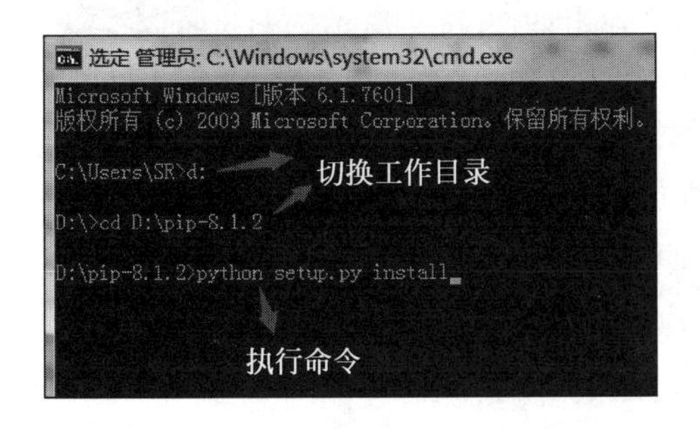

图　4-11

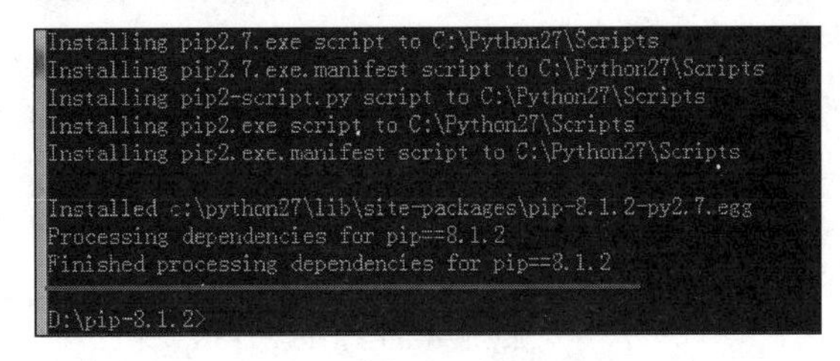

图　4-12

(5) 将 C:\Python35\Scripts 路径添加到系统的 Path 环境变量中,添加方法参照第 4.1.2 节"配置 Python 3 环境",兹不赘述。

4.2　安装 Python 集成开发环境 PyCharm

PyCharm 是比较好用的 Python 编辑器,不仅功能强大,而且还可以跨平台。本书所有实例默认都使用该软件开发,本小节主要讲解此软件的安装步骤。

(1) 访问网址 http://www.jetbrains.com/pycharm/download/#section=windows。

(2) 在打开的网页中,选择社区版本进行下载(免费),如图 4-13 所示。

(3) 等待几分钟,下载完成后会得到一个 exe 安装文件 pycharm-community-2016.2.3.exe。

(4) 双击 pycharm-community-2016.2.3.exe 离线安装文件进行安装,然后只需要一路单击 Next 按钮,期间需要根据读者自己的操作系统位数完成如图 4-14 的操作。安装成功后,会在操作系统桌面上出现一个如图 4-15 所示的快捷启动图标。

图　4-13

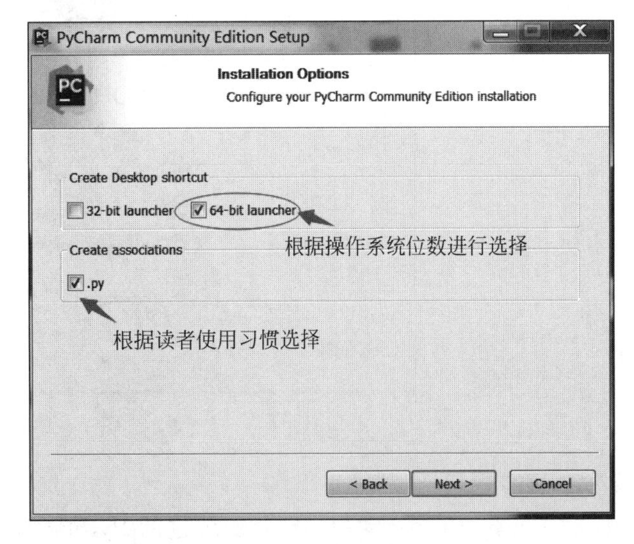

图　4-14

(5) 双击图 4-15 所示的图标，启动 PyCharm，第一次启动会出现如图 4-16 所示的界面，直接单击 OK 按钮即可。等待几秒后，会弹出如图 4-17 所示的对话框，根据读者喜好自行修改或者直接单击 Skip 按钮跳过此步。

图 4-15

(6) 首次使用，需要先进行新工程的创建，如图 4-18 所示的界面，单击 Create New Project 图标，出现如图 4-19 所示的界面。

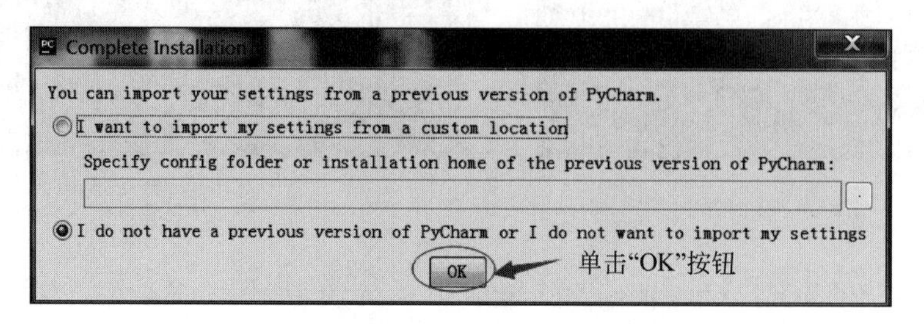

图 4-16

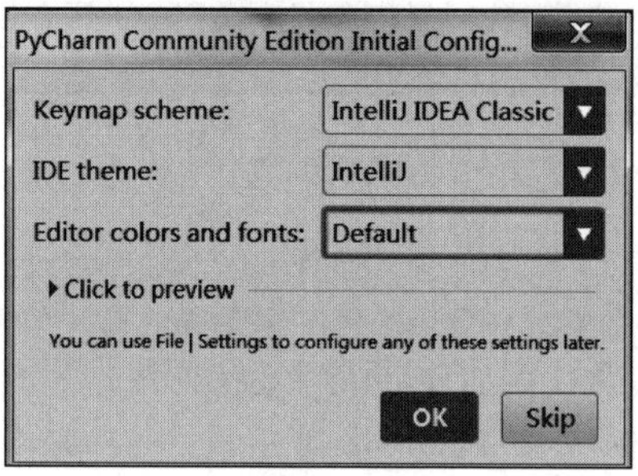

PyCharm Community Edition Initial Config...
Keymap scheme: IntelliJ IDEA Classic
IDE theme: IntelliJ
Editor colors and fonts: Default
Click to preview
You can use File | Settings to configure any of these settings later.
OK Skip

图 4-17

图 4-18

更多说明：

PyCharm 会自动扫描已经安装过的 Python 解释器，然后显示在 Interpreter 下拉选择框中，如果没有显示，读者可以自行单击后面的设置按钮 ，进行 Python 解释器的选择。

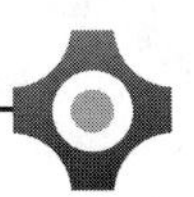

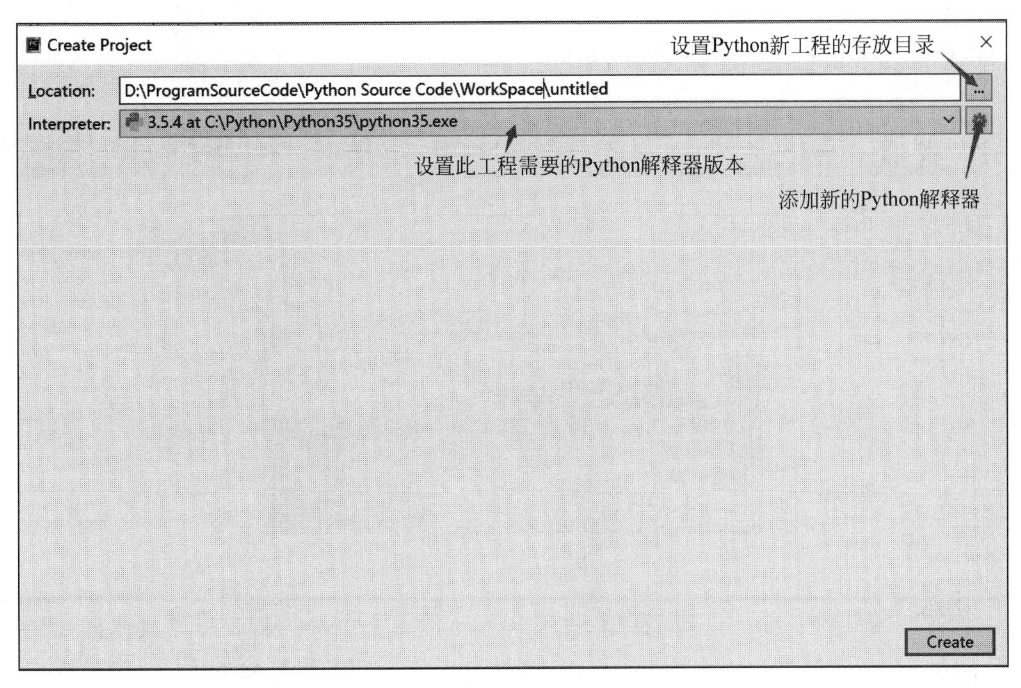

图　4-19

（7）在新工程编辑界面设置好 Python 新工程存放路径以及新工程需要的 Python 解释器，如图 4-19 所示，然后单击 Create 按钮，等待几分钟后，即可进入 PyCharm 集成开发环境界面，如图 4-20 所示。

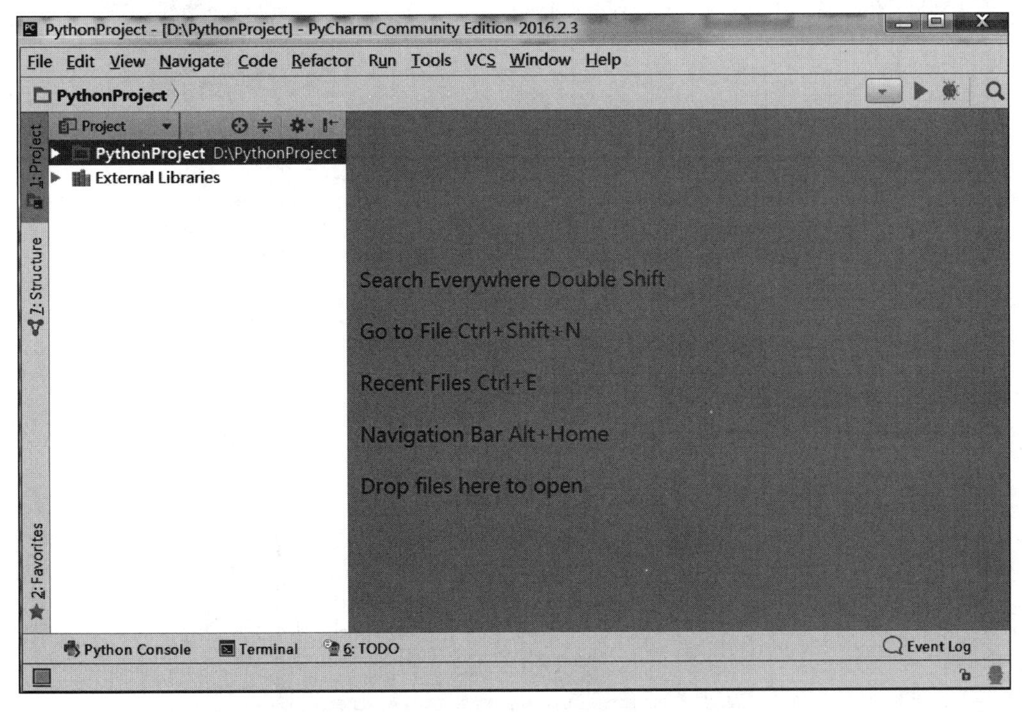

图　4-20

说明：

PyCharm 工具的默认程序文件或工程的编码为 UTF-8。

4.3 新建一个 Python 工程

具体操作步骤如下。

(1) 选择 File→New Project...命令，如图 4-21 所示。

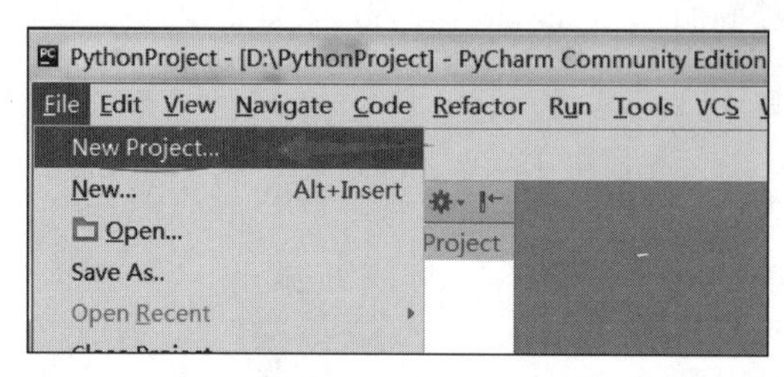

图 4-21

(2) 弹出新建 Python 工程对话框，在 Location 输入框中输入新工程存放路径，或者直接单击浏览按钮选择路径。在 Interpreter 选择框中选择新工程需要的 Python 解释器，或者单击后面的设置按钮添加新的 Python 解释器，单击 Create 按钮。

(3) 在工程名上右击，在弹出的菜单列表中选择 New→File(也可以选择创建目录或 Python 包等)，如图 4-22 所示。

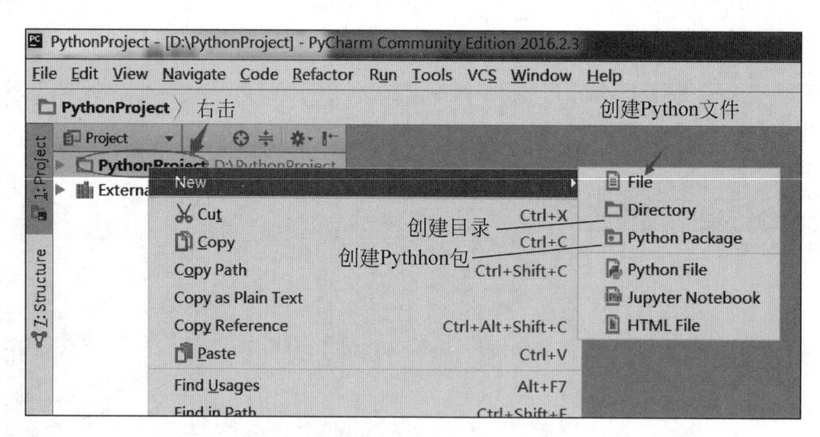

图 4-22

(4) 在弹出的文件编辑对话框中输入文件名，如 test. py，单击 OK 按钮，完成 Python 文件创建，如图 4-23 所示。

图 4-23

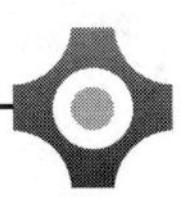

(5) 完成以上步骤以后，在新建的 Python 工程下会看到新创建的文件 test.py，单击选中这个文件，可以将光标切换到该文件编辑区域，插入一行 Hello World 代码 print("Hello World!")，如图 4-24 所示。

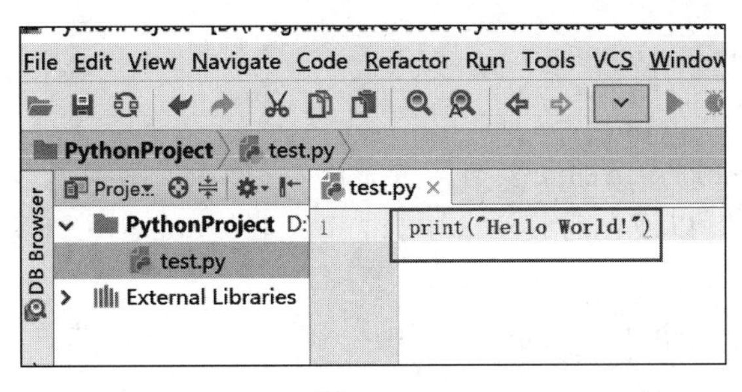

图 4-24

(6) 在 PyCharm 工具栏中，单击 Run，在弹出的菜单列表中单击 Run 命令，弹出一个运行文件选择界面，选择刚创建的 Python 文件，如图 4-25 所示。

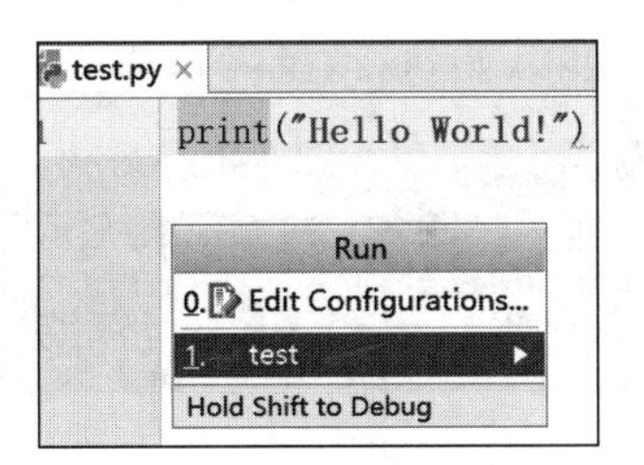

图 4-25

(7) 在 PyCharm 的 Console 窗口中可看到程序的输出结果为 Hello World!，如图 4-26 所示。

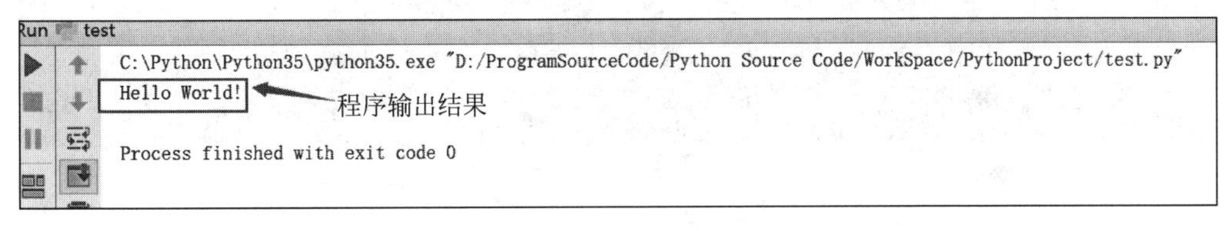

图 4-26

第5章 Selenium 3(WebDriver)的安装

本章主要讲解 WebDriver 的安装与配置方法，请读者按照本章节的内容安装和配置好 WebDriver 的运行环境，以便在后续章节中，讲解基于 WebDriver 实例时，能及时执行及查看结果。

本书以笔者撰稿时最新的 Selenium(3.0.2)版本为例，介绍其安装及使用方法。

5.1 在 Python 中安装 WebDriver

(1) 安装好 Python、pip 工具后，尝试直接在 CMD 下输入 pip install selenium，如图 5-1 所示，如果成功直接跳转到第(5)步开始执行，笔者选择的是 selenium==3.14.0 版本。

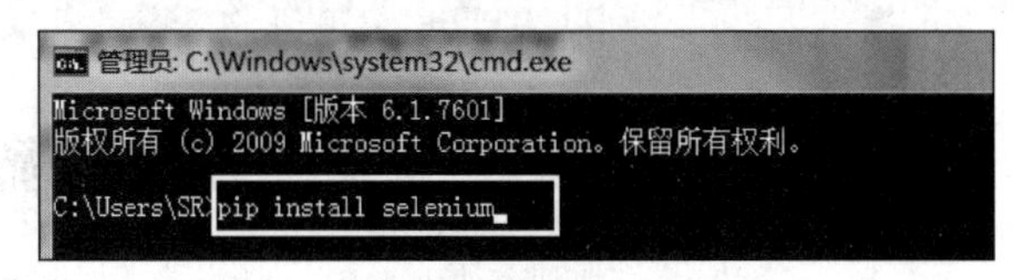

图 5-1

(2) 下载 Selenium 离线安装包，访问网址：https://pypi.python.org/pypi/selenium，选择扩展名为 gz 的源码包进行下载，如图 5-2 所示。

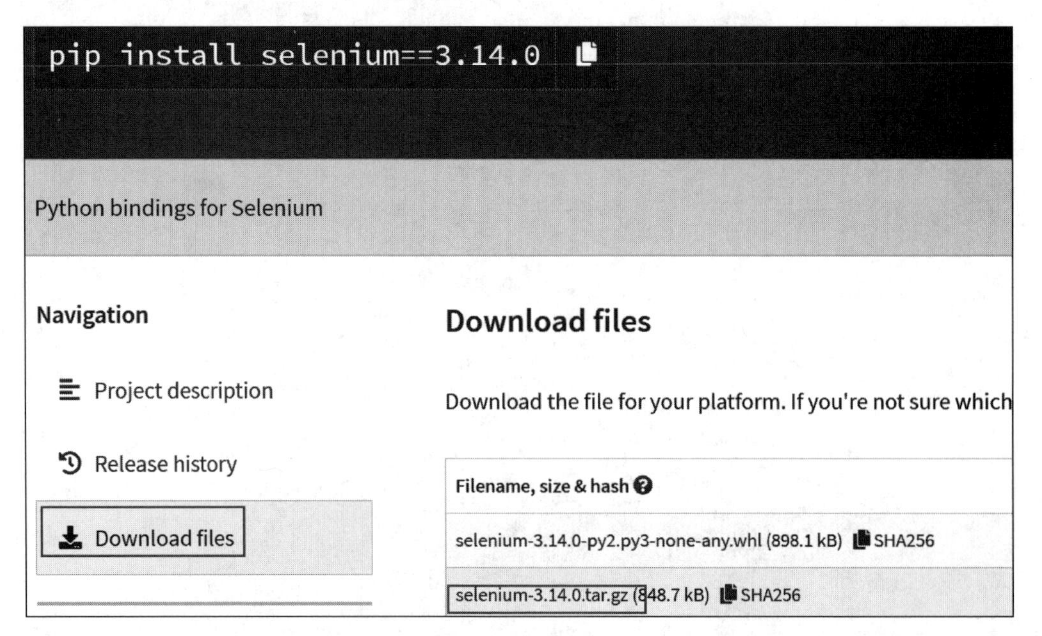

图 5-2

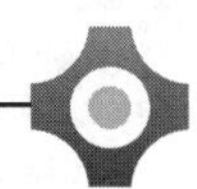

(3) 下载完以后会得到一个 selenium-3.14.0.tar.gz 离线安装文件，解压该文件到任意目录。CMD 下通过 cd 命令将当前的工作目录切换到 setup.py 文件所在的目录，如图 5-3 所示。

```
D:\>cd D:\selenium-3.14.0

D:\selenium-3.14.0>dir
 驱动器 D 中的卷是 Study
 卷的序列号是 6A05-836C

 D:\selenium-3.14.0 的目录

2018/12/22  15:57    <DIR>          .
2018/12/22  15:57    <DIR>          ..
2018/08/02  21:06            25,201 CHANGES
2017/03/08  20:00               915 MANIFEST.in
2018/08/02  21:22             7,737 PKG-INFO
2018/07/26  17:59             5,650 README.rst
2018/12/22  15:57    <DIR>          selenium
2018/12/22  15:57    <DIR>          selenium.egg-info
2018/08/02  21:22               209 setup.cfg
2018/08/02  21:18             3,220 setup.py
               6 个文件         42,932 字节
               4 个目录 85,147,713,536 可用字节

D:\selenium-3.14.0>
```

图　5-3

(4) 然后在 CMD 下执行 python setup.py install 命令进行安装，安装成功后的界面如图 5-4 所示。

```
Finished processing dependencies for selenium==3.14.0

D:\selenium-3.14.0>
```

图　5-4

(5) 安装成功后，CMD 下输入 python 回车，进入 Python 交互模式，执行 import selenium 命令，如果没有报错，说明 Selenium 已经安装成功，如图 5-5 所示。

```
C:\Python\Python35\python35.exe
Python 3.5.4 (v3.5.4:3f56838, Aug  8 2017, 02:17:05) [MSC v.1900 64 bit (AMD64)] on win32
Type "help", "copyright", "credits" or "license" for more information.
>>> import selenium
>>> _   <———— 没有报错，说明安装成功了
```

图　5-5

5.2　第一个 WebDriver 脚本

测试目标：

使用 Firefox 浏览器验证 WebDriver 是否可用。

测试用例步骤：

(1) 在 Firefox 浏览器中打开搜狗浏览器首页。

(2) 在搜索输入框中输入“光荣之路自动化测试”。

(3) 单击“搜索”按钮。

(4) 页面显示搜索结果。

环境准备：

（1）使用 Firefox 浏览器执行 Selenium 3 编写的自动化测试脚本时，需要从 https://github.com/mozilla/geckodriver/releases 网址根据读者操作系统类型及浏览器位数（64 位驱动兼容 32 位 Firefox 浏览器）下载对应的 WebDriver 操作 Firefox 浏览器的驱动程序，如图 5-6 所示。

▼ Assets 7

geckodriver-v0.26.0-linux32.tar.gz	2.22 MB
geckodriver-v0.26.0-linux64.tar.gz	2.28 MB
geckodriver-v0.26.0-macos.tar.gz	1.91 MB
geckodriver-v0.26.0-win32.zip	1.37 MB
geckodriver-v0.26.0-win64.zip	1.46 MB
Source code (zip)	
Source code (tar.gz)	

图 5-6

（2）下载并解压得到 geckodriver.exe 文件，将该文件保存到本地硬盘任意位置，比如 D:\下。

测试脚本程序：

```
from selenium import webdriver
import time

# 通过 executable_path 参数指明 Firefox 驱动文件所在路径
driver = webdriver.Firefox(executable_path = "c:\\geckodriver")
# driver = webdriver.Chrome(executable_path = "c:\\chromedriver")
# 打开搜狗首页
driver.get("http://www.sogou.com")
# 清空搜索输入框默认内容
driver.find_element_by_id("query").clear()
# 在搜索输入框中输入"光荣之路自动化测试"
driver.find_element_by_id("query").send_keys("光荣之路自动化测试")
# 单击"搜索"按钮
driver.find_element_by_id("stb").click()
# 等待 3 秒
time.sleep(3)
# 退出浏览器
driver.quit()
```

PyCharm 中执行该脚本，会看到程序自动启动浏览器，访问搜狗首页，并在搜索输入框中输入搜索关键内容“光荣之路自动化测试”，单击搜索按钮后展示搜索结果，3 秒后自动退出浏览器。

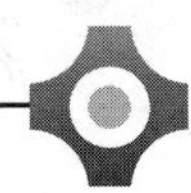

更多说明：

从 Selenium 3 版本开始，webdriver/firefox/webdriver.py 程序文件中的__init__.py 文件中，设置 executable_path="geckodriver"，而 Selenium 2 是 executable_path="wires"，所以使用 Selenium 3 编写的自动化测试脚本，使用 Firefox 浏览器测试时需要指明 Firefox 浏览器驱动程序 geckodriver.exe 文件所在路径。

问题及建议：

建议 Firefox 浏览器在安装时使用默认安装路径。如果使用了自定义安装路径，可能无法找到 Firefox.exe 文件来启动执行此测试脚本的 Firefox 浏览器。

报错 1：

```
Exception in thread "main" org.openqa.selenium.WebDriverException: Cannot find firefox binary
in PATH. Make sure firefox is installed.
```

解决办法：

在 driver = webdriver.Firefox(executable_path = "c:\geckodriver")这行代码前一行增加如下代码：

```
os.environ["webdriver.firefox.driver"] = "C:\Program Files (x86)\Mozilla Firefox\firefox.
exe"
```

其中"C:\ProgramFiles (x86)\Mozilla Firefox\firefox.exe"代表 firefox.exe 文件所在的路径，读者须根据自己机器上 firefox.exe 文件所在的路径进行修改。

报错 2：

```
WebDriverException: Message: Expected browser binary location, but unable to find binary in
default location, no 'moz:firefoxOptions.binary' capability provided, and no binary flag set on
the command line
```

解决方法：

脚本顶部导入 FirefoxBinary 模块：

```
from selenium.webdriver.firefox.firefox_binary import FirefoxBinary
```

修改启动浏览器的代码如下：

```
binary = FirefoxBinary('D:\\FirefoxPortable\\Firefox.exe')
driver = webdriver.Firefox(firefox_binary = binary,
    executable_path = r"c:\geckodriver")
```

5.3 各浏览器驱动的使用方法

Selenium 3 版本开始不再提供默认浏览器支持，所有浏览器都是通过各个浏览器官方提供相应的浏览器驱动进行支持，这使得运行在各种浏览器上的自动化测试更稳定。

除了本章 5.2 小节中直接通过 executable_path 参数指明支持对应浏览器的驱动程序文件方法外，读者还可以创建一个目录(比如 D:\driver\目录)，把不同浏览器的驱动文件均放到该目录中(比如：geckodriver.exe、chromedriver.exe、MicrosoftWebDriver.exe、

IEDriverServer.exe、operadriver.exe 等)，然后将该目录(比如 D:\driver\目录)添加到系统环境变量 path 中，WebDriver 在启动浏览器时，会自动到环境变量中设定的路径中寻找相应的驱动文件。

在本书中，笔者将采用 5.2 小节实例程序中介绍的添加浏览器驱动的方式，仅是为了更灵活。

第6章 pytest单元测试框架

Python 语言编写的 WebDriver 测试脚本通常使用单元测试框架来运行，所以了解单元测试框架的基本方法和单元测试框架的使用技巧是很必要的。

6.1 单元测试的定义

单元测试(Unit Testing)是指在计算机编程中，针对程序模块(软件设计的最小单位)来进行正确性检验的测试工作。

单元测试的特点如下：

- 程序单元是应用的最小可测试部件，通常基于类或者类的方法进行测试。
- 程序单元和其他单元是相互独立的。
- 单元测试的执行速度很快。
- 单元测试发现的问题，相对容易定位。
- 单元测试通常由开发人员来完成。
- 通过了解代码的实现逻辑进行测试，通常称之为白盒测试。

6.2 pytest 单元测试框架

pytest 是 Python 的第三方单元测试框架，与 Python 自带的 unittest 测试框架类似，但是比 unittest 框架使用起来更简洁，效率更高。

根据 pytest 官方网站介绍，它具有如下特点：

(1) 入门简单，非常容易上手，文档丰富，并且文档中有很多实例可以参考。

(2) 能够支持简单的单元测试和复杂的功能测试。

(3) 支持参数化。

(4) 执行测试过程中支持跳过某些测试，或者对某些预期失败的用例标记成失败。

(5) 支持执行失败的用例的重试。

(6) 支持运行由 nose、unittest 编写的测试用例。

(7) pytest 具有很多第三方插件，并且可以自定义扩展，比较好用的如 pytest-selenium(集成 selenium)、pytest-html(生成完美 HTML 测试报告)、pytest-rerunfailures(失败用例重复执行)、pytest-xdist(多 CPU 分发)等。

(8) 方便和 CI 工具集成，比如 Jenkins。

6.3 安装 pytest 测试框架

pip install pytest 或者 pip install -U pytest

测试是否安装成功,直接在 Python 中导入,如果未报错说明安装成功。

```
>>> import pytest
>>>
```

6.4 pytest 用例编写规则

使用 pytest 单元测试框架进行单元测试,必须遵循其用例编写规则。

(1) 测试文件以 test_开头或者以_test 结尾。

(2) 测试类以 Test 开头,并且不能带有__ init __方法。

(3) 测试函数以 test_开头。

(4) 断言使用基本的 assert 即可。

6.5 pytest 单元测试框架初体验

测试脚本文件 test_try.py 内容如下。

```
import pytest

def test_case_1():
    print(" --- test_case_1 --- ")
    assert 2 == 2

def test_case_2():
    print(" --- test_case_2 --- ")
    assert "s" in "str"
```

CMD 下或者 PyCharm 下进入该文件所在目录,执行命令 pytest test_try.py 或者 py.test test_try.py,执行后的结果如图 6-1 所示。

```
============================= test session starts =============================
platform win32 -- Python 3.7.0, pytest-5.3.2, py-1.8.0, pluggy-0.13.1
rootdir: D:\ProgramSourceCode\pytest-self
plugins: cov-2.8.1
collected 2 items

test_try.py ..                                                           [100%]
============================== 2 passed in 0.04s ==============================
```

图 6-1

6.6 如何执行 pytest 测试用例

在上一小节中,传递了测试文件给 pytest 或 py.test,其实 pytest 也提供了多种方法执行测试,下面通过一个实例来呈现,在同一个目录中创建如下两个脚本文件。

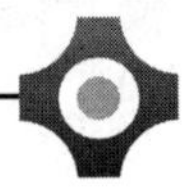

Calc. py 文件内容如下：

```
class Calc(object):
    @classmethod
    def add(cls, x, y, *d):
        # 加法计算
        result = x + y
        for i in d:
            result += i
        return result

    @classmethod
    def sub(cls, x, y, *d):
        # 减法计算
        result = x - y
        for i in d:
            result -= i
        return result

    @classmethod
    def mul(cls, x, y, *d):
        # 乘法计算
        result = x * y
        for i in d:
            result *= i
        return result

    @staticmethod
    def div(x, y, *d):
        # 除法计算
        if y != 0:
            result = x / y
        else:
            return -1
        for i in d:
            if i != 0:
                result /= i
            else:
                return -1
        return result
```

test_calc. py 文件内容如下：

```
from Calc import Calc
def test_add():
    assert Calc.add(1, 2, 3) == 6

def test_sub():
    assert Calc.sub(100, 20, 30) == 50

class TestCalc():
```

```
    def test_mul(self):
        assert Calc.mul(2, 3, 4) == 24

    def test_div(self):
        assert Calc.div(32, 8, 4) == 2
```

(1) pytest 或 py. test：默认执行当前目录及子目录下所有以 test 开头的. py 文件。

(2) pytest/py. test test_xx. py：执行指定的. py 文件。

执行 pytest test_calc. py，结果如图 6-2 所示。

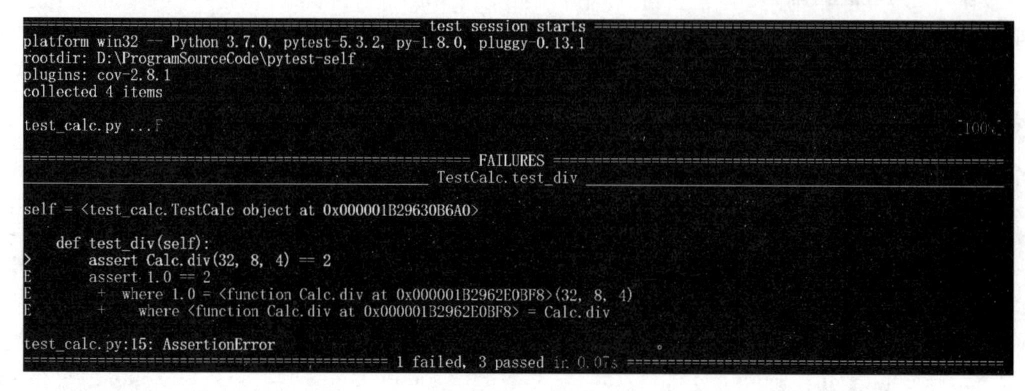

图 6-2

(3) pytest/py. test somepath：执行指定路径 somepath 下所有的以 test 开头的. py 文件。

(4) pytest/py. test -k stringexpr：会执行. py 文件中所有能匹配字符串 stringexpr 的方法或测试类中的方法。比如：py. test -k "TestClass and not method" -v test_x. py，只会执行 test_x. py 文件中 TestClass 类中方法名中不包含"method"字符串的方法。

执行 pytest -k "TestCalc and not div" -v test_calc. py 命令，结果如图 6-3 所示。

```
============================= test session starts =============================
platform win32 -- Python 3.7.0, pytest-5.3.2, py-1.8.0, pluggy-0.13.1 -- c:\python\python37\python3.exe
cachedir: .pytest_cache
rootdir: D:\ProgramSourceCode\pytest-self
plugins: cov-2.8.1
collected 4 items / 3 deselected / 1 selected

test_calc.py::TestCalc::test_mul PASSED                                  [100%]

======================= 1 passed, 3 deselected in 0.02s =======================
```

图 6-3

(5) pytest/py. test test_mod. py::test_func：显示指定执行 test_mod. py 文件中方法名为 test_func 的方法。

执行 py. test test_calc. py::test_add -v 命令，结果如图 6-4 所示。

```
============================= test session starts =============================
platform win32 -- Python 3.7.0, pytest-5.3.2, py-1.8.0, pluggy-0.13.1 -- c:\python\python37\python3.exe
cachedir: .pytest_cache
rootdir: D:\ProgramSourceCode\pytest-self
plugins: cov-2.8.1
collected 1 item

test_calc.py::test_add PASSED                                            [100%]

============================== 1 passed in 0.03s ==============================
```

图 6-4

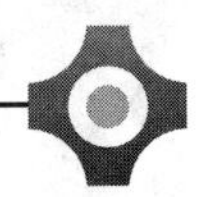

有时还需要配合 Console 参数，查看执行过程信息或调试测试脚本，常用参数如下：

(1) -v：打印出每个测试函数的执行结果，如 pytest test * . py -v。

(2) -q：只打印整体测试结果，如 pytest test * . py -q。

(3) -s：打印出函数中 print()函数输出结果。

(4) -x：在发生第一次错误或失败测试后立即退出。

(5) -h：查看帮助信息。

6.7　setup 和 teardown 函数

运行一次测试函数就会运行一次 setup 和 teardown，类似于 unittest 的 setup 和 teardown。

- setup，在测试函数或类之前执行，完成准备工作，例如数据库连接、准备测试数据、打开文件等。
- teardown，在测试函数或类之后执行，完成收尾工作，例如断开数据库连接、回收内存资源等。

实例代码：

```
class Test_ST():
    def setup(self):
        print("------setup------")

    def teardown(self):
        print("------teardown------")

    def test_001(self):
        print("test_001...")
        assert 1 == 1
```

执行 pytest test_t2. py -s，结果如图 6-5 所示。

```
=============================== test session starts ===============================
platform win32 -- Python 3.7.0, pytest-5.3.2, py-1.8.0, pluggy-0.13.1
rootdir: D:\ProgramSourceCode\pytest-self
plugins: cov-2.8.1
collected 1 item

test_t2.py ------setup------
test_001...
.------teardown------

=============================== 1 passed in 0.05s ===============================
```

图　6-5

6.8　失败重试

pytest 框架要实现重试功能需要安装插件，直接执行如下命令进行安装即可。

pip install pytest - rerunfailures 或 pip install　- U pytest - rerunfailures

命令语法：运行命令 pytest test * . py --reruns n --reruns-delay m，其中 n 表示重试次

数,m 表示失败后延迟多少秒后重试,可以不传。test_t2.py 文件的示例代码如下。

```
class Test_ST():
    def test_001(self):
        print("test_001...")
        assert 1 == 1

    def test_002(self):
        print("test_002...")
        assert False
```

执行 py.test test_t2.py -v --reruns 2,结果如图 6-6 所示。

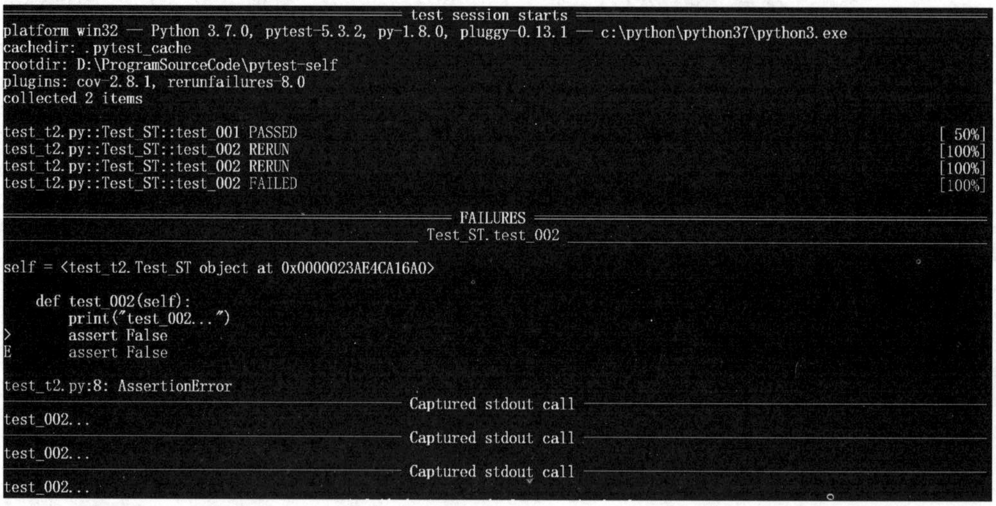

```
============================= test session starts =============================
platform win32 -- Python 3.7.0, pytest-5.3.2, py-1.8.0, pluggy-0.13.1 -- c:\python\python37\python3.exe
cachedir: .pytest_cache
rootdir: D:\ProgramSourceCode\pytest-self
plugins: cov-2.8.1, rerunfailures-8.0
collected 2 items

test_t2.py::Test_ST::test_001 PASSED                                     [ 50%]
test_t2.py::Test_ST::test_002 RERUN                                      [100%]
test_t2.py::Test_ST::test_002 RERUN                                      [100%]
test_t2.py::Test_ST::test_002 FAILED                                     [100%]

================================== FAILURES ===================================
_____________________________ Test_ST.test_002 ________________________________

self = <test_t2.Test_ST object at 0x0000023AE4CA16A0>

    def test_002(self):
        print("test_002...")
>       assert False
E       assert False

test_t2.py:8: AssertionError
---------------------------- Captured stdout call -----------------------------
test_002...
---------------------------- Captured stdout call -----------------------------
test_002...
---------------------------- Captured stdout call -----------------------------
test_002...
```

图 6-6

6.9 控制测试函数运行顺序

运行命令安装控制运行顺序的插件:pip install pytest-ordering。该插件通过提供的标记方法来控制函数的执行顺序,具体的使用方法如下。

(1) 使用@pytest.mark.run(order=n)标记被测试函数。

(2) 函数的运行顺序由 order 传入的参数决定,按 order 从小到大的顺序执行。

示例代码如下。

```
import pytest

class Test_Order():
    @pytest.mark.run(order = 3)
    def test_001(self):
        print("001...")
        assert True

    @pytest.mark.run(order = 2)
    def test_002(self):
```

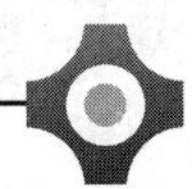

```
        print("002...")
        assert True

    @pytest.mark.run(order = 1)
    def test_003(self):
        print("003...")
        assert True
```

执行结果如图 6-7 所示。

```
============================== test session starts ==============================
platform win32 -- Python 3.7.0, pytest-5.3.2, py-1.8.0, pluggy-0.13.1
rootdir: D:\ProgramSourceCode\pytest-self
plugins: cov-2.8.1, ordering-0.6, rerunfailures-8.0
collected 3 items

test_t2.py 003...
.002...
.001...
.

=============================== 3 passed in 0.03s ===============================
```

图　6-7

6.10 生成 HTML 测试报告

pytest 也提供了生成测试报告的功能，但需要先安装插件，直接运行安装命令 pip install pytest-html 即可。

以本章的 6.9 小节的实例 test_t1. py 为例，运行测试命令 pytest test_t1. py -sv --html=report. html，该命令会在 test_t1. py 文件所在的目录中生成一个 report. html 文件，直接在浏览器查看，效果如图 6-8 所示。

report.html

Report generated on 18-Jan-2020 at 17:42:44 by pytest-html v2.0.1

Environment

JAVA_HOME	C:\Program Files\Java\jdk1.8.0_181
Packages	{'pytest': '5.3.2', 'py': '1.8.0', 'pluggy': '0.13.1'}
Platform	Windows-10-10.0.18362-SP0
Plugins	{'cov': '2.8.1', 'html': '2.0.1', 'metadata': '1.8.0', 'ordering': '0.6', 'rerunfailures': '8.0'}
Python	3.7.0

Summary

3 tests ran in 0.10 seconds.

(Un)check the boxes to filter the results.

3 passed, 0 skipped, 0 failed, 0 errors, 0 expected failures, 0 unexpected passes, 0 rerun

Results

Show all details / Hide all details

Result	Test	Duration	Links
Passed (show details)	test_t2.py::Test_Order::test_003	0.00	
Passed (show details)	test_t2.py::Test_Order::test_002	0.00	
Passed (show details)	test_t2.py::Test_Order::test_001	0.00	

图　6-8

除了 pytest-html 插件可以生成测试报告外，pytest 还有另外一个很好用的插件 Allure，有需要的读者可以自己下载来学习一下，这里就不介绍了。

6.11 通过配置文件配置要执行的测试用例

pytest 的配置文件名叫 pytest. ini，是 pytest 的主配置文件，里面的选项可以改变 pytest 的默认行为，通常放在测试目录下，命令行运行时会使用该配置文件中的配置选项。

(1) 配置 pytest 命令行运行参数,可添加多个命令参数,以空格分隔,所有参数均为插件包的参数。

```
[pytest]
addopts = -sx
```

(2) 配置测试搜索的路径。

```
[pytest]
testpaths = D:\scripts          # 表示D盘下的scripts目录,srcipts目录名可以自定义
```

(3) 配置测试搜索的文件名。

```
[pytest]
python_files = test_*.py        # 表示当前目录下的scripts目录下,以test_开头,.py结尾的所
                                # 有文件,文件名可自定义
```

(4) 配置测试的测试类名。

```
[pytest]
python_classes = Test_*
#表示运行当前目录下的scripts目录下,以test_开头,以.py结尾的所有文件中,以Test_开头的
类,名字可以自定义
```

(5) 配置测试的测试函数名。

```
[pytest]
python_functions = test_*
#表示运行当前目录下的scripts目录下,以test_开头,以.py结尾的所有文件中,以Test_开头的
类内,以test_开头的方法,名字可以自定义
```

以上配置项中参数名字固定不可更改,这并不是全部的可用参数,其他的参数可以通过 pytest --help 命令查看。

如下实例:

新建目录 pytest_ini_test,进入该目录,新建存放测试报告的目录 report,同时新建 pytest.ini 文件,具体内容如下。

```
[pytest]
addopts = -sv --html=./report/report.html
testpaths = ./scripts
python_files = test_*.py
python_classes = Test_*
python_functions = test_*
```

新建存放测试脚本的目录 scripts,并新建 test_case_1.py、test_case_2.py、test_case_3.py 文件,内容分别如下。

```
test_case_1.py
import pytest

def test_case1():
    print("test_case1")
```

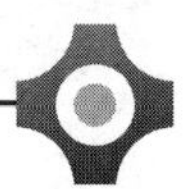

```
        assert True
test_case_2.py
import pytest

def test_case2():
    print("test_case2")
    assert True
test_case_3.py
import pytest

class Test_Case():
    def test_case_1(self):
        print("Test_Case->test_case_1")
        assert True

    def test_case_2(self):
        print("Test_Case->test_case_2")
        assert True

    def t_case_3(self):
        print("Test_Case->t_case_3")
        assert True
```

在 pytest.ini 文件所在的目录下执行命令 pytest，结果如图 6-9 所示。执行后会在 report 目录中生成 report.html 测试报告，具体内容可以自行用浏览器查看。

```
============================= test session starts =============================
platform win32 -- Python 3.7.0, pytest-5.3.2, py-1.8.0, pluggy-0.13.1 -- c:\python\python37\python3.exe
cachedir: .pytest_cache
metadata: {'Python': '3.7.0', 'Platform': 'Windows-10-10.0.18362-SP0', 'Packages': {'pytest': '5.3.2', 'py': '1.8.0', 'pluggy': '0.13.1'}, 'Plugins': {'cov': '2.8.1', 'html': '2.0.1', 'metadata': '1.8.0', 'ordering': '0.6', 'rerunfailures': '8.0'}, 'JAVA_HOME': 'C:\\Program Files\\Java\\jdk1.8.0_161'}
rootdir: D:\ProgramSourceCode\pytest-self\pytest_ini, inifile: pytest.ini, testpaths: ./scripts
plugins: cov-2.8.1, html-2.0.1, metadata-1.8.0, ordering-0.6, rerunfailures-8.0
collected 4 items

scripts/test_case_2.py::test_case2 test_case2
PASSED
scripts/test_case_3.py::Test_Case::test_case_1 Test_Case->test_case_1
PASSED
scripts/test_case_3.py::Test_Case::test_case_2 Test_Case->test_case_2
PASSED
scripts/test_csae_1.py::test_case1 test_case1
PASSED

------------ generated html file: file://D:\ProgramSourceCode\pytest-self\pytest_ini\report\report.html ------------
============================== 4 passed in 0.15s ==============================
```

图 6-9

6.12 捕获异常

在测试过程中，经常需要测试是否如期抛出预期的异常，以确定异常处理模块生效。在 pytest 中使用 pytest.raises()方法进行异常捕获。

get_catch.py 文件内容如下。

```
import pytest


def test_raises():
```

```
    with pytest.raises(ZeroDivisionError) as e:
        2/0
    exec_msg = e.value.args[0]
    assert exec_msg == 'division by zero'
```

执行后的结果如下。

```
============================ test session starts ============================
platform win32 -- Python 3.7.0, pytest-5.3.2, py-1.8.0, pluggy-0.13.1
rootdir: D:\ProgramSourceCode\pytest-self
plugins: cov-2.8.1, html-2.0.1, metadata-1.8.0, ordering-0.6, rerunfailures-8.0
collected 1 item

get_catch.py                                                           [100%]
============================ 1 passed in 0.03s ============================
```

6.13 标记函数

在实际工作当中，要编写的自动化用例一般都是一个大型项目的所有测试用例，不会都放在一个 py 文件里，有可能还会根据模块划分测试用例，但在这些测试用例中，可能存在未开发完成的测试用例，或者并不想每次运行时执行所有的测试用例，但代码还想保留，为解决这些问题，pytest 给我们提供了标记函数，通过提供的标记或自定义的标记，实现灵活控制测试用例的运行。

6.13.1 过滤测试函数

默认情况下，pytest 会递归查找当前目录下所有以 test 开头或结尾的 Python 脚本，并执行文件内的所有以 test 开头的函数和类。但由于某些原因，比如 某些方法的功能尚未开发完成无法实时执行，但又不想删除代码，想跳过某些方法的执行，我们只想执行指定并且开发完成的测试函数，为了解决这类问题，pytest 给我们提供了几种解决方式。

第一种，执行测试命令时，通过::显示指定要执行的函数。比如 pytest get_catch.py::test_raises，这种方法只能指定一个测试函数，当要进行批量测试时，这种方法就不适用了。

第二种，使用-k 参数，通过模糊匹配。比如 py.test -k "TestClass and not method" -v test_x.py，这种方法可以批量操作，但需要所有测试的函数名包含相同的模式，也不够灵活。

第三种，就是我们这节要讲的使用 pytest.mark 标记函数，测试时使用 -m 选择被标记的测试函数，就可以实现灵活执行测试函数了。

新建 test_with_mark.py 文件，内容如下。

```
import pytest

@pytest.mark.finished
def test_func1():
    print("marked finished")
    assert 1 == 1
```

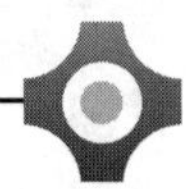

```
@pytest.mark.finished
def test_add():
    print("add...")
    assert 2 == (1+1)

@pytest.mark.unfinished
def test_func2():
    print("marked unfinished")
    with pytest.raises(AssertionError) as e:
        assert 1 != 1
    exec_msg = e.value.args[0]
    print(exec_msg)

if __name__ == '__main__':
    pytest.main(["-s","test_with_mark.py","-m=finished"])
```

pytest 提供了 main()方法代替 CMD 中的命令执行方式，将要传的命令通过列表的方式传给 main()方法，实例中的运行命令等价于 pytest -m finished test_with_mark.py -s，表示指定只运行通过 pytest.mark.finished 标记的方法，结果如下。

```
====== 2 passed, 1 deselected, 2 warnings in 0.08s ======
```

6.13.2　跳过测试

上一节提到使用 pytest 的标记过滤测试函数，是按正向思维，只要通过标记指定要测试的就可以了。但有时候我们也可以进行反向的操作去解决这个问题，即通过标记指定要跳过的测试。pytest 还支持使用 pytest.mark.skipif 为测试函数指定被忽略的条件。修改 6.13.1 小节代码如下。

```
import pytest

def test_func1():
    print("marked finished")
    assert 1 == 1

def test_add():
    print("add...")
    assert 2 == (1+1)

@pytest.mark.skip(reason="故意的")
def test_func2():
    print("marked unfinished")
    with pytest.raises(AssertionError) as e:
        assert 1 != 1
    exec_msg = e.value.args[0]
    print(exec_msg)


if __name__ == '__main__':
    # 添加-rs 控制台打印出跳过原因
```

```
    pytest.main(["-rs", "test_with_mark.py"])
```

运行 python test_with_mark.py,执行结果如下。

```
============================ test session starts ============================
platform win32 -- Python 3.7.0, pytest-5.3.2, py-1.8.0, pluggy-0.13.1
rootdir: D:\ProgramSourceCode\pytest-self
plugins: cov-2.8.1, html-2.0.1, metadata-1.8.0, ordering-0.6, rerunfailures-8.0
collected 3 items

test_with_mark.py ..s                                                [100%]

=========================== short test summary info ==========================
SKIPPED [1] test_with_mark.py:11: 故意的
========================== 2 passed, 1 skipped in 0.10s ======================
```

上面给的实例是无条件跳过用例的执行,下面我们看一个有条件跳过执行,修改上例代码如下。

```
import pytest
import sys

class TestMark():
    def setup(self):
        self.num = 0

    def test_func1(self):
        print("marked finished")
        assert 1 == 1

    @pytest.mark.run(order=1)
    def test_add(self):
        print("add...")
        self.num = self.num + 1
        assert 1 == self.num

    # 当条件为 True 时,跳过测试函数
    @pytest.mark.skipif(sys.version_info > (3,7), reason="requires python3.6 or lower")
    @pytest.mark.run(order=2)
    def test_func2(self):
        print("marked unfinished")
        with pytest.raises(AssertionError) as e:
            assert 1 != 1
        exec_msg = e.value.args[0]
        print(exec_msg)

if __name__ == '__main__':
    # 添加-rs 控制台打印出跳过原因
    pytest.main(["-rs", "test_with_mark.py"])
```

运行 python test_with_mark.py 命令,执行结果如下。

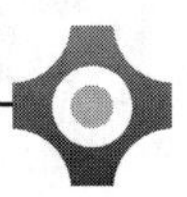

```
======================== test session starts ================================
platform win32 -- Python 3.7.0, pytest-5.3.2, py-1.8.0, pluggy-0.13.1
rootdir: D:\ProgramSourceCode\pytest-self
plugins: cov-2.8.1, html-2.0.1, metadata-1.8.0, ordering-0.6, rerunfailures-8.0
collected 3 items

test_with_mark.py .s.                                                    [100%]

============================ short test summary info ========================
SKIPPED [1] test_with_mark.py:19: requires python3.6 or lower
========================== 2 passed, 1 skipped in 0.09s =====================
```

6.13.3　预期失败

如果我们事先知道测试函数会执行失败，但又不想直接跳过，而是希望显示预期失败的提示。pytest 使用 pytest.mark.xfail 实现预期失败的功能。如下示例。

```
import pytest
import sys

class Test_xfail:
    @pytest.mark.xfail(sys.platform == "win32",reason = "only for linux")
    def test_a(self):
        assert 2 == 1

    # 条件为 boolean 值,True(标记失败)/False(正常执行)
    @pytest.mark.xfail(sys.platform == "win32",reason = "only for linux")
    def test_b(self):
        assert True

if __name__ == "__main__":
    pytest.main([" -v","test_xfail.py"])
```

执行结果如下。

```
================================ test session starts ======================
collected 2 items

test_xfail.py::Test_xfail::test_a XFAIL                                  [ 50%]
test_xfail.py::Test_xfail::test_b XPASS                                  [100%]

============================= 1 xfailed, 1 xpassed in 0.12s =================
```

说明：XFAIL(同 x)，表示预期失败，实际执行也失败，XPASS(同 X)，预期失败，实际执行没有失败。

6.13.4　参数化

测试时，通常需要给测试函数传递一组或多组参数。比如测试登录功能时，我们需要模拟各种千奇百怪的账号和密码。当然，我们可以把这些需要的参数写在测试函数内部然后进行遍历，当某一组参数导致断言失败，测试就会终止。此时就需要通过捕获异常，来保证所有参数完整被执行，但要分析测试结果就需要做不少额外的工作。pytest 已经提供了解决这类问题的方法，通过使用@pytest. mark. parametrize(argnames, argvalues)标记方法，可以实现向测试函数传递一组或多组参数，同时完成测试结果统计。

```
@pytest.mark.parametrize(argnames,argvalues,indirect=False,ids=None,scope=None)
```

常用参数说明：

```
argnames:参数名称
argvalues:参数对应的值,必须 list 类型
```

（1）单个参数时，示例代码如下。

```
import pytest

class TestParam:
    @pytest.mark.parametrize('name', ["lily", "lisa", "lucy"])
    def test_p1(self, name):
        # 要传参数的名称,且和参数化中定义的一致
        print("name:", name)
        assert name != "haha"

if __name__ == "__main__":
    pytest.main(["test_parametrize.py", "-s"])
```

执行结果如下所示。

```
================================= test session starts =======================
collected 3 items

test_parametrize.py name: lily
.name: lisa
.name: lucy
.

================================= 3 passed in 0.06s ==========================
```

（2）多个参数时，示例代码如下所示。

```
import pytest

def login_params():
    data = [
        ('lily','lily-test'),
        ('lucy','lucy-test'),
        ('anne','anne-test')
    ]
    return data

class TestParam:
    @pytest.mark.parametrize("username,password",login_params())
```

```
    def test_login(self,username,password):
        print("用户名:%s,密码:%s,登录成功!" %(username,password))
        assert True

if __name__ == "__main__":
    pytest.main(["test_parametrize.py", "-s"])
```

执行结果如下所示。

```
============================== test session starts ========================
collected 3 items

test_parametrize.py 用户名:lily,密码:lily-test,登录成功!
.用户名:lucy,密码:lucy-test,登录成功!
.用户名:anne,密码:anne-test,登录成功!
.

================================ 3 passed in 0.06s ========================
```

(3) 参数组合。

示例代码如下。

```
import pytest

@pytest.mark.parametrize("x", [1, 2])
@pytest.mark.parametrize("y", [3, 3])
def test_foo(x, y):
    print("测试数据组合:x->%s, y->%s" % (x, y))
    with pytest.raises(AssertionError) as e:
        assert x + y == 4
    print(e.value.args[0])

if __name__ == "__main__":
    pytest.main(["test_parametrize.py","-s"])
```

测试数据组合测试结果情况:

```
x->1, y->3 passed
x->2, y->3 failed
x->1, y->3 passed
x->2, y->3 failed
```

6.13.5　超时时间

pytest 支持给测试函数设定超时时间,使用前需要安装支持的插件,安装命令 pip3 install pytest-timeout。

标记超时时间方法原型 pytest.mark.timeout(timeout=0, method=DEFAULT_METHOD)

参数说明:

timeout:超时时间,单位秒,默认为 0。

method：超时方法，两个可选值 thread 和 signal。pytest-timeout 插件会根据平台选择最合适的方法，但有时并不完全可行，可能需要显式地指定特定的超时方法。

thread 选项值是最可靠、最方便的方法。它也是不支持 signal 方法的系统的默认值。对于每个测试项，pytest-timeout 插件将启动一个计时器线程，该线程将在指定超时后终止整个进程。当一个测试项完成时，这个计时器线程将被取消，测试运行将继续。这种方法的缺点是，运行每个测试的开销相对较大，而且测试运行没有完成。这也意味着 py. test 的其他特性，如 JUnit XML 输出或 fixture teardown，将不能正常工作。第二个问题可以通过使用 pytest-xdist 插件的--boxed 选项来缓解。这种方法的好处是它总是有效的。此外，它还将通过将应用程序中所有线程的堆栈打印到 stderr 来提供调试信息。

signal 选项值，如果系统支持 SIGALRM 信号，则默认使用 signal 方法。此方法在测试项启动时调度警报，并在完成时取消警报。如果在测试期间警报过期，则信号处理程序将转储任何其他运行到 stderr 的线程的堆栈，并使用 pytest. fail()中断测试。这种方法的优点是 py. test 过程没有终止，测试运行可以正常完成。这个方法要注意的主要问题是它可能会干扰被测试的代码。如果被测试的代码本身使用的 SIGALRM，那么就会出错，此时将不得不选择 thread 方法。

示例代码如下。

```
import pytest
import time

@pytest.mark.timeout(timeout = 2, method = 'thread')
def test_foo():
    time.sleep(3)
    print("timeout")
```

也可以通过命令行参数--timeout＝2 设定所有用例超时时间。

6.13.6 失败重跑

在本章的 6.8 小节中提到的失败重跑方法是通过命令行参数设定的，对所有的测试函数生效，本小节将通过使用 pytest. mark. flaky (reruns, reruns_delay)标记方法实现的失败重跑，从而实现针对某些测试函数实现灵活的失败后重跑设置。示例代码如下。

```
import pytest

class TestReruns():
    # 如果失败则延迟 1s 后重跑
    # 最多重跑 2 次, reruns_delay 可以不传
    @pytest.mark.flaky(reruns = 2, reruns_delay = 1)
    def test_001(self):
        print("test_001...")
        assert 1 == 1

    @pytest.mark.flaky(reruns = 2, reruns_delay = 1)
    def test_002(self):
        print("test_002...")
        assert 1 == 2
```

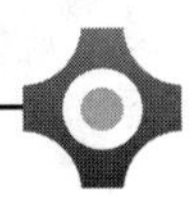

执行后的部分结果如下所示。

```
============================== test session starts ==========================
test_reruns.py test_001...
.test_002...
Rtest_002...
Rtest_002...
F

========================== 1 failed, 1 passed, 2 rerun in 2.19s =================
```

从执行结果可以看出，test_002 方法在执行失败后，重试了两次。

6.13.7　自定义标记

pytest 支持自定义标记，自定义标记可以把一个项目划分成多个模块，然后指定模块名称执行。比如一个 Web 项目的测试用例，可以分成网页模块、Android 模块及 iOS 模块，测试时只需要根据自定义的标记进行分模块执行测试用例即可。

第一步，注册标签名。新建测试目录 custom_mark，在目录中新建 pytest.ini 配置文件，在该文件中注册 3 个自定义标签 webtest、android 以及 ios，具体内容如下。

```
[pytest]
markers =
    # tag name: tag description,
    webtest: this is a web tag name.
    android: this is a android tag name.
    ios: this is a ios tag name.
```

第二步，在测试用例或测试类中给用例打上已注册的自定义标签。标签格式@pytest.mark.已注册标签名。在 custom_mark 目录中新建 test_custom_mark.py 文件，具体内容如下。

```
import pytest

@pytest.mark.webtest
def test_login_web():
    print("web test.")

@pytest.mark.android
def test_login_android():
    print("android test.")

@pytest.mark.ios
def test_login_ios():
    print("ios test.")
```

运行时，根据用例标签过滤参数(-m 标签名)过滤执行满足条件的测试用例即可。比如 pytest -m webtest test_custom_mark.py -s，只会执行 custom_mark 目录下所有通过@pytest.mark.webtest 标签标记过的函数或类。执行命令 pytest -m webtest test_custom_ mark. py -s -v,结果如下所示。

```
===================== test session starts =========================
collected 4 items / 2 deselected / 2 selected

test_custom_mark.py::test_login_web web test.
PASSED
test_custom_mark.py::Test_tt::test_try try...
PASSED

================= 2 passed, 2 deselected in 0.09s ==================
```

6.14 固件

fixture(固件)是 pytest 单元测试框架中特有的功能,是一些函数通过 pytest. fixture 标记,定义在函数前面,使测试能够可靠、重复地执行。比如测试前数据库的连接或测试后断开数据库连接。和 unitest 单元测试框架中的 setup、teardown 方法类似,但又有很大的改进。比如 fixture 可以仅在执行某一个或某几个特定用例执行前调用。

(1) fixture 具有明确的名称,在测试函数、模块、类或整个项目中声明时会被激活。

(2) fixture 是基于模块化的方式,因为每个 fixture 的名称都会触发一个 fixture 函数,它自身也可以调用其他的 fixture。

(3) fixture 管理从简单的单元扩展到复杂的功能测试,允许根据配置和组件选项对 fixture 和测试进行参数化,或者在跨功能、类、模块或整个测试会话范围内重复使用 fixture。pytest. fixture()函数原型定义以及常用参数说明如下。

```
fixture(callable_or_scope = None, * args, scope = 'function', params = None, autouse = False, ids =
None, name = None)
```

scope:有四个级别参数"function"(默认)、"class"、"module"、"session"。

scope='function',函数级别的 fixture,每个测试函数只运行一次。配置代码在测试用例运行之前运行,销毁代码在测试用例运行之后执行。function 是 fixture 的默认值参数。

scope='class',类级别的 fixture,每个测试类只运行一次,不管测试类中有多少个类方法都可以共享这个 fixture,并且只在第一个方法调用时执行。

scope='module',模块级别的 fixture,每个模块只运行一次,不管模块里有多少个测试函数,类方法或其他 fixture,都共享这一个 fixture,比如一个. py 文件调用一次。

scope='session',会话级别的 fixture,每次会话只运行一次。一次 pytest 会话中的所有测试函数、方法都可以共享这个 fixture,比如多个文件调用一次,可以跨. py 文件。

params:一个可选参数列表,供调用标记方法的函数使用,它将导致对 fixture 函数和使用它的所有测试的多次调用。当前参数在 request. param 中可用。

autouse:如果为 True,则对所有可以看到它的测试自动使用 fixture。如果为 False(默认值),则需要显式声明函数参数值或使用 usefixture 装饰器。

ids:每个参数对应的字符串 id 的列表,因此它们是测试 id 的一部分。如果没有提供 id,它们将从 params 自动生成。

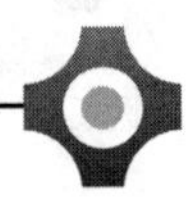

name：fixture 的名称。这默认为装饰函数的名称。如果 fixture 在定义它的统一模块中使用，则该 fixture 的函数名将被请求该 fixture 的函数名隐藏，解决这个问题的一种方法是将装饰的函数命令为"fixture_<fixturename>"，然后使用"@pytest.fixture(name='<fixturename>')"。

下面看一个固件的简单使用示例。

```
import pytest

@pytest.fixture(scope = 'function')
def login():
    print("测前登录")

def test_1():
    print('运行测试用例 1')

def test_2(login):
    print('运行测试用例 2')

if __name__ == "__main__":
    pytest.main(['test_fixture.py','-s'])
```

运行结果如下。

```
========================== test session starts ==========================

test_fixture.py 运行测试用例 1
.测前登录
运行测试用例 2
.

========================== 2 passed in 0.03s ==========================
```

在特性的函数请求中，高范围(如会话)的 fixture 比低范围 fixture(如函数或类)先被实例化。相同范围内的固定装置的相对顺序遵循测试功能中声明的顺序，并尊重固定装置之间的依赖关系。示例如下。

```
import pytest

@pytest.fixture(scope = "module")
def m():
    print("m...module...")

@pytest.fixture(scope = "session")
def s():
    print("\ns...session....")

@pytest.fixture
def f1():
    print("f1...function...")

@pytest.fixture
```

```
def f2():
    print("f2...function...")

def test_foo(f1, m, f2, s):
    #分别执行上面的 fixture
    f1 = f1
    m = m
    f2 = f2
    s = s
```

通过添加-s -v 参数执行结果如下所示。

```
========================== test session starts =============================
collected 1 item

test_fixture.py::test_foo
s...session....
m...module...
f1...function...
f2...function...
PASSED

============================== 1 passed in 0.07s =========================
```

6.14.1 作为参数引用

示例代码如下。

```
import requests
import pytest

@pytest.fixture
def baidu_response():
    print("... 执行 fixture ...")
    return requests.get("http://www.baidu.com")

# 将 fixture 作为函数参数
def test_baidu(baidu_response):
    response = baidu_response
    # 断言响应状态码
    assert response.status_code == 200

if __name__ == "__main__":
    pytest.main(['test_fixture.py','-s'])
```

执行结果如下。

```
============================== test session starts ========================
collected 1 item

test_fixture.py 执行 fixture
.

============================== 1 passed in 0.10s ===========================
```

6.14.2　作为函数引用

示例如下。

```
import pytest
# fixture 作为函数引用

@pytest.fixture()
def before():
    print("run before()...")

# 每个测试方法执行前执行,不包括 setup、teardown
@pytest.mark.usefixtures("before")
class Test_before():
    def setup(self):
        print("run setup()...")

    def test_a(self):
        print("run test_a()...")

    def test_b(self):
        print("run test_b()...")

    def teardown(self):
        print("run teardown()...")

# 函数执行前执行
@pytest.mark.usefixtures("before")
def test_c():
    print("run test_c()...")

if __name__ == "__main__":
    pytest.main(['test_fixture.py','-s'])
```

运行结果如下。

```
============================ test session starts ============================
collected 3 items

run before()...
run setup()...
run test_a()...
.run teardown()...
run before()...
run setup()...)
run test_b()...
.run teardown()...
run before()...
run test_c()...
.

============================ 3 passed in 0.07s ============================
```

从结果可以看出，在执行类中的方法时，只调用了一次定义的固件方法，但 setup()和 teardown()方法在每个类方法执行时都调用了。

6.14.3 设置自动使用 fixture

示例如下。

```
import pytest

@pytest.fixture(autouse = True)
def before():
    print("run before()...")

class Test_before():
    def setup(self):
        print("run setup()...")

    def test_a(self):
        print("run test_a()...")

    def test_b(self):
        print("run test_b()...")

    def teardown(self):
        print("run teardown()...")

def test_c():
    print("run test_c()...")

if __name__ == "__main__":
    pytest.main(['test_fixture.py','-s'])
```

运行结果如下。

```
============================ test session starts ============================
collected 3 items

run before()...
run setup()...
run test_a()...
.run teardown()...
run before()...)
run setup()...
run test_b()...
.run teardown()...
run before()...)
run test_c()...
.

================================ 3 passed in 0.06s ==========================
```

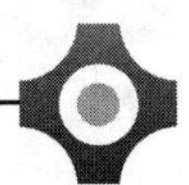

在其他范围内使用自动 fixture，需要注意如下几点。

(1) 自动 fixture 遵循 scope=关键字参数，如果一个自动 fixture 的 scope='session'，它将只运行一次，不管它在哪里定义。scope='class'表示每个类会运行一次，遵循 scope 参数取值的运行规则。

(2) 如果在一个测试模块中定义一个自动 fixture，所有的测试函数都会自动使用它。

(3) 如果在 conftest.py 文件中定义了自动 fixture，那么在其目录下的所有测试模块中的所有测试都将调用 fixture。

(4) 最后，要小心使用自动 fixture，如果我们在插件中定义了一个自动 fixture，它将在插件安装的所有项目中的所有测试中被调用。例如，在 pytest.ini 文件中，这样一个全局性的 fixture 应该全局适配，避免无用的导入或计算。规范的方法是将 fixture 的定义放在 conftest.py 文件而不使用自动运行。

6.14.4 设置作用域为 function

每个函数、类中的方法都将运行一次。示例代码如下。

```
import pytest

# 如果不设置自动运行将不会被调用
@pytest.fixture(scope = "function", autouse = True)
def before():
    print("run before()...")

class Test_before():
    def test_a(self):
        print("run test_a()...")

    def test_b(self):
        print("run test_b()...")

def test_c():
    print("run test_c()...")

if __name__ == "__main__":
    pytest.main(['test_fixture.py','-s'])
```

运行结果如下。

```
============================== test session starts ========================
collected 3 items

run before()...
run test_a()...
.run before()...
run test_b()...
.run before()...
run test_c()...
.

================================== 3 passed in 0.05s ======================
```

6.14.5 设置作用域为 class

一个类只运行一次。示例代码如下。

```
import pytest

# 如果不设置自动运行将不会被调用
@pytest.fixture(scope="class", autouse=True)
def before():
    print("run before()...")

class Test_before():
    def test_a(self):
        print("run test_a()...")

    def test_b(self):
        print("run test_b()...")

def test_c():
    print("run test_c()...")

if __name__ == "__main__":
    pytest.main(['test_fixture.py','-s'])
```

运行结果如下。

```
========================== test session starts ==============================
collected 3 items

run before()...
run test_a()...
.run test_b()...
.run before()...
run test_c()...
.

================================ 3 passed in 0.06s ========================
```

6.14.6 设置作用域为 module

一个模块运行一次。示例代码如下。

```
import pytest

# 如果不设置自动运行将不会被调用
@pytest.fixture(scope="module",autouse=True)
def before():
    print("run before()...")

class Test_before():
    def test_a(self):
```

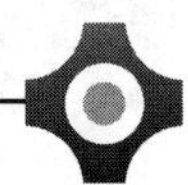

```
        print("run test_a()...")

    def test_b(self):
        print("run test_b()...")

def test_c():
    print("run test_c()...")

if __name__ == "__main__":
    pytest.main(['test_fixture.py','-s'])
```

运行结果如下。

```
========================== test session starts ==========================
collected 3 items

run before()...
run test_a()...
.run test_b()...
.run test_c()...
.

================================ 3 passed in 0.06s ======================
```

从结果可以看出，整个.py 文件只调用了一次。

6.14.7　设置作用域为 session

声明为 session 级别的 fixture，将会在每个测试类中调用一次，所以使用需谨慎。在同一个目录中新建 conftest.py 和 test_session.py 文件，内容分别如下。

conftest.py 内容如下。

```
import pytest

@pytest.fixture(scope="session")
def login():
    print("\n--购物前先登录")
    yield
    print(" --购物后退出登录")
```

test_session.py 文件内容如下。

```
def test_add_shopping_cart(login):
    print('用例 1,登录后执行添加购物车操作')

def test_search():
    print('用例 2,不登录操作查询')

def test_pay(login):
    print("用例 3,登录后执行支付操作")
```

将工作目录切到该目录下，执行 pytest -s -v 命令，输入结果如下所示。

```
=============================== test session starts ==========================
collected 3 items

test_session.py::test_add_shopping_cart
--购物前先登录
用例 1,登录后执行添加购物车操作
PASSED
test_session.py::test_search 用例 2,不登录操作查询
PASSED
test_session.py::test_pay 用例 3,登录后执行支付操作
PASSED--购物后退出登录

================================= 3 passed in 0.06s ==========================
```

将 fixture 函数放入单独的 conftest.py 文件中,以便目录中多个测试模块的测试可以访问 fixture 函数,使用时不需要导入,pytest 会自动发现它。fixture 函数的发现规则从测试类开始,然后是测试模块,再然后是 conftest.py 文件,最后是内置插件和第三方插件。将在类、模块或会话中跨测试共享一个 fixture 示例。

使用 conftest.py 的规则:

(1) conftest.py 这个文件名是固定的,不可以更改。

(2) conftest.py 与运行用例在同一个目录下,并且该目录中必须有__init__.py 文件。

(3) 使用时无须导入 conftest.py,会自动寻找,所有同目录测试文件运行前都会执行 conftest.py 文件。

(4) conftest.py 文件的作用域是当前包内(包括子包);如果函数被调用,会优先从当前测试类中寻找,然后是模块(.py 文件)中,接着是当前包中寻找(conftest.py 中),如果没有再找父包直至根目录;如果要声明全局的 conftest.py 文件,则可以将其放在根目录下。

6.14.8 使用 fixture 返回值

当 fixture 作为函数参数引用时,可以使用其返回值。示例如下。

```
import requests
import pytest

@pytest.fixture
def baidu_response():
    response = requests.get("http://www.baidu.com")
    return response

# 将 fixture 作为函数参数
def test_baidu(baidu_response):
    response = baidu_response
    # 使用 fixture 函数的返回值
    assert response.status_code == 200

if __name__ == "__main__":
    pytest.main(['test_fixture.py','-s'])
```

执行结果如下。

```
============================ test session starts ============================
collected 1 item

test_fixture.py .

=============================== 1 passed in 0.11s ========================
```

6.14.9　参数化

示例代码如下。

```
import pytest

custom_info = [
  {"name": "lily","password": "123lily"},
  {"name": "jack","password": "123jack"},
  {"name": "tom","password": "123tom"}
]

@pytest.fixture(params = custom_info)
def init_params(request):                    # 参数为固定用法
    # 取出单个参数,固定用法
    return request.param

class Test_Params:
    def setup_class(self):
        print("setup_class...")

    def teardown_class(self):
        print("teardown_class...")

    def test_params(self,init_params):
        # 每个参数执行一遍
        print("获取到的参数:\n")
        print("name:", init_params["name"], ",password:", init_params["password"])

    def test_a(self):
        # 只执行一次
        print("test_a...")
        assert True

if __name__ == "__main__":
    pytest.main(['test_fixture.py','-s'])
```

执行结果如下所示。

```
=========================== test session starts ============================
collected 4 items

test_fixture.py setup_class...
获取到的参数:
```

```
name: lily ,password: 123lily
.获取到的参数:

name: jack ,password: 123jack
.获取到的参数:

name: tom ,password: 123tom
.test_a...
.teardown_class...


============================== 4 passed in 0.06s ========================
```

6.14.10 yield 与 addfinalizer

pytest 支持执行 fixture 特定的最终代码,通过使用 yield 语句,yield 语句之后的所有代码都用作拆卸代码,会在最后一次测试完成后执行,不管测试是否为异常状态,比如清除脏数据等。在同一目录下新建 conftest.py 和 test_fixture.py 两文件。

conftest.py 文件内容如下。

```
import smtplib
import pytest

@pytest.fixture(scope="module")
def smtp():
    smtp = smtplib.SMTP("smtp.qq.com", 587, timeout=5)
    yield smtp
    print("teardown smtp...")
    smtp.close()
```

test_fixture.py 文件内容如下。

```
def test_ehlo(smtp):
    response, msg = smtp.ehlo()
    assert response == 250
    assert b"qq.com" in msg

def test_noop(smtp):
    response, msg = smtp.noop()
    assert response == 250
```

添加-s -v 参数执行结果如下所示。

```
========================== test session starts ==========================
collected 2 items

test_fixture.py::test_ehlo PASSED
test_fixture.py::test_noop PASSEDteardown smtp...


============================= 2 passed in 0.36s ========================
```

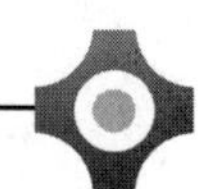

注意　如果我们用 scope='function'修饰 fixture 函数,那么每个测试都会发生 fixture 设置和清理。

执行拆卸代码的另一种选择是使用请求上下文对象的 addFinalizer 方法来注册完成函数。修改 conftest.py 文件内容如下。

```
import smtplib
import pytest

@pytest.fixture(scope = "module")
def smtp(request):
    smtp  = smtplib.SMTP("smtp.qq.com", 587, timeout = 5)
    def fin():
        print("teardown smtp...")
        smtp.close()
    request.addfinalizer(fin)
    return smtp
```

执行结果同 yield 示例一致。

yield 和 addfinalizer 方法在测试结束后调用其代码的工作方式相似,但 addfinalizer 与 yield 有两个关键区别:

(1) addfinalizer 可以注册多个终结器函数。

(2) addfinalizer 无论 fixture 设置代码是否引发异常,都将始终调用终结器。即使其中一个无法创建/获取,也可以方便地正确关闭设备创建的所有资源。

第7章 unittest单元测试框架

7.1 关于 unittest

unittest 被称作 Python 版本的 JUnit，由 Kent Beck 和 Erich Gamma 开发，有时也被称为 PyUnit。是 Python 语言自带的单元测试框架。

unittest 框架拥有支持自动化测试、测试用例间共享 setUp(实现测试前的初始化工作)和 shutDown(实现测试结束后的清理工作)代码块，集合所有的测试用例并且将测试结果独立的展示在报告框架中的特性，在一组测试中，通过 unittest 框架提供的类很容易支持它的这些特性。

7.2 unittest 框架四个重要概念

官方文档给出了 unittest 框架中 4 个重要的概念，介绍如下。

(1) test fixture(测试固件)。一个 test fixture 代表一个或多个测试执行前的准备动作和测试结束后的清理动作，比如，创建数据库连接、启动服务进程、测试环境的清理或者关闭数据库连接等。

(2) test case(测试用例)。一个 test case 就是一个最小测试单元，也就是一个完整的测试流程。针对一组特殊的输入进行特殊的验证与响应。通过继承 unittest 提供的测试基类(TestCase)，可以创建新的测试用例。

(3) test suite(测试套件)。一个 test suite 就是一组测试用例，一组测试套件或者两者共同组成的集合。它的作用是将测试用例集合到一起，然后一次性执行集合中所有的测试用例。

(4) test runner(测试运行器)。一个 test runner 由执行设定的测试用例和将测试结果提供给用户两部分功能组成。

7.3 单元测试加载方法

在 unittest 单元测试框架中，提供了两种单元测试的加载方法：

(1) 直接通过 unittest.main()方法加载单元测试的测试模块，这是一种最简单的加载方法，所有的测试方法执行顺序都是按照方法名的字符串表示的 ASCII 码升序排序。

(2) 将所有的单元测试用例(test case)添加到测试套件(test suite)集合中，然后一次性加载所有测试对象。

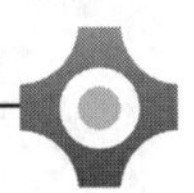

7.4　测试用例

软件测试中最基本组成单元是测试用例，unittest 框架通过 TestCase 类来构建测试用例，并要求所有自定义的测试类都必须继承该类，它是所有测试用例的基类，传入一个测试方法名，将返回一个测试用例实例。TestCase 的子类中实现测试用例的代码既可以单独运行，也可以和其他测试用例构成测试用例集，然后批量执行。

TestCase 作为 unittest 单元测试框架中测试单元的运行实体，单元测试脚本编写员可以通过它派生出自定义的测试过程与方法。TestCase 子类从父类继承的几个特殊的方法，在测试用例执行时均会被依次执行。

TestCase 类中定义的几个特殊方法：

（1）setUp()：每个测试方法运行前运行，测试前的初始化工作。

（2）tearDown()：每个测试方法运行结束后运行，测试后的清理工作。

（3）setUpClass()：所有的测试方法运行前运行，单元测试前期准备，必须使用@classmethod 装饰器进行修饰，setUp()函数之前执行，整个测试过程只执行一次。

（4）tearDownClass()：所有的测试方法运行结束后执行，单元测试后期清理，必须使用@classmethod 装饰器进行修饰，tearDown()之后执行，整个测试过程只执行一次。

最简单的测试用例只需要通过覆盖 runTest()方法来执行自定义的测试代码，这种称为静态方法，如下示例：

```
import unittest
import random

class TestSequenceFunctions(unittest.TestCase):
    def setUp(self):
        # 初始化一个递增序列
        self.seq = list(range(10))

    def run_test(self):
        # 从序列 seq 中随机选取一个元素
        element = random.choice(self.seq)
        # 验证随机元素确实属于列表中
        self.assertTrue(element in self.seq)

class TestDictValueFormatFunctions(unittest.TestCase):
    def setUp(self):
        self.seq = list(range(10))

    def test_shuffle(self):
        # 随机打乱原 seq 的顺序
        random.shuffle(self.seq)
        self.seq.sort()
        self.assertEqual(self.seq, list(range(10)))
        # 验证执行函数时抛出了 TypeError 异常
        self.assertRaises(TypeError, random.shuffle, (1, 2, 3))
```

```
if __name__ == '__main__':
    unittest.main()
```

如果要在 unittest 单元测试框架中构造上述测试类的一个实例，需要按 testCase = TestSequenceFunctions()这行代码实现，并且一个测试用例通常只能对测试模块中一个方法进行单元测试，如果要对测试模块中多个方法进行单元测试，就需要构造多个执行测试类，就如上例中的 TestSequenceFunctions 类和 TestDictValueFormatFunctions 类，而对于同一测试模块，测试用例间可能有着相同的初始状态，如果还是采用上述方法就会出现很多冗余代码，并且还是一项费时的枯燥工作。

unittest 框架针对这一问题，给出了一种动态的解决办法，脚本编写人员只需要写一个测试类来完成对整个测试模块的单元测试，而初始化工作直接在 setUp()方法中完成，资源的释放等清理工作直接在 tearDown()方法中完成即可，这种方法规定所有需要被执行的测试方法都必须以 test 开头，具体示例如下。

示例脚本程序：

```
import unittest


class MyClass:
    """被测试类"""

    @classmethod
    def sum(cls, a, b):
        """将两个传入参数进行相加操作"""
        return a + b #

    @classmethod
    def sub(cls, a, b):
        """将两个传入参数进行相减操作"""
        return a - b


class MyTest(unittest.TestCase):
    @classmethod
    def setUpClass(cls):
        """初始化类固件"""
        print(" ---- setUpClass")

    @classmethod
    def tearDownClass(cls):
        """清理类固件"""
        print(" ---- tearDownClass")

    def setUp(self):
        """初始化工作"""
        self.a = 3
        self.b = 1
        print(" -- setUp")

    def tearDown(self):
```

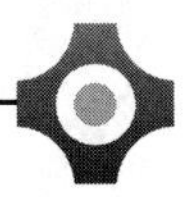

```
            """退出清理工作"""
            print(" -- tearDown")

        def test_sum(self):
            """具体的测试用例,一定要以 test 开头"""
            # 断言两数之和的结果是否是 4
            self.assertEqual(MyClass.sum(self.a, self.b), 4, 'test sum fail')

        def test_sub(self):
            # 断言两数之差的结果是否是 2
            self.assertEqual(MyClass.sub(self.a, self.b), 2, 'test sub fail')


if __name__ == '__main__':
    unittest.main()                          # 启动单元测试
```

在 PyCharm 工具中执行该测试脚本。

测试结果：

```
..
----------------------------------------------------------------------
Ran 2 tests in 0.000s

OK
---- setUpClass
-- setUp
-- tearDown
-- setUp
-- tearDown
---- tearDownClass
```

测试结果说明：

从输出结果看出，setUpClass()和 tearDownClass()方法在整个测试类运行过程中只被执行了一次，而 setUp()和 tearDown()方法在每个测试方法执行前和执行后均被调用，被执行了多次。在测试结果输出中，出现了两个点(.)，这表示测试类 mytest 中两个测试方法 testsum()和 testsub()均被成功执行，一个点表示一个测试方法。

说明：

在 PyCharm 工具中执行单元测试代码，输出结果信息有可能会出现乱序，如图 7-1 所示，此时直接在 CMD 中执行.py 文件即可解决此问题。

更多说明：

(1) 有时在测试结果输出中会出现 E 和 F 字符的情况，这说明测试用例执行失败或发生了异常，下面分别介绍一下这两字符出现的情况。按如图 7-2 所示内容修改上面脚本程序，然后再次执行该脚本程序，执行结果如图 7-3 所示。

测试结果中只出现了一个点(.)，但出现了一个“F”，这表示 MyTest 类中只有一个测试方法执行通过了，另一个失败了，因为 3 加 1 不等于 3，所以断言失败了。

再次按如图 7-4 所示的代码修改脚本程序，修改后再次执行该脚本，测试结果如图 7-5 所示。

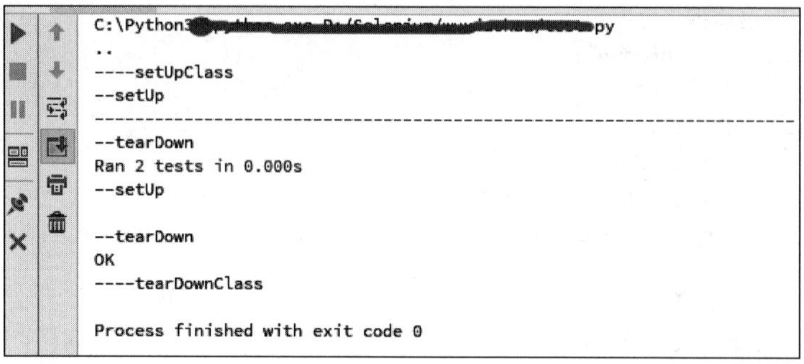

```
C:\Python3...py
..
----setUpClass
--setUp
----------------------------------------------------------------------
--tearDown
Ran 2 tests in 0.000s
--setUp

--tearDown
OK
----tearDownClass

Process finished with exit code 0
```

图 7-1

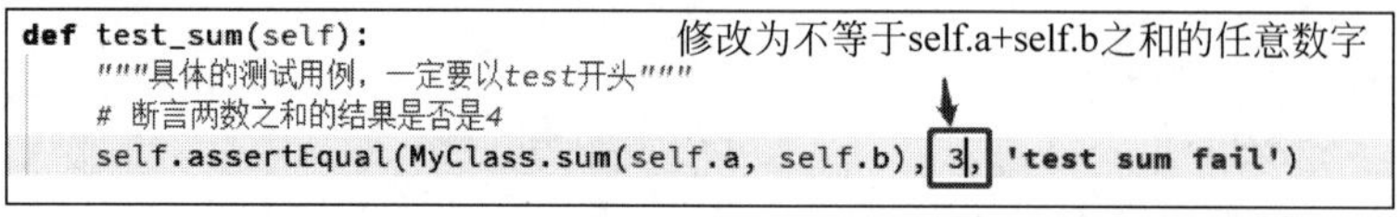

```
def test_sum(self):                       修改为不等于self.a+self.b之和的任意数字
    """具体的测试用例，一定要以test开头"""
    # 断言两数之和的结果是否是4
    self.assertEqual(MyClass.sum(self.a, self.b), 3, 'test sum fail')
```

图 7-2

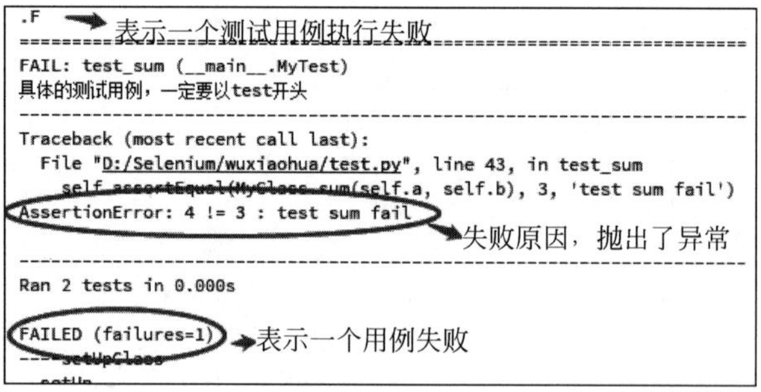

```
.F  → 表示一个测试用例执行失败
======================================================================
FAIL: test_sum (__main__.MyTest)
具体的测试用例，一定要以test开头
----------------------------------------------------------------------
Traceback (most recent call last):
  File "D:/Selenium/wuxiaohua/test.py", line 43, in test_sum
    self.assertEqual(MyClass.sum(self.a, self.b), 3, 'test sum fail')
AssertionError: 4 != 3 : test sum fail  → 失败原因，抛出了异常

----------------------------------------------------------------------
Ran 2 tests in 0.000s

FAILED (failures=1)  → 表示一个用例失败
----setUpClass
--setUp
```

图 7-3

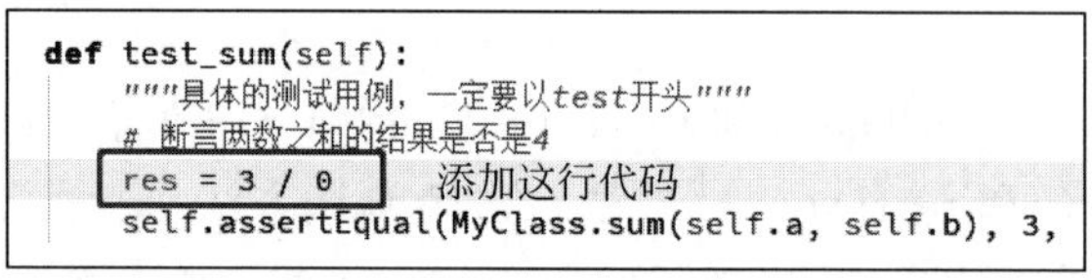

```
def test_sum(self):
    """具体的测试用例，一定要以test开头"""
    # 断言两数之和的结果是否是4
    res = 3 / 0      添加这行代码
    self.assertEqual(MyClass.sum(self.a, self.b), 3,
```

图 7-4

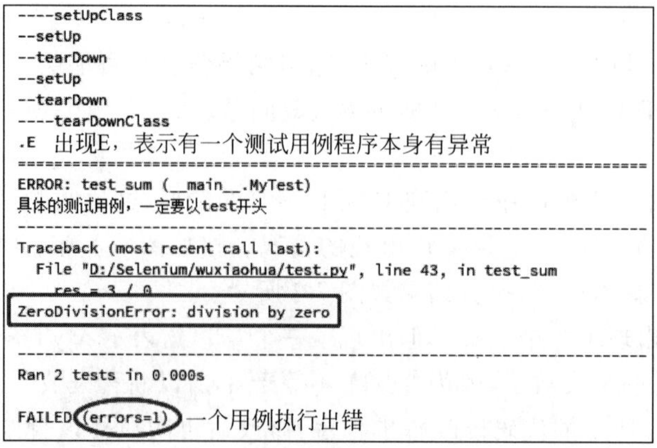

```
----setUpClass
--setUp
--tearDown
--setUp
--tearDown
----tearDownClass
.E  出现E，表示有一个测试用例程序本身有异常
======================================================================
ERROR: test_sum (__main__.MyTest)
具体的测试用例，一定要以test开头
----------------------------------------------------------------------
Traceback (most recent call last):
  File "D:/Selenium/wuxiaohua/test.py", line 43, in test_sum
    res = 3 / 0
ZeroDivisionError: division by zero

----------------------------------------------------------------------
Ran 2 tests in 0.000s

FAILED (errors=1) 一个用例执行出错
```

图 7-5

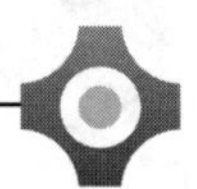

因为在测试方法 test_sum()函数中，增加了一行 res=3/0，由于除数不能为零，所以程序执行到这行代码时，抛出 ZeroDivisionError 异常，但并没有捕获该异常，导致脚本执行中断。

(2) 动态方法不再覆盖 runTest()方法，而是直接在一个测试类中编写多个测试方法。

mytest 类继承自 unittest. TestCase 类，同时重写了 setUp()、setUpClass()、tearDown()、tearDownClass()方法，并且定义了两个以 test 开头的方法(注意必须以 test 开头，中间可以插入_、字母等字符)。当然也可以同时在一个. py 文件中编写多个自定义测试类，这些自定义测试类都必须继承 unittest. TestCase 类。

7.5 测试集合

在自动化测试的执行过程中，通常会有批量运行多个测试用例的需求，此需求称为运行测试集合(test suite)。将功能相关的测试用例组合到一起称为一个测试用例集，unittest 框架中通过 TestSuite 类来组装所有的测试用例集。也就是说使用测试集合可以同时执行同一. py 文件中的多个测试用例类。

加载测试集合步骤：

第一步：TestLoader(测试用例加载器)根据传入的参数获取相应的测试用例的测试方法。

第二步：然后 makeSuite(通常由单元测试框架调用，用于生产 test suite 对象的实例)。把所有的测试用例组装成 test suite 集合。

第三步：最后将 test suite 集合传给 test runner 进行执行。

示例代码：

```
import random
import unittest

class TestSequenceFunctions(unittest.TestCase):
    def setUp(self):
        self.seq = list(range(10))

    def tearDown(self):
        pass

    def test_choice(self):
        # 从序列 seq 中随机选取一个元素
        element = random.choice(self.seq)
        # 验证随机元素确实属于列表中
        self.assertTrue(element in self.seq)

    def test_sample(self):
        # 验证执行的语句是否抛出了异常
        with self.assertRaises(ValueError):
            random.sample(self.seq, 20)
        for element in random.sample(self.seq, 5):
```

```
            self.assertTrue(element in self.seq)


class TestDictValueFormatFunctions(unittest.TestCase):
    def setUp(self):
        self.seq = list(range(10))

    def tearDown(self):
        pass

    def test_shuffle(self):
        # 随机打乱原 seq 的顺序
        random.shuffle(self.seq)
        self.seq.sort()
        self.assertEqual(self.seq, list(range(10)))
        # 验证执行函数时抛出了 TypeError 异常
        self.assertRaises(TypeError, random.shuffle, (1, 2, 3))

if __name__ == '__main__':
    # 根据给定的测试类,获取其中的所有以 test 开头的测试方法,并返回一个测试套件
    testCase1 = unittest.TestLoader().loadTestsFromTestCase(TestSequenceFunctions)
    testCase2 = unittest.TestLoader().loadTestsFromTestCase(TestDictValueFormatFunctions)
    # 将多个测试类加载到测试套件中
    suite = unittest.TestSuite([testCase1, testCase2])
    # 设置 verbosity = 2,可以打印出更详细的执行信息
    unittest.TextTestRunner(verbosity = 2).run(suite)
```

测试结果:

```
C:\Python3.5\python.exe D:/PythonProject/test/a.py
test_choice (__main__.TestSequenceFunctions) ... ok
test_sample (__main__.TestSequenceFunctions) ... ok
test_shuffle (__main__.TestDictValueFormatFunctions) ... ok

----------------------------------------------------------------------
Ran 3 tests in 0.000s

OK
```

代码解释:

(1) TestLoader 类:测试用例加载器,返回一个测试用例集合。

(2) loadTestsFromTestCase 类:根据给定的测试类,获取其中的所有以 test 开头的测试方法,并返回一个测试集合。

(3) TestSuite 类:组装测试用例的实例,支持添加和删除用例,最后将传递给 test runner 进行测试执行。

(4) TextTestRunner 类:测试用例执行类,其中 Text 的表示以文本形式输出测试结果。

更多说明:

(1) 设置 verbosity <=0 的数字,输出结果中不提示执行成功的用例数。

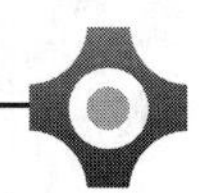

(2) 设置 verbosity=1,输出结果中仅以点(.)表示执行成功的用例数。

(3) 设置 verbosity>=2 的数字,可以输出每个用例执行的详细信息,特别是在大批量执行测试用例时,可以根据这些信息判断哪些用例执行失败。

(4) TestRunner.run()方法会返回一个 TestResult 实例对象,该实例对象里存储着所有测试用例执行过程的详细信息,有需要的读者可以使用 Python 的 dir()方法查看,这里不详细介绍。

7.6 按照特定顺序执行测试用例

通过 TestSuite 类可以改变测试用例执行顺序。在介绍按照特定顺序执行测试用例之前,让我们先看看 unittest 框架在使用 unittest.main()方法启动单元测试时,测试用例的执行顺序。

测试代码:

在 PyCharm 工具中新建 Python 工程 Calc,在该工程下新建两 Python 文件,即 Calc.py 和 MyTest.py,工程结构如图 7-6 所示。

Python 文件 Calc.py 详细代码如下。

```
class Calc:

    def add(self, x, y, *d):
        # 加法计算
        result = x + y
        for i in d:
            result += i
        return result

    def sub(self, x, y, *d):
        # 减法计算
        result = x - y
        for i in d:
            result -= i
        return result

    @classmethod
    def mul(cls, x, y, *d):
        # 乘法计算
        result = x * y
        for i in d:
            result *= i
        return result

    @staticmethod
    def div(x, y, *d):
        # 除法计算
        if y != 0:
            result = x / y
```

图 7-6

```
        else:
            return -1
        for i in d:
            if i != 0:
                result /= i
            else:
                return -1
        return result
```

Python 文件 MyTest.py 详细代码如下。

```
import unittest
from Calc import Calc

class MyTest(unittest.TestCase):

    @classmethod
    def setUpClass(cls):
        print("单元测试前,创建 Calc 类的实例")
        cls.c = Calc()

    # 具体的测试用例,一定要以 test 开头
    def test_add(self):
        print("run add()")
        self.assertEqual(MyTest.c.add(1, 2, 12), 15, 'test add fail')

    def test_sub(self):
        print("run sub()")
        self.assertEqual(MyTest.c.sub(2, 1, 3), -2, 'test sub fail')

    def test_mul(self):
        print("run mul()")
        self.assertEqual(Calc.mul(2, 3, 5), 30, 'test mul fail')

    def test_div(self):
        print("run div()")
        self.assertEqual(Calc.div(8, 2, 4), 1, 'test div fail')


if __name__ == '__main__':
    unittest.main()                       # 启动单元测试
```

CMD 下执行 MyTest.py 文件。

测试结果：

```
单元测试前,创建 Calc 类的实例
run add()
.run div()
.run mul()
.run sub()
.
```

```
----------------------------------------
Ran 4 tests in 0.007s

OK
```

结果说明：

从输出结果可以看出，以 unittest.main()这种方式启动的单元测试，各测试方法的执行顺序是所有方法名的字符串按照 ASCII 码排序后的顺序。但如果我们想让 test_mul()这个测试用例在 test_add()方法之前执行，该怎么办？可以按以下方式修改 MyTest.py 文件中(如图 7-7 所示)。

```
if __name__ == '__main__':
# unittest.main()                          # 启动单元测试
# 获取 TestSuite 的实例对象
suite = unittest.TestSuite()
# 将测试用例添加到测试容器中
suite.addTest(MyTest("test_mul"))
suite.addTest(MyTest("test_div"))
suite.addTest(MyTest("test_add"))
suite.addTest(MyTest("test_sub"))
# 创建 TextTestRunner 类的实例对象
runner = unittest.TextTestRunner()
runner.run(suite)
```

```
if __name__ == '__main__':
    unittest.main() # 启动单元测试
```

图 7-7

再次执行 MyTest.py 文件，可以看到 test_mul()方法最先被执行，结果如下：

```
单元测试前,创建 Calc 类的实例
run mul()
.run div()
.run add()
.run sub()
.
----------------------------------------
Ran 4 tests in 0.007s

OK
```

从测试结果可以看出，测试用例已经按照我们设定的特定顺序进行执行。

代码解释：

首先是获得一个单元测试中用例集 TestSuite 的实例对象 suite，然后按照我们设定的顺序将相应的用例添加到集合对象的实例中，然后创建一个用单元测试中的 TextTestRunner 类的实例对象 runner，最后将集合对象 suite 添加到执行对象实例 runner 中，并调用它的 run 方法开始执行这些测试用例。

使用这种方法，我们就可以添加多个用例到用例集中，组成一组用例，然后再按照设定的顺序去执行这些用例，特别是在某些复杂的依赖测试场景中，只有当某个或某几个测试用例被执行后才能执行其他的测试用例，这种方法给我们提供了极大的方便，比如只有在登录 126 邮箱后，才能发邮件或添加新联系人等场景。但这种方法只能线性执行，如果想实现并发的执行用例，读者可以自行编写多线程去实现。

更多说明：

在 MyTest 类中，每一个以 test 字符串开头的方法都是一个测试用例，具体从 TestLoader 类加载测试用例原理上来解释。在 unittest. TestLoader 类（文件路径为\Python3. 5\Lib\unittest\loader. py）中有一个 loadTestsFromTestCase()方法，其具体代码如下：

```
def loadTestsFromTestCase(self, testCaseClass):
"""Return a suite of all tests cases contained in testCaseClass"""
if issubclass(testCaseClass, suite.TestSuite):
    raise TypeError("Test cases should not be derived from TestSuite." \
                    " Maybe you meant to derive from TestCase?")
testCaseNames = self.getTestCaseNames(testCaseClass)
if not testCaseNames and hasattr(testCaseClass, 'runTest'):
    testCaseNames = ['runTest']
loaded_suite = self.suiteClass(map(testCaseClass, testCaseNames))
return loaded_suite
```

testCaseNames=self. getTestCaseNames(testCaseClass)这行代码中的 getTestCaseNames()（具体代码也在 loader. py 文件中）方法是从 testCaseClass 这个类中寻找所有以 test 开头的方法（所以，不以 test 开头的方法将会被忽略执行），然后将其赋给 testCaseNames 变量。loaded_suite = self. suiteClass(map(testCaseClass, testCaseNames))这行代码用于创建 TestSuite 类对象，其中使用 Python 内建的 map()方法，为 testCaseNames 变量中每一个元素都构建 testCaseClass 类对象示例，最后使用 suiteClass()方法构造成一个 TestSuite 对象集合并赋给 loaded_suite 变量，如下代码更能详细说明其过程：

```
testCasesList = []
for caseName in testCaeNames:
    testCaseList.append(TestCase(caseName))
loaded_suite = self.suiteClass(tuple(testCasesList))
```

由此可见，MyTest 类中每一个以 test 开头的方法，都是一个 TestCase 对象，也就是说是一个测试用例。

7.7 忽略某个测试方法

在批量执行测试用例时，可能会遇到某些测试用例不需要执行，但又想保留测试代码，除了可以注释掉代码以外，unittest 框架提供了一个更简便的注解方法用来忽略那些暂时不需要执行的测试用例，单元测试框架在执行过程中遇到被标上忽略的注解的用例时，就会自动跳过。忽略测试用例分为：无条件忽略和有条件忽略，详见如下代码介绍。

示例代码：

```
import random
import unittest
import sys

class TestSequenceFunctions(unittest.TestCase):
```

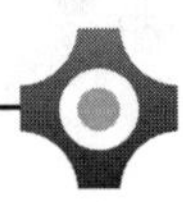

```
    a = 1

    def setUp(self):
        self.seq = list(range(10))

    @unittest.skip("skipping")              # 无条件忽略该测试方法
    def test_shuffle(self):
        random.shuffle(self.seq)
        self.seq.sort()
        self.assertEqual(self.seq, list(range(10)))
        self.assertRaises(TypeError, random.shuffle, (1, 2, 3))

    # 如果变量 a > 5,则忽略该测试方法
    @unittest.skipIf(a > 5, "condition is not satisfied!")
    def test_choice(self):
        element = random.choice(self.seq)
        self.assertTrue(element in self.seq)

    # 除非执行测试用例的平台是 Windows 平台,否则忽略该测试方法
    @unittest.skipUnless(sys.platform.startswith("linux"), "requires Windows")
    def test_sample(self):
        with self.assertRaises(ValueError):
            random.sample(self.seq, 20)
        for element in random.sample(self.seq, 5):
            self.assertTrue(element in self.seq)


if __name__ == '__main__':
    # unittest.main()
    testCases = unittest.TestLoader().loadTestsFromTestCase(TestSequenceFunctions)
    suite = unittest.TestSuite(testCases)
    unittest.TextTestRunner(verbosity = 2).run(suite)
```

执行结果：

```
test_choice (__main__.TestSequenceFunctions) ... ok
test_sample (__main__.TestSequenceFunctions) ... skipped 'requires Windows'
test_shuffle (__main__.TestSequenceFunctions) ... skipped 'skipping'

----------------------------------------------------------------------
Ran 3 tests in 0.006s

OK (skipped = 2)
```

结果说明：

从测试结果可以分析出，test_sample 和 test_shuffle 方法被忽略执行了，test_choice 方法被成功执行了，skipped=2 表示有两个测试方法被跳过执行。

7.8 命令行模式执行测试用例(x)

unittest 框架支持命令行模式运行测试模块、类，甚至单独有效的测试方法。通过命令行模式，可以传入任何模块名组合、有效的测试类或者测试方法的参数列表。详细使用方法

参看下面的实例(本小节测试示例使用前面小节中编写的 Python 工程 Calc 实例)。

1. 通过命令直接运行整个测试模块

命令格式:

```
python -m unittest test_module1 test_module2 ...
```

实战步骤:

(1) CMD 下切换当前工作目录到 Python 工程 Calc 目录下,如图 7-8 所示。

图 7-8

(2) CMD 下执行 python -m unittest -v MyTest 命令,输出结果如图 7-9 所示。

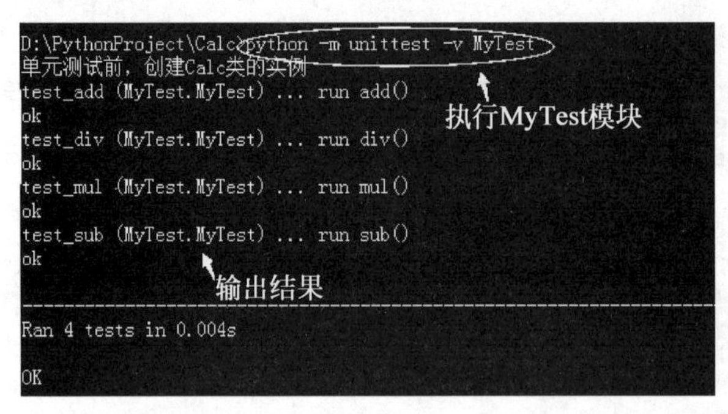

图 7-9

2. 执行测试模块中某个测试类

命令格式:

```
python -m unittest test_module.TestClass
```

实战步骤:

CMD 下将当前工作目录切换到 Python 工程 Calc 目录下,然后执行命令:python -m unittest -v MyTest. MyTest,运行结果如图 7-10 所示。

```
D:\PythonProject\Calc>python -m unittest -v MyTest.MyTest
单元测试前,创建Calc类的实例
test_add (MyTest.MyTest) ... run add()
ok
test_div (MyTest.MyTest) ... run div()
ok
test_mul (MyTest.MyTest) ... run mul()
ok
test_sub (MyTest.MyTest) ... run sub()
ok
----------------------------------------------------------------------
Ran 4 tests in 0.005s

OK
```

执行MyTest模块下的MyTest类

详细输出结果

图 7-10

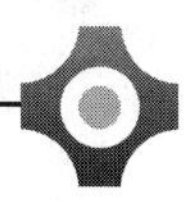

3. 执行测试模块中某个测试类下的某个测试方法

命令格式：

```
python -m unittest test_module.TestClass.test_method
```

实战步骤：

CMD 下将当前工作目录切换到 Python 工程 Calc 目录下，然后执行命令：python -m unittest -v MyTest.MyTest.test_add MyTest.MyTest.test_sub，运行结果如图 7-11 所示。

图　7-11

更多说明：

(1) 使用命令执行测试用例前，必须将 CMD 当前的工作目录切换到测试脚本文件存放目录，如果想直接指定脚本文件所在路径去运行，会抛出 ImportError 异常。

(2) 这种命令执行方式对脚本所在目录名以及测试脚本文件名都没有特殊要求。

(3) 命令中-v 参数表示输出测试用例执行的详细信息，等价于 verbosity=2。

7.9　批量执行测试模块

之前我们了解的都是针对同一测试模块来展开的测试运行，本小节介绍一下 unittest 单元测试框架提供的批量执行测试模块方法，官方称之为测试发现。unittest 单元测试框架支持简单的测试发现，即可以自动发现并执行给定目录下满足规则的测试模块。为了更好地匹配测试模块，给定目录下所有的测试文件都必须是模块或者是能从工程的顶层目录导入的包，也就意味着所有的文件名必须是有效的标识符，同时目录下需要被执行的测试脚本文件名都必须以 test 字符串开头，比如 testEqual.py。文件发现分两种形式：程序文件模式和命令行模式。

1. 程序文件模式

程序文件模式，就是将测试发现代码编写在测试脚本中，然后直接执行脚本文件即可，具体由 TestLoader.discover()方法实现。

实例代码：

PyCharm 工具中新建一名叫 project_discover（工程名可以随便取）的 Python 工程，在该工程下新建 5 个 Python 文件 Calc.py、testCalc.py、testFact.py、testSeqSum.py 和 run.py，各文件详细代码如下。

```
Calc.py(供其他模块调用)
class Calc:
```

```
    def add(self, x, y, *d):
        # 加法计算
        result = x + y
        for i in d:
            result += i
        return result

    def mul(self, a, b):
        # 返回两数之积
        return a * b
```

testCalc.py(测试模块)

```
import unittest
from Calc import Calc

class MyTest(unittest.TestCase):
    c = None

    @classmethod
    def setUpClass(cls):
        print("单元测试前,创建 Calc 类的实例")
        cls.c = Calc()

    # 具体的测试用例,一定要以 test 开头
    def test_add(self):
        print("run add()")
        self.assertEqual(MyTest.c.add(1, 2, 12), 15, 'test add fail')
```

testFact.py(测试模块)

```
    # encoding = utf - 8
import unittest
from functools import reduce

class MyTestCase(unittest.TestCase):
    def setUp(self):
        self.num = 5

    def testFactorial(self):
        # 生成一个递增序列
        seq = range(1, self.num + 1)
        # 求阶乘
        res = reduce(lambda x, y: x * y, seq)
        # 断言阶乘结果
        self.assertEqual(res, 120, "断言阶乘结果错误!")
```

testSeqSum.py(测试模块)

```
import unittest

class MyTestCase(unittest.TestCase):
    def testEqual(self):
        seq = range(11)
```

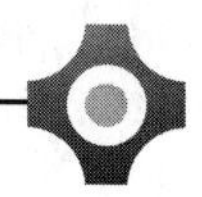

```
        self.assertEqual(sum(seq), 55, "断言列表元素之和结果错误!")
```

run.py(用于发现目录下的测试模块并执行)

```
    #coding = utf - 8
import unittest

if __name__ == '__main__':
    # 加载当前目录下所有有效的测试模块,"."表示当前目录
    testSuite = unittest.TestLoader().discover('.')
    unittest.TextTestRunner(verbosity = 2).run(testSuite)
```

拥有这段代码的 Python 文件,必须放在需要被执行的测试脚本所在目录。

进入 run.py 文件中,鼠标移到最后一行代码的 run()方法上,右击,选择 ▶ Run 'run' 命令运行 run.py 文件。

执行结果:

```
单元测试前,创建 Calc 类的实例
test_add (testCalc.MyTest) ... ok
run add()
testFactorial (testFact.MyTestCase) ... ok
testEqual (testSeqSum.MyTestCase) ... ok

----------------------------------------------------------------------
Ran 3 tests in 0.000s

OK
```

结果说明:

从执行结果分析,工程 project_discover 下 test_add、testFactorial 和 testEqual 模块均被执行了。

2. 命令行模式

命令行批量执行某个目录下的测试脚本,通过 unittest 单元测试框提供的 discover 命令实现。

实战步骤:

(1) CMD 下切换当前工作目录到上一小节创建的 Python 工程 project_discover 目录下。

(2) 然后 CMD 下执行命令 python -m unittest discover,执行结果如图 7-12 所示。

更多说明:

(1) 上述执行命令的顺序不能更改。

(2) 在执行批量执行测试脚本命令前,必须将 CMD 当前的工作目录切换到存放测试脚本所在目录。

(3) discover 命令还有一些其他参数,说明如下:

-v: 输出详细测试信息,比如: python -m unittest discover -v;

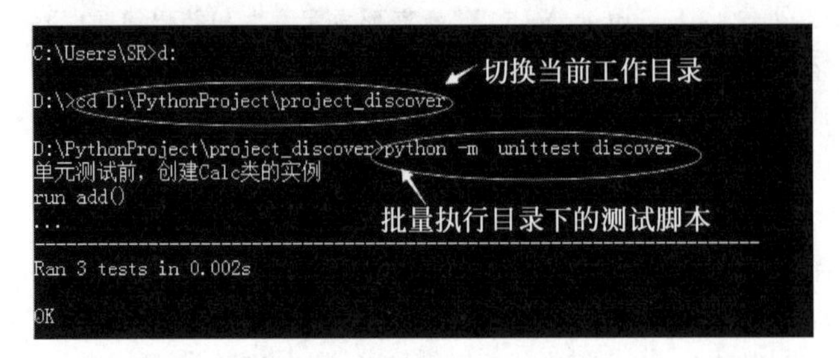

图 7-12

-s：执行发现测试脚本的目录，默认为当前目录(.)，比如：python -m unittest discover -v -s D:\PythonProject\project_discover；

-p：模式匹配测试文件，比如：python -m unittest discover -p "test *.py"；

-t directory：工程的根目录下搜索可执行的测试脚本，默认是当前目录，比如：python -m unittest discover -v -t D:\PythonProject\project_discover；

(4) discover 命令添加-s、-p、-t 参数时，由于可以人为指定搜索测试脚本目录，所以执行测试脚本命令也等价于下面两命令。

```
python -m unittest discover -s project_directory -p "*_test.py"
python -m unittest discover project_directory "*_test.py"
```

7.10 常用的断言方法

断言表示为一些布尔表达式，编写代码时，程序员总是会在某些特定点做出一些假设，来判断程序是否达到预期。断言为真时，表示达到预期，否则未达到预期。而对于自动化测试人员来说，借助断言能更好地检测被测试对象是否满足测试期望。

在单元测试过程中必须使用断言。unittest 单元测试框架中的 TestCase 类提供了很多断言方法，便于检验测试是否满足预期结果，并能在断言失败后抛出失败的原因。本节只介绍其中一部分常用的断言方法，如表 7-1 所示。

表 7-1

断言方法	检测
assertEqual(first, second, msg=None)	测试 first==second，否则抛出断言异常信息 msg
assertNotEqual(first, second, msg=None)	测试 first!=second，否则抛出断言异常信息 msg
assertTrue(expr, msg=None)	测试表达式 expr 为 True，否则抛出断言异常信息 msg
assertFalse(expr, msg=None)	测试表达式 expr 为 False，否则抛出断言异常信息 msg
assertIs(a, b, msg=None)	测试 a 和 b 是同一对象，否则抛出断言异常信息 msg
assertIsNot(a, b, msg=None)	测试 a 和 b 不是同一对象，否则抛出断言异常信息 msg
assertIsNone(expr, msg=None)	测试表达式 expr 结果为 None，否则抛出断言异常信息 msg
assertIsNotNone(expr, msg=None)	测试表达式 expr 结果不为 None，否则抛出断言异常信息 msg
assertIn(a, b, msg=None)	测试 a 包含在 b 中，否则抛出断言异常信息 msg

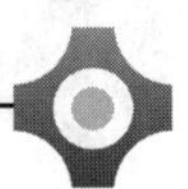

续表

断 言 方 法	检　　测
assertNotIn(a, b, msg=None)	测试 a 不包含在 b 中,否则抛出断言异常信息 msg
assertIsInstance(obj, cls, msg=None)	断言 obj 为 cls 类型,否则抛出断言异常信息 msg。可以用 isinstance(obj,cls)或者 assertIs(type(obj), cls)代替
assertNotIsInstance(obj, cls, msg=None)	断言 obj 不为 cls 类型,否则抛出断言异常信息 msg。可以用 not isinstance(obj,cls)或者 assertIsNot(type(obj), cls)代替
assertRaises(exc[,fun, * args, ** kwds])	测试函数 fun(* args, ** kwds)抛出 exc 异常,否则抛出断言异常
assertRaisesRegexp(exc,r[,fun, * args, ** kwds])	测试函数 fun(* args, ** kwds)抛出 exc 异常,同时可用正则 r 去匹配异常信息 exc,否则抛出断言异常

各方法的实例脚本:

```
# encoding = utf - 8
import unittest
import random

# 被测试类
class MyClass:
    @classmethod
    def sum(cls, a, b):
        return a + b

    @classmethod
    def div(cls, a, b):
        return a / b

    @classmethod
    def retrun_None(cls):
        return None

# 单元测试类
class MyTest(unittest.TestCase):

    # assertEqual()方法实例
    def test_assertEqual(self):
        # 断言两数之和的结果
        try:
            a, b = 1, 2
            sum = 13
            self.assertEqual(a + b, sum, '断言失败, %s + %s != %s' % (a, b, sum))
        except AssertionError as e:
            print(e)

    # assertNotEqual()方法实例
    def test_assertNotEqual(self):
        # 断言两数之差的结果
```

```
        try:
            a, b = 5, 2
            res = 3
            self.assertNotEqual(a - b, res, '断言失败,%s - %s = %s' % (a, b, res))
        except AssertionError as e:
            print(e)

    # assertTrue()方法实例
    def test_assertTrue(self):
        # 断言表达式的为真
        try:
            self.assertTrue(1 == 1, "表达式为假")
        except AssertionError as e:
            print(e)

    # assertFalse()方法实例
    def test_assertFalse(self):
        # 断言表达式为假
        try:
            self.assertFalse(3 == 2, "表达式为真")
        except AssertionError as e:
            print(e)

    # assertIs()方法实例
    def test_assertIs(self):
        # 断言两变量类型属于同一对象
        try:
            a = 12
            b = a
            self.assertIs(a, b, "%s与%s不属于同一对象" % (a, b))
        except AssertionError as e:
            print(e)

    # test_assertIsNot()方法实例
    def test_assertIsNot(self):
        # 断言两变量类型不属于同一对象
        try:
            a = 12
            b = "test"
            self.assertIsNot(a, b, "%s与%s属于同一对象" % (a, b))
        except AssertionError as e:
            print(e)

    # assertIsNone()方法实例
    def test_assertIsNone(self):
        # 断言表达式结果为 None
        try:
            result = MyClass.retrun_None()
            self.assertIsNone(result, "not is None")
        except AssertionError as e:
            print(e)
```

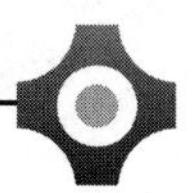

```
# assertIsNotNone()方法实例
def test_assertIsNotNone(self):
    # 断言表达式结果不为 None
    try:
        result = MyClass.sum(2, 5)
        self.assertIsNotNone(result, "is None")
    except AssertionError as e:
        print(e)

# assertIn()方法实例
def test_assertIn(self):
    # 断言对象 A 是否包含在对象 B 中
    try:
        strA = "this is a test"
        strB = "is"
        self.assertIn(strB, strA, "%s 不包含在 %s 中" % (strB, strA))
    except AssertionError as e:
        print(e)

# assertNotIn()方法实例
def test_assertNotIn(self):
    # 断言对象 A 不包含在对象 B 中
    try:
        strA = "this is a test"
        strB = "Selenium"
        self.assertNotIn(strB, strA, "%s 包含在 %s 中" % (strB, strA))
    except AssertionError as e:
        print(e)

# assertIsInstance()方法实例
def test_assertIsInstance(self):
    # 测试对象 A 的类型是否值指定的类型
    try:
        x = MyClass
        y = object
        self.assertIsInstance(x, y, "%s 的类型不是 %s" % (x, y))
    except AssertionError as e:
        print(e)

# assertNotIsInstance()方法实例
def test_assertNotIsInstance(self):
    # 测试对象 A 的类型不是指定的类型
    try:
        a = 123
        b = str
        self.assertNotIsInstance(a, b, "%s 的类型是 %s" % (a, b))
    except AssertionError as e:
        print(e)

# assertRaises()方法实例
```

```
    def test_assertRaises(self):
        # 测试抛出的指定的异常类型
        # assertRaises(exception)
        with self.assertRaises(TypeError) as cm:
            random.sample([1,2,3,4,5], "j")
        # 打印详细的异常信息
        print("===", cm.exception)

        # assertRaises(exception, callable, *args, **kwds)
        try:
            self.assertRaises(ZeroDivisionError, MyClass.div, 3, 0)
        except ZeroDivisionError as e:
            print(e)

    # assertRaisesRegexp()方法实例
    def test_assertRaisesRegexp(self):
        # 测试抛出的指定异常类型,并用正则表达式具体验证
        # assertRaisesRegexp(exception, regexp)
        with self.assertRaisesRegex(ValueError, 'literal') as ar:
            int("xyz")
        # 打印详细的异常信息
        print(ar.exception)
        # 打印正则表达式
        print(ar.expected_regex)

        # assertRaisesRegexp(exception, regexp, callable, *args, **kwds)
        try:
            self.assertRaisesRegex(ValueError, "invalid literal for.*XYZ'$", int, 'XYZ')
        except AssertionError as e:
            print(e)


if __name__ == '__main__':
    # 执行单元测试
    unittest.main()
```

PyCharm 执行该脚本程序,输出结果如图 7-13 所示。

```
3 != 13 : 断言失败,1 + 2 != 13
3 == 3 : 断言失败,5 - 2 = 3
=== '<=' not supported between instances of 'int' and 'str'
invalid literal for int() with base 10: 'xyz'
re.compile('literal')
..............
----------------------------------------------------------------------
Ran 14 tests in 0.001s

OK
```

图 7-13

从 PyCharm 的 Console 中看出,所有的单元测试方法都成功执行。读者可以自行修改断言函数中的参数值,来查看不同的断言结果。

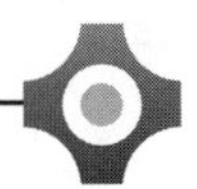

更多说明：

对于 assertRaises(exc, fun, * args, ** kwds)，assertRaisesRegexp(exc, r, fun, * args, ** kwds)断言异常的函数来说，如果只给出异常类型参数 exc，将返回一个上下文管理器，以便测试的代码可以直接内联而不是写入函数中。上下文管理器将在其异常属性中存储被捕获的异常对象。如果意图是在引发的异常上执行额外的检查，这些信息就非常有用了。

7.11 在 unittest 中运行第一个 WebDriver 测试用例

本小节将在 unittest 单元测试框架中实现第一个 WebDriver 测试用例，其具体实现步骤如下。

(1) 在 PyCharm 工具中新建一个 Python 工程 UnitTestProj。

(2) 在 UnitTestProj 工程下新建一个 Python 文件 gloryroad.py，并在该文件中编写如下代码。

```
#encoding = utf - 8
import time
import unittest

from selenium import webdriver

class GloryRoad(unittest.TestCase):
    def setUp(self):
        # 启动 Firefox 浏览器
        self.driver = webdriver.Firefox(executable_path = "c:\\geckodriver")

    def testSoGou(self):
        # 访问搜狗首页
        self.driver.get("http://sogou.com")
        # 清空搜索输入框默认内容
        self.driver.find_element_by_id("query").clear()
        # 在搜索输入框中输入"光荣之路自动化测试"
        self.driver.find_element_by_id("query").send_keys(u"WebDriver 实战宝典")
        # 单击"搜索"按钮
        self.driver.find_element_by_id("stb").click()
        # 等待 3 秒
        time.sleep(3)
        assert u"吴晓华" in self.driver.page_source, "页面中不存在要寻找的关键字!"

    def tearDown(self):
        # 退出浏览器
        self.driver.quit()

if __name__ == '__main__':
    unittest.main()
```

(3) 在 main()函数上右击，选择 ▶ Run 'gloryroad' 命令开始执行测试用例，待所有操作步骤都完成后，浏览器就会自动退出，同时在 PyCharm 结果输出区域也可以看到用例执行结果信息。

注意，WebDriver 启动浏览器过程可能会比较慢，请耐心等待一会。

第8章 页面元素定位方法

Selenium WebDriver 可以根据网页中页面元素拥有不同的标签名和属性值等特征来定位不同的页面元素，并完成对已定位到的页面元素的各种操作。

在自动化测试实施过程中，测试脚本中常用的页面元素操作步骤如下：

(1) 定位网页上的页面元素，并获得该页面元素对象。

(2) 通过获取的页面元素对象拥有的属性操作该页面元素。例如，单击、拖曳页面元素或在输入框中输入内容等。

(3) 设定页面元素的操作值。例如，设定输入框中输入的内容或指定下拉选项框中哪个选项等。

通过以上 3 步，可以完成对页面元素的自动化操作，但在此之前必须要获取到要操作的页面元素对象，否则接下来的两步将是空谈。由于网页技术的实现过于复杂，在自动化测试实践过程中，经常出现各种页面元素难以定位的难题，常常有人绞尽脑汁也无法成功完成某些页面元素的定位。为了更好地解决页面元素的定位难题，本章将根据笔者多年的成功实践经验，详细介绍定位页面元素的常用方法。

8.1 定位页面元素方法汇总

在低版本的 Selenium WebDriver 中，通常使用 find_element 和 find_elements 方法，再结合 By 类来实现定位页面元素，但在高版本的 Selenium WebDriver 官方文档中，已经开始建议使用 find_element_by_ * 和 find_elements_by_ * 方法代替前两种方法。WebDriver 对象内置的 find_element_by_ * 方法仅用于定位单个页面元素，find_elements_by_ * 方法可用于同时定位多个页面元素，以列表形式返回所有定位到的页面元素。定位到的页面元素需要进行存储，以便在以后的测试过程中能随时取用。

WebDriver 提供了 8 种不同的定位方法，分别为 id、name、xpath、class name、tag name、link text、partial link text 以及 css selector，其详细使用方法如表 8-1 所示。

表 8-1

定位方法	定位方法的 Python 语言实现实例	
	定位单个元素	定位多个元素
使用 ID 定位	find_element_by_id("ID 值") find_element(by="id",value="ID 值")	因为 ID 唯一，所以不能定位多个
使用 name 定位	find_element_by_name("name 值") find_element(by="name",value="name 值")	find_elements_by_name("name 值") find_elements(by="name",value="name 值")

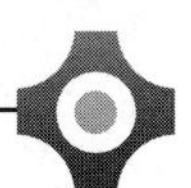

续表

定位方法	定位方法的 Python 语言实现实例	
	定位单个元素	定位多个元素
使用 class name 定位	find_element_by_class_name("页面元素的 Class 属性值") find_element(by="classname",value="页面元素的 Class 属性值")	find_elements_by_class_name("页面元素的 Class 属性值") find_elements(by="classname",value="页面元素的 Class 属性值")
使用标签名称定位	find_element_by_tag_name("页面中的 HTML 标签名称") find_element(by="tag name",value="页面中的 HTML 标签名称")	find_elements_by_tag_name("页面中的 HTML 标签名称") find_elements(by="tag name",value="页面中的 HTML 标签名称")
使用链接的全部文字定位	find_element_by_link_text("链接的全部文字内容") find_element(by="link text",value="链接的全部文字内容")	find_elements_by_link_text("链接的全部文字内容") find_elements(by="link text",value="链接的全部文字内容")
使用部分链接文字定位	find_element_by_partial_link_text("链接的部分文字内容") find_element(by="partial link text",value="链接的部分文字内容")	find_elements_by_partial_link_text("链接的部分文字内容") find_elements(by="partial link text",value="链接的部分文字内容")
使用 XPath 定位	find_element_by_xpath("XPath 定位表达式") find_element(by="xpath",value="XPath 定位表达式")	find_elements_by_xpath("XPath 定位表达式") find_elements(by="xpath",value="XPath 定位表达式")
使用 CSS 方式定位	find_element_by_css_selector("CSS 定位表达式") find_element(by="css selector",value="CSS 定位表达式")	find_elements_by_css_selector("CSS 定位表达式") find_elements(by="css selector",value="CSS 定位表达式")

8.2 使用 ID 定位

被测试网页的 HTML 代码：

```
<html>
    <body>
        <label>用户名</label>
        <input id="username"></input>
        <label>密码</label>
        <input id="password"></input>
        <br>
        <button id="submit">登录</button>
    </body>
</html>
```

生成可见网页方法，在硬盘某路径下新建一个扩展名为 html 的文件，将上面代码复制到该文件并保存，然后将该文件直接拖曳到浏览器中，就可以看到生成的 HTML 页面。

定位语句代码：

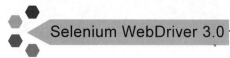

```
password = driver.find_element_by_id("password")
username = driver.find_element(by = "id", value = "username")
```

代码解释：

由于页面元素的ID属性都是唯一的，所以不存在根据ID属性值同时定位多个页面元素的情况，由此上面只给出了两种定位页面元素的方法。语句1使用driver对象的find_element_by_id方法通过元素ID属性进行页面元素定位，所传的参数值password为将要被定位的页面元素的ID属性值，查看被测试网页的HTML代码可以找到密码输入框元素的ID属性值为password。

同理，语句2用username作为ID属性值，此时的定位方式是通过参数by传入。

由于页面元素的ID属性都是唯一的，所以使用ID值定位页面元素可以保证定位的唯一性，不会像其他定位方式可能会同时定位到多个页面元素。但在自动化测试实施过程中，很多核心页面元素均无ID属性值，导致无法使用ID值进行定位操作。

建议：

可与Web开发工程师约定，所有核心页面元素都添加ID属性，以提高网页程序的可测试性，降低自动化测试的实施难度，提升效率。

8.3 使用 name 定位

被测试网页的HTML代码：

```
<html>
    <body>
        <label>用户名</label>
        <input name = "username"></input>
        <label>密码</label>
        <input name = "password"></input>
        <br>
        <button name = "submit">登录</button>
    </body>
</html>
```

定位语句代码：

```
username = driver.find_element_by_name("username")
password = driver.find_element(by = "name", value = "password")
pwdList = driver.find_elements_by_name("password")
butList = driver.find_elements(by = "name", value = "submit")
```

代码解释：

语句1和语句2均是用于定位单个页面元素，均通过name定位方式进行定位，定位需要的name值均通过参数传入，比如username和password，它对应于被测试HTML代码中用户名输入框和密码输入框页面元素的name属性值。

语句3和语句4均通过name定位方式同时定位页面上拥有相同name属性值（比如

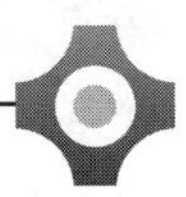

name="password"或 name="submit")的所有页面元素,每个定位到的页面元素都将作为列表的一个元素,然后返回这个列表给调用者。

更多说明:

页面元素的 name 属性和 ID 属性有所区别,name 属性值在当前网页可以不唯一,因而使用单个页面元素定位方法时通过 name 方式定位时可能会定位到多个元素,此时需要进一步定位以保证操作元素的唯一性。

8.4 使用链接的全部文字定位

被测试网页的 HTML 代码:

```
<html>
    <body>
        <a href = "http://www.sogou.com"> sogou 搜索</a><br>
        <a href = "http://www.baidu.com"> baidu 搜索</a>
    </body>
</html>
```

定位语句:

```
link = driver.find_element_by_link_text("sogou 搜索")
link = driver.find_element(by = "link text", value = "sogou 搜索")
linkList = driver.find_elements_by_link_text("baidu 搜索")
linkList = driver.find_elements(by = "link text", value = "baidu 搜索")
```

代码解释:

以上语句都通过 link text 定位方式进行定位。定位需要的链接文字均通过参数传入,比如"sogou 搜索",它对应于被测试 HTML 代码中超链接标签的文本内容,链接文字需要完全匹配"sogou 搜索"或"baidu 搜索"这几个关键字,否则将无法找到链接页面元素。

更多说明:

使用此方式定位页面链接元素,需要完全匹配链接标签中的文本内容,常用于页面中存在多个链接文字高度相似,且无法使用部分链接文字进行定位的情况。

8.5 使用部分链接文字定位

被测试网页的 HTML 代码:

```
<html>
    <body>
        <a href = "http://www.sogou.com"> sogou 搜索</a><br>
        <a href = "http://www.baidu.com"> baidu 搜索</a>
    </body>
</html>
```

定位语句:

```
partialLink = driver.find_element_by_partial_link_text("sogou")
```

```
partialLink = driver.find_element(by = "partial link text", value = "sogou")
partialLinkList = driver.find_elements_by_partial_link_text("搜索")
partialLinkList = driver.find_elements(by = "partial link text", value = "搜索")
```

代码解释：

语句 1 和语句 2 都使用 partial link text 方式查找包含“sogou”关键字符串的链接元素。若匹配到了多个包含“sogou”关键字符串的链接元素，将返回第一个匹配到的页面元素对象，并将其赋值给 partialLink 变量。

语句 3 和语句 4 将匹配页面中所有包含“搜索”两关键字的超链接页面元素，在被测试网页的 HTML 代码中可看到有两个包含“搜索”关键字的链接元素，这两个链接元素都会被抓取到，并存于列表中，然后将该列表对象赋值给 partialLinkList 变量。

更多说明：

使用此方式定位页面链接只需要模糊匹配链接文字即可，常用于匹配页面链接文字不定期发生少量变化的情况。使用模糊匹配的方式可以提高链接定位的准确率，也可以用于模糊匹配一组链接的情况。

8.6 使用 HTML 标签名定位

被测试网页的 HTML 代码：

```
<html>
    <body>
        关键字:<input id = "keyword"></input><br />
        <a href = "http://www.sogou.com"> sogou 搜索</a>
        <a href = "http://www.baidu.com"> baidu 搜索</a>
    </body>
</html>
```

定位语句：

```
input = driver.find_element_by_tag_name("input")
input = driver.find_element(by = "tag name", value = "input")
aList = driver.find_elements_by_tag_name("a")
aList = driver.find_elements(by = "tag name", value = "a")
```

代码解释：

语句 1 和语句 2 使用 tag name 方式定位页面中 HTML 标签名为“input”的页面元素，并将查找到的页面对象赋值给 input 变量。

语句 3 和语句 4 将定位页面中所有 HTML 标签名为“a”的页面元素，在被测试网页的 HTML 代码中可看到有两个 HTML 标签名为“a”的页面元素，这两个链接元素都会被定位到，并存于列表中，然后将该列表对象赋值给 aList 变量。

更多说明：

HTML 标签名称的定位方式主要用于匹配多个页面元素的情况，将查找到的网页元素对象进行计数、遍历、修改属性等操作。

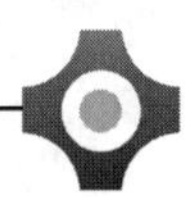

8.7 使用 Class 名称定位

被测试网页的 HTML 代码：

```
<html>
    <head>
        <style type = "text/css">
            input.spread { FONT - SIZE: 20pt;}
            input.tight { FONT - SIZE: 10pt;}
        </style>
    </head>
    <body>
        <input class = "spread" type = text></input>
        <input class = "tight" type = text></input>
    </body>
</html>
```

定位语句：

```
spread = driver.find_element_by_class_name("spread")
spread = driver.find_element(by = "class name", value = "spread")
tightList = driver.find_elements_by_class_name("tight")
tightList = driver.find_elements(by = "class name", value = "tight")
```

代码解释：

查看被测试网页的 HTML 代码，可以看到两个输入框均有 class 属性，spread 类定义的字体比 tight 类大 10pt。上面 4 条语句都是使用 class 属性名称来查找页面元素。

更多说明：

可以根据 class 属性值来查找一个或者一组显示效果相同的页面元素。

8.8 使用 XPath 定位

XPath 定位方式是自动化测试定位技术中的必杀技，几乎可以解决所有的定位难题，就算 HTML 标签没有 id、name 等属性，使用 XPath 也能轻松解决这些页面元素定位问题。强烈建议读者深入掌握该定位方式的详细使用方法。

8.8.1 关于 XPath

XPath 是 XML Path 语言的缩写，是一门在 XML 文档中查找信息的语言，它在 XML 文档中通过元素和属性进行导航，主要用于在 XML 文档中选择节点。基于 XML 树状文档结构，XPath 语言可以用于在整棵树中沿着路径或 step(步)来寻找指定的节点。XPath 定位和即将讲到的 CSS 定位相比具备更大的灵活性，在 XML 文档树中的某个节点既可以向前搜索，也可以向后搜索，而 CSS 定位只能在 XML 文档树中向前搜索，但 XPath 的定位速度比 CSS 稍慢。

XPath 使用路径表达式来选取 XML 文档中的节点或节点集。节点是通过沿着路径或者步来选取的。

8.8.2 XPath 节点

XPath 语言中提供了七种节点：文档节点（根节点）、元素、属性、文本、命名空间、处理指令以及注释。XML 文档被作为节点树对待，树的根被称作文档节点或根节点。

1. 节点

XML 实例文档：

```
<?xml version = "1.0" encoding = "utf - 8" ?>
<! -- 这是一个注释节点 -- >
< booklist type = "science and engineering">
    < book category = "Selenium">
        < title > WebDriver 实战宝典</title >
        < author >吴晓华</author >
        < pageNumber > 400 </pageNumber >
    </book >
</booklist >
```

在上面的 XML 文档实例中展示的节点如下：

```
< booklist >:文档节点
< title >:元素节点
type = "science and engineering":属性节点
```

2. 节点间关系

（1）父节点（Parent）。每个元素以及属性都有一个父节点。上面的 XML 文档实例中，book 元素是 title、author 以及 pageNumber 元素的父节点。

（2）子节点（Children）。一个元素节点可有零个、一个或多个子节点。上面的 XML 文档实例中，title、author 以及 pageNumber 元素是 book 元素的子节点。

（3）同胞节点（Sibling）。同胞节点表示拥有相同父节点的节点。上面的 XML 文档实例中，title、author 以及 pageNumber 元素都是同胞节点。

（4）先辈节点（Ancestor）。先辈节点表示的是某节点的父节点、父节点的父节点，以及父节点的所有祖先节点。上面的 XML 文档实例中，title 元素的先辈节点有 book 和 booklist。

（5）后代节点（Descendant）。后代节点表示某个节点的子节点、子节点的子节点，以及子节点的所有后代节点。上面的 XML 文档实例中，booklist 元素的后代节点有 book、title、author 以及 pageNumber 元素。

8.8.3 XPath 定位语法

由于用于 Web 开发的 HTML 语言的语法结构跟 XML 很相似，所以 XPath 也支持在 HTML 代码中定位 HTML 树状文档结构中的节点。后续小节均使用 HTML 代码实例来支持对 XPath 语法的讲解。

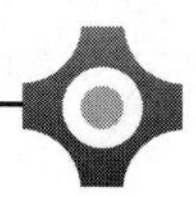

被测试网页的 HTML 代码：

```
<html>
    <body>
        <div id="div1" style="text-align:center">
            <img alt="div1-img1"
            src="http://www.sogou.com/images/logo/new/sogou.png"
            href="http://www.sogou.com">搜狗图片</img><br />
            <input name="div1input">
            <a href="http://www.sogou.com">搜狗搜索</a>
            <input type="button" value="查询">
        </div>
        <br>
        <div name="div2" style="text-align:center">
            <img alt="div2-img2" src="http://www.baidu.com/img/bdlogo.png"
            href="http://www.baidu.com">百度图片</img><br />
            <input name="div2input">
            <a href="http://www.baidu.com">百度搜索</a>
        </div>
    </body>
</html>
```

使用上面 HTML 代码生成被测试网页，基于此网页来实践各种不同的页面元素的 XPath 定位方法。

1. 使用绝对路径来定位元素

绝对路径表示页面元素在被测网页的 HTML 代码结构中，从根节点一层层地搜索到需要被定位的页面元素，绝对路径起始于正斜杠(/)，每一步均被斜杠分割。

目的：

在被测试网页中，查找第一个 div 标签下的“查询”按钮。

XPath 定位表达式：

```
/html/body/div/input[@value="查询"]
```

Python 定位语句：

```
query = driver.find_element_by_xpath('/html/body/div/input[@value="查询"]')
```

代码解释：

上述 XPath 定位表达式从 HTML DOM 树的根节点(HTML 节点)开始逐层查找，最后定位到“查询”按钮节点。路径表达式“/”表示根节点。

更多说明：

使用绝对路径定位页面元素的好处在于可以验证页面是否发生变化。如果页面结构发生变化，可能会造成原先有效的 XPath 表达式失效。使用绝对路径定位是十分脆弱的，因为即便页面代码结构只发生了微小的变化，也可能会造成原先有效的 XPath 定位表达式定位失败。因此，建议在自动化测试的定位实施环节中，优先考虑使用后面将要介绍的相对路径进行页面元素的定位。

后续无特殊要求，定位元素的方法默认均使用 find_element_by_*或 find_elements_by_*。

2. 使用相对路径定位元素

相对路径的每一步都根据当前节点集之中的节点来进行计算，起始于双正斜杠(//)。

目的：

在被测试网页中，查找第一个 div 标签下的“查询”按钮。

XPath 定位表达式：

```
//input[@value = "查询"]
```

Python 定位语句：

```
query = driver.find_element_by_xpath('//input[@value = "查询"]')
```

代码解释：

上述 XPath 定位表达式中的“//”表示从匹配选择的当前节点开始选择文档中的节点，而不考虑它们的位置。input[@value="查询"]表示定位 value 值为“查询”两个字的 input 页面元素。

更多说明：

相对路径的 XPath 定位表达式更加简洁，不管页面发生了何种变化，只要 input 标签的 value 属性值没变，始终都可以定位到。推荐使用相对路径的 XPath 表达式，并且越简洁越好，这样可大大降低测试脚本中定位表达式的维护成本。

1. 使用索引号定位元素

索引号表示某个被定位的页面元素在其父元素节点下的同名元素中的位置序号，需要从 1 开始。

目的：

在被测试网页中，查找第一个 div 标签下的“查询”按钮。

XPath 定位表达式：

```
//input[2]
Python 定位语句：
query = driver.find_element_by_xpath("//input[2]")
```

代码解释：

索引号定位方式是根据该页面元素在页面中相同标签名之间出现的索引位置来进行定位。上述 XPath 定位表达式表示查找页面中第二个出现的 input 元素，即被测试页面上的“查询”按钮。

更多说明：

若在 Firefox 浏览器的 Firepath 插件中使用“//input[1]”定位表达式进行页面元素定位，可以发现在被测试网页的 HTML 代码区域高亮显示了两行代码，两个 div 标签下的第一个 input 标签都被定位到，这和只查找第一个 input 元素相冲突，这是由于被测试网页中两个 div 标签下都包含了 input 标签，XPath 在查找的时候把每个 div 节点都当作相同的起始层级开始查找，所以用“//input[1]”表达式会同时查找到两个 div 节点下的第一个 input 元素。如果在两 div 标签下还有嵌套的 div，并且嵌套的 div 下也有 input 标签，使用“//input[1]”定位表达式，也会定位到嵌套 div 下的 input 标签，也就是说，无论嵌套多少层

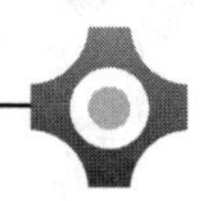

HTML标签，只要这些HTML标签的子标签里有input标签，第一个input标签都会被定位到。因此在使用索引号定位页面元素的时候，需要注意网页HTML代码中是否包含了多个层级完全相同的代码结构(比如本例中的两个div层级)，若出现了这种情况，就需要修改定位表达式，以确保自动化测试脚本中使用的定位表达式能唯一定位所需要的页面元素。

但如果想同时定位多个相同input页面元素，则可以使用如下Python语句：

```
inputList = driver.find_elements_by_xpath("//input[1]")
```

将定位的多个元素存储到list对象中，然后根据list对象的索引号获取想要的页面元素。但如果发现页面元素会经常增加或减少，就不建议使用索引号定位的方式，因为页面变化很可能会让使用索引号的XPath定位表达式定位失败。

基于实例中的被测试网页，下面给出更多的通过索引号定位的实例，如表8-2所示。

表 8-2

预期定位的页面元素	定位表达式实例	使用的属性值
定位第二个div下的超链接	//div[last()]/a	div[last()]表示最后一个div元素，last()函数获取的是指定元素的最后的索引号
定位第一个div中的超链接	//div[last()－1]/a	div[last()－1]表示倒数第二个div元素
定位最前面一个属于div元素的子元素中的input元素	//div/input[position()<2]	position()函数获取当前元素input的位置序列号

2. 使用页面元素的属性值定位元素

在定位页面元素的时候，经常会遇到各种复杂结构的被测试网页，并且很多页面元素也没有设计ID、Name等属性，同时又不想使用绝对路径或索引号来定位页面元素，但是发现要被定位的页面元素拥有某些固定不变的属性及属性值，此时推荐使用属性定位方式来定位页面元素。

目的：

定位被测试网页中的第一个img元素。

XPath定位表达式：

```
//img[@alt = "div1 - img1"]
```

Python定位语句：

```
img = driver.find_element_by_xpath('//img[@alt = "div1 - img1"]')
```

代码解释：

表达式使用了相对路径再结合元素拥有的特定属性的方法进行定位，定位元素img的属性是alt，其属性值为div1-img1，使用@符号指明后面接的是属性，并同属性及属性值一起写到元素后的方括号中。

更多解释：

被测试网页的元素通常会包含各种各样的属性值，并且很多属性值具有唯一性。若能确认属性值不常变并且唯一，强烈建议使用相对路径再结合属性的定位方式来编写XPath定位表达式，使用此方法可以解决99%的页面元素定位难题。基于实例中的被测试网页，下面给出更多的属性值定位实例，如表 8-3 所示。

表 8-3

预期定位的页面元素	定位表达式实例	使用的属性值
定位页面的第一张图片	//img[@href='http://www.sogou.com']	使用 img 标签的 href 属性值
定位第二个 div 中第一个 input 输入框	//div[@name='div2']/input[@name='div2input']	使用 div 标签的 name 属性值 使用 input 标签的 name 属性值
定位第一个 div 中的第一个链接	//div[@id='div1']/a[@href='http://www.sogou.com']	使用 div 标签的 ID 属性值 使用 a 标签的 href 属性值
定位页面的查询按钮	//input[@type='button']	使用 type 属性值

3. 使用模糊属性值定位元素

模糊属性值定位方式表示使用属性值的一部分内容进行定位。在自动化测试的实施过程中，常常会遇到页面元素的属性值是动态生成的，也就是说每次访问属性值都不一样，此类页面元素会加大定位难度，使用模糊属性值定位方式可以解决一部分此类难题，但前提是属性值中有一部分内容保持不变。XPath 提供了一些可实现模糊属性值的定位需求的函数，如表 8-4 所示。

表 8-4

XPath 函数	定位表达式实例	表达式解释
starts-with(str1, str2)	//img[starts-with(@alt,'div1')]	查找属性 alt 的属性值以"div1"关键字开始的页面元素
contains((str1, str2)	//img[contains(@alt,'img')]	查找 alt 属性的属性值包含"img"关键字的页面元素，只要包含即可，无须考虑位置

contains()函数属于 XPath 的高级用法，使用场景比较多，尽管页面元素的属性值经常发生变化，但只要其属性值有几个固定不变的关键词，就可以使用 contains()函数进行定位。

4. 使用 XPath 轴(Axes)定位元素

轴可以定义相对于当前节点的节点集。使用 XPath(Axes)定位方式可根据在文档树中的元素相对位置关系进行页面元素定位。先找到一个相对好定位的元素，让它作为轴，根据它和要定位元素间的相对位置关系进行定位，可解决一些元素难以定位的问题。

根据本节提供的被测试网页 HTML 代码，画出一棵图形化文档树状图，如图 8-1 所示。

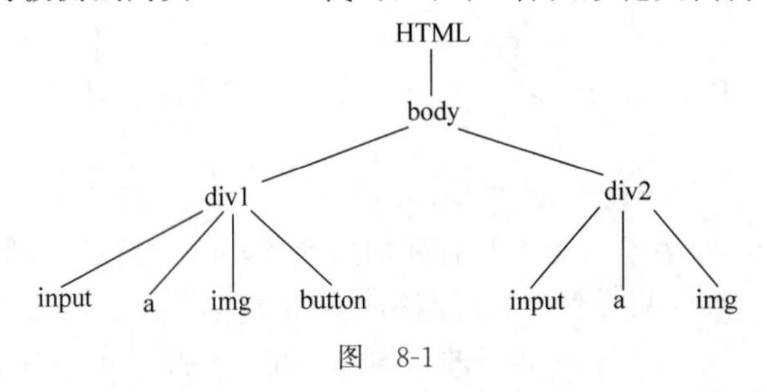

图 8-1

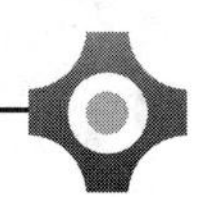

XPath 常用轴关键字如表 8-5 所示。

表　8-5

XPath 轴关键字	轴的含义说明	定位表达式实例	表达式解释
parent	选择当前节点的上层父节点	//img[@alt='div2-img2']/parent::div	查找到属性 alt 的属性值为 div2-img2 的 img 元素，并基于该 img 元素的位置找到它上一级的 div 页面元素
child	选择当前节点的下层所有子节点	//div[@id='div1']/child::img	查找到 ID 属性值为 div1 的 div 元素，并基于 div 的位置找到它下层节点中的 img 页面元素
ancestor	选择当前节点所有上层的节点	//img[@alt='div2-img2']/ancestor::div	查找到属性 alt 的属性值为 div2-img2 的 img 元素，并基于该 img 元素的位置找到它上级的 div 页面元素
descendant	选择当前节点所有下层的节点(子、孙等)	//div[@name='div2']/descendant::img	查找到属性 name 的属性值为 div2 的 div 页面元素，并基于该元素的位置找到它下级所有节点中的 img 页面元素
following	选择在当前节点之后显示的所有节点	//div[@id='div1']/following::img	查找到 ID 属性值为 div1 的 div 页面元素，并基于 div 的位置找到它后面节点中的 img 页面元素
following-sibling	选择当前节点后续所有兄弟节点	//a[@href='http://www.sogou.com']/following-sibling::input	查找到链接地址为 http://www.sogou.com 的链接页面元素 a，并基于链接的位置找到它后续兄弟节点中的 input 页面元素
preceding	选择当前节点前面的所有节点	//img[@alt='div2-img2']/preceding::div	查找到属性 alt 的属性值为 div2-img2 的图片页面元素 img，并基于图片的位置找到它前面节点中的 div 页面元素
preceding-sibling	选择当前节点前面的所有兄弟节点	//input[@value='查询']/preceding-sibling::a[1]	查找到 value 属性值为“查询”的输入框页面元素，并基于该输入框的位置找到它前面同级节点中的第一个链接页面元素

更多说明：

有时候我们会在轴后面加一个星号(*)，表示通配符，比如//input[@value='查询']/preceding-sibling::*，它表示查找属性 value 的值为“查询”的输入框 input 元素前面所有的同级元素，但不包括 input 元素本身。

5. 使用页面元素的文本定位元素

通过 text()函数可以定位到元素文本包含某些关键内容的页面元素。

XPath 表达式：

(1) //a[text()="搜狗搜索"]

(2) //a[.="搜狗搜索"]

(3) //a[contains(., "百度")]

(4) //a[contains(text(), "百度")]

(5) //a[contains(text(), "百度")]/preceding::div

(6) //a[contains(., "百度")]/..

Python 定位语句：

```
sogou_a = driver.find_element_by_xpath('//a[text() = "搜狗搜索"]')
sogou_a = driver.find_element_by_xpath('//a[. = "搜狗搜索"]')
baidu_a = driver.find_element_by_xpath('//a[contains(., "百度")]')
baidu_a = driver.find_element_by_xpath('//a[contains(text(), "百度")]')
div = driver.find_element_by_xpath('//a[contains(text(), "百度")]/preceding::div')
div = driver.find_element_by_xpath('//a[contains(., "百度")]/..')
```

代码解释：

- XPath 表达式 1 和表达式 2 等价，都是查找文本内容为“搜狗搜索”的链接页面元素，使用的是精准匹配方式，也就是说文本内容必须完全匹配，不能多一个字也不能少一个字。第二个 XPath 语句中使用了一个点(.)，这里的点等价于 text()函数，都指代的是当前节点的文本内容。
- XPath 表达式 3 和表达式 4 等价，都是查找文本内容包含“百度”关键字的链接页面元素，使用的是模糊匹配方式，即可以根据部分文本关键字进行匹配。
- XPath 表达式 5 和表达式 6 等价，都是查找文本内容包含“百度”关键字的链接页面元素 a 的上层父元素 div，第 6 句 XPath 表达式最后使用了两个点(..)，它表示选取当前节点的父节点，等价于 preceding::div。

更多说明：

使用文本内容匹配模式进行定位，为定位复杂的页面元素又提供了一种强大的定位模式，在遇到定位困难时，可以优先考虑使用此方式进行定位。建议读者对此定位方式进行大量练习，以便做到可随意定位页面元素中的任意元素。

8.8.4 XPath 运算符

XPath 也提供了一下运算符来更好地支持定位页面元素，XPath 表达式可返回节点集、字符串、逻辑值以及数字等，读者可以根据自己的需求选择使用。

被测试网页的 HTML 代码：

```
<html>
    <body>
        <div id="div1" style="text-align:center">
            <img alt="div1-img1"
            src="http://www.sogou.com/images/logo/new/sogou.png"
            href="http://www.sogou.com">搜狗图片</img><br />
            <a>10</a>
            <span>13</span>
            <input name="div1input">
            <a href="http://www.sogou.com">搜狗搜索</a>
            <input type="button" value="查询">
        </div>
        <br>
        <div name="div2" style="text-align:center">
            <img alt="div2-img2" src="http://www.baidu.com/img/bdlogo.png"
            href="http://www.baidu.com">百度图片</img><br />
```

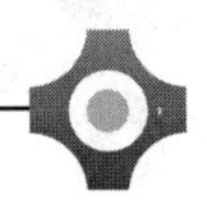

```
            <input name="div2input">
            <a href="http://www.baidu.com">百度搜索</a>
        </div>
    </body>
</html>
```

XPath 运算符如表 8-6 所示。

表　8-6

XPath 运算符	含　义	实　例	表达式解释
|	获取两个或多个节点集	//div | //a //div | //a | //input	返回所有拥有 div 和 a 元素的节点集 返回所有拥有 div 和 a 和 input 元素的节点集
+	两数或多数连加	1+ 1 + 2 //div[1 + 1]	返回数字 4 查找页面中第二个 div 元素
−	两数或多数连减	10−1−3	返回数字 6
*	两数或多数连乘	2 * 1 * 4	返回数字 8
div	两数或多数连除	10 div 2	返回数字 5
=	等于	//span = 10 //div[a=10]	如果 span 是 10,返回 True,否则返回 False 查找子元素中有 a 元素,并且其文本内容为数字且为 10 的 div 元素
!=	不等于	//span! =12 //div[a! =10]	如果 span 不是 12,返回 True,否则返回 False 查找子元素中有 a 元素,并且其文本内容为数字且不为 10 的 div 元素
<	小于	//span < 12 //div[a < 11]	如果 span 小于 12,返回 True,否则返回 False 查找子元素中有 a 元素,并且其文本内容为数字且小于 11 的 div 元素
<=	小于等于	//span <=12 //div[a <=10]	如果 span 小于等于 12,返回 True,否则返回 False 查找子元素中有 a 元素,并且其文本内容为数字且小于等于 10 的 div 元素
>	大于	//span > 12 //div[a > 9]	如果 span 大于 12,返回 True,否则返回 False 查找子元素中有 a 元素,并且其文本内容为数字且大于 9 的 div 元素
>=	大于等于	//span >=12 //div[a >=10]	如果 span 大于等于 12,返回 True,否则返回 False 查找子元素中有 a 元素,并且其文本内容为数字且大于等于 10 的 div 元素
and	与	//span > 1 and span <= 3 //div[a <=10 and span=13]	如果 span 大于 1 并且 span 小于等于 3,返回 True,否则返回 False 查找子元素中有 a 元素,并且其文本内容为数字且小于等于 10,同时满足存在子元素 span,其文本内容也为数字且等于 13 的 div 元素

续表

XPath 运算符	含　义	实　例	表达式解释
or	或	//span =10 or span =20 //div[a<=10 or span=13]	如果 span 等于 10 或者 span 等于 20，返回 True，否则返回 False 查找子元素中有 a 元素，并且其文本内容为数字且小于等于 10，或者存在子元素 span，其文本内容也为数字且等于 13 的 div 元素
mod	取余	7 mod 3	返回数字 1

8.9 CSS 定位

CSS 定位方式和 XPath 定位方式很类似，能够解决大部分常见的定位难题。如果读者已经深入掌握了 XPath 定位方式的使用方法，可以选择跳过此节。

8.9.1 关于 CSS

CSS 是英文 Cascading Style Sheets 的缩写，中文意思指层叠样式表。CSS 是一种用于表现 HTML 或 XML 等文件样式的前端页面语言，主要用于描述页面元素的展现和样式的定义。

8.9.2 CSS 定位语法

CSS 定位方式和 XPath 定位方式基本相同，只是 CSS 定位表达式有其自己的格式。CSS 定位方式拥有比 XPath 定位速度快，且比 XPath 稳定的特性。下面详细介绍一下 CSS 定位方式的使用方法。

被测试网页 HTML 代码：

```
<html>
<head>
    <style type="text/css">
        input.spread { FONT-SIZE: 20pt;}
        input.tight { FONT-SIZE: 10pt;}
    </style>
</head>
<body onload="document.getElementById('div1input').focus()">
    <div id="div1" style="text-align:center">
        <input id="div1input" class="spread"></input>
        <a href="http://www.sogou.com">搜狗搜索</a>
        <img alt="div1-img1"src=http://www.sogou.com/images/logo/new/sogou.png
        href="http://www.sogou.com">搜狗图片</img>
        <input type="button" value="查询"></input>
    </div>
    <br>
    <p>第一段文字:时间管理好,每天学习 1 小时,改变只会手工测试的命运</p>
```

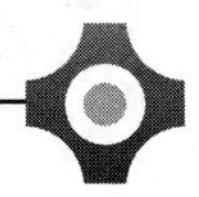

```
    <p>第二段文字:现在不努力,老大搞 IT</p>
    <p>第三段文字:1 万小时理论,1 万小时的努力和积累让你与众不同</p>
    <input type = "checkbox" >学习</input>
    <div name = "div2" style = "text - align:center">
        <input name = "div2input" class = "tight"></input>
        <a href = "http://www.baidu.com">百度搜索</a>
        <img alt = "div2 - img2" src = "http://www.baidu.com/img/bdlogo.png"
        href = "http://www.baidu.com">百度图片</img>
    </div>
    <div class = "foodDiv">
        <ul id = "recordlist">
            <p>土豆</p>
            <li>黄瓜</li>
            <li>西红柿</li>
            <li>冬瓜</li>
            <li>茄子</li>
        </ul>
    </div>
</body>
</html>
```

使用上面 HTML 代码生成被测试网页,基于此网页来实践各种不同的页面元素的 CSS 定位方法。

1. 使用绝对路径定位元素

目的:

在被测试网页中,查找第一个 div 元素中的"查询"按钮。

CSS 定位表达式:

```
html > body > div > input[value = "查询"]
```

Python 定位语句:

```
query = driver.find_element_by_css_selector('html > body > div > input[value = "查询"]')
```

代码解释:

上述 CSS 定位表达式使用绝对路径定位属性 value 的值为"查询"的页面元素。从 CSS 定位表达式可以看出,步间通过">"符号分割,区别于 XPath 路径中的正斜杠(/),并且也不再使用@符号选择属性。

提示:

不推荐在频繁变化的被测试页面上使用绝对路径方式定位页面元素。

2. 使用相对路径定位元素

目的:

在被测试网页中,查找第一个 div 元素中的"查询"按钮。

CSS 定位表达式:

```
input[value = "查询"]
```

Python 定位语句：

```
query = driver.find_element_by_css_selector('input[value="查询"]')
```

代码解释：

上述 CSS 表达式通过相对路径使用元素名称和元素的属性及属性值进行页面元素的定位。

3. 使用 class 名称定位元素

目的：

在被测试网页中，查找第一个 div 元素下的 input 输入框。

CSS 定位表达式：

```
input.spread
```

Python 定位语句：

```
input = driver.find_element_by_css_selector('input.spread')
```

代码解释：

上述 CSS 定位表达式使用 input 页面元素的 class 属性名称 spread 来进行定位，用点(.)分隔元素名与 class 属性名，点后面是 class 属性名称。

4. 使用 ID 属性值定位元素

目的：

在被测试网页中，查找第一个 div 元素下 ID 属性值为 div1input 的 input 页面元素。

CSS 定位表达式：

```
input#div1input
```

Python 定位语句：

```
input = driver.find_element_by_css_selector('input.spread')
```

代码解释：

上述 CSS 定位表达式使用 input 页面元素的 ID 属性值 div1input 进行定位，使用#分割元素名和 ID 属性值，#后面是 ID 属性值。

5. 使用页面其他属性值定位

目的：

在被测试网页中，查找 div 元素下的第一个图片元素 img。

CSS 定位表达式：

```
表达式 1:img[alt="div1-img1"]
表达式 2:img[alt="div1-img1"][href="http://www.sogou.com"]
```

Python 定位语句：

```
img = driver.find_element_by_css_selector('img[alt="div1-img1"]')
img = driver.find_element_by_css_selector('img[alt="div1-img1"][href="
```

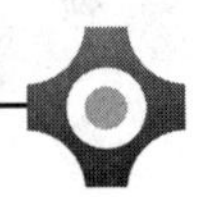

```
http://www.sogou.com"]')
```

代码解释：

- CSS 表达式 1 和 CSS 表达式 2 在本章节提供的被测试网页中是等价的，都是定位的第一个 div 元素下的 img 元素。
- CSS 表达式 1：表示使用 img 页面元素的 alt 属性值 div1-img1 进行定位。若想定位的页面元素始终具有唯一的属性值，此定位方法可解决页面频繁变化的部分定位难题。
- CSS 表达式 2：表示同时使用了 alt 属性和 href 属性进行页面元素的定位。在某些复杂的定位场景，可使用多个属性来确保定位元素的唯一性。

6. 使用属性值的一部分内容定位元素

目的：

在被测试网页中，查找"搜狗搜索"链接。

CSS 定位表达式：

```
表达式 1:a[href^ = "http://www.so"]
表达式 2:a[href $ = "gou.com"]
表达式 3:a[href * = "so"]
```

Python 定位语句：

```
a = driver.find_element_by_css_selector('a[href^ = "http://www.so"]')
a = driver.find_element_by_css_selector('a[href $ = "gou.com"]')
a = driver.find_element_by_css_selector('a[href * = "so"]')
```

代码解释：

- 上述 3 个 CSS 定位表达式在本章提供的被测试网页中是等价的，都是查找"搜狗搜索"链接。
- CSS 表达式 1：表示匹配链接地址开始包含 http://www.so 关键字串的链接元素，以字符"^"指明从字符串的开始匹配。
- CSS 表达式 2：表示匹配链接地址结尾包含 gou.com 关键字串的链接元素，以字符"$"指明在字符串的结尾匹配。
- CSS 表达式 3：表示匹配链接地址包含 so 关键字串的链接元素，以符号"*"指明需要进行模糊匹配。

更多说明：

使用此模糊定位方式，可匹配动态变化的属性值的页面元素，只要找到属性值固定不变的关键部分，就可以进行模糊匹配定位。此方法可以解决大部分复杂定位的难题。

7. 使用页面元素进行子页面元素的查找

目的：

在被测试网页中，查找第一个 div 下的第一个 input 页面元素。

CSS 定位表达式：

```
表达式 1:div#div1 > input#div1input
```

```
表达式 2:div input
```

Python 定位语句：

```
input = driver.find_element_by_css_selector('div#div1 > input#div1input')
inputList = driver.find_elements_by_css_selector('div input')
```

代码解释：

- CSS 表达式 1 中的 div#div1，表示在被测试网页上定位到 ID 属性值为 div1 的 div 页面元素，“>”表示在以查找到的 div 元素的子页面元素中进行查找，input#div1input 表示查找 ID 属性值为 div1input 的 input 页面元素。此方法可实现查找 div 下子页面元素的目的。
- 表达式 2 表示匹配所有属于 div 元素后代的 input 元素，父元素 div 和子元素 input 间必须用空格分隔。

8. 使用伪类定位元素

目的：

在被测试网页中，查找第一个 div 下的指定子页面元素。

CSS 定位表达式：

```
表达式 1:div#div1 :first-child
表达式 2:div#div1 :nth-child(2)
表达式 3:div#div1 :last-child
表达式 4:input:focus
表达式 5:input:enabled
表达式 6:input:checked
表达式 7:input:not([id])
```

Python 定位语句：

```
frist_elem = driver.find_element_by_css_selector('div#div1 :first-child')
second_elem = driver.find_element_by_css_selector('div#div1 :nth-child(2)')
last_elem = driver.find_element_by_css_selector('div#div1 :last-child')
focus = driver.find_element_by_css_selector('input:focus')
enabled = driver.find_elements_by_css_selector('input:enabled')
checked = driver.find_elements_by_css_selector('input:checked')
inputList = driver.find_elements_by_css_selector('input:not([id])')
```

代码解释：

伪类表达式是 CSS 语法支持的定位方式，前 3 个 CSS 定位表达式要特别注意的是，在冒号(:)前一定要有一个空格，否则就会定位不到期望的页面元素。

- CSS 表达式 1：表示查找 ID 属性值为 div1 的 div 页面元素下的第一个子元素，参考被测试网页的 HTML 可以看到定位到的页面元素是 input 元素，:first-child 表示查找某个页面元素下的第一个子页面元素。
- CSS 表达式 2：表示查找 ID 属性值为 div1 的 div 页面元素下的第二个子元素，参考被测试网页的 HTML 可看到定位到的页面元素是一个链接元素，:nth-child(2)表示查找某个页面元素下的第二个子页面元素，如果改成:nth-child(3)则表示某个页

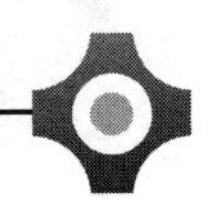

面元素下的第三个子元素，以此类推。

- CSS 表达式 3：表示查找 ID 属性值为 div1 的 div 页面元素下的最后一个子元素，参考被测试网页的 HTML 可看到定位到的页面元素是按钮元素，:last-child 表示查找某个页面元素下的最后一个子页面元素。
- CSS 表达式 4：表示查找当前获取焦点的 input 页面元素。
- CSS 表达式 5：表示查找可操作的 input 页面元素。
- CSS 表达式 6：表示查找处于勾选状态的 checkbox 页面元素。
- CSS 表达式 7：表示查找所有无 id 属性的 input 页面元素。

更多说明：

伪类定位方式可基于子元素的相对位置和元素的状态进行定位，此定位方式可解决自动化测试中一部分页面元素定位难的问题。

9. 查找同级兄弟页面元素

目的：

在被测试网页中，查找第一个 div 下第一个 input 子页面元素的同级兄弟页面元素。

CSS 定位表达式：

```
表达式 1:div#div1 > input + a
表达式 2:div#div1 > input + a+ img
表达式 3:div#div1 > input + * + img
表达式 4:ul#recordlist > p~li
```

Python 定位语句：

```
a = driver.find_element_by_css_selector('div#div1 > input + a')
img = driver.find_element_by_css_selector('div#div1 > input + a+ img')
img = driver.find_element_by_css_selector('div#div1 > input + * + img')
tagList = driver.find_elements_by_css_selector('ul#recordlist > p~li')
```

代码解释：

- CSS 表达式 1：表示在 ID 属性值为 div1 的 div 页面元素下，查找 input 页面元素后面的同级且临近的链接元素 a。
- CSS 表达式 2：表示在 ID 属性值为 div1 的 div 页面元素下，查找 input 页面元素和链接元素后面的同级且临近的图片元素 img。
- CSS 表达式 3：表示在 ID 属性值为 div1 的 div 页面元素下，查找 input 页面元素和任一种页面元素后面的同级且临近的图片元素 img，* 表示任意类型的一个页面元素，只能表示一个页面元素，如果想用此种方法查找第一个 div 下的最后一个 input 元素，表达式写法为 div#div1 > input+ * + * +input 或 div#div1 > input+a+ * +input 或 div#div1 > input+a+img+input 等。
- CSS 表示式 4：表示在 ID 属性值为 recordlist 的 ul 页面元素下，查找 p 页面元素以后所有的 li 元素。

更多说明：

使用此方法可基于相对位置和页面元素类型来定位页面元素，星号(*)在此处表示通配符，可以表示任意类型的页面元素名。

10. 多元素选择器

CSS定位方式支持多元素选择器，也就是一次可以同时选择多个相同的标签，也可以同时选择多个不同的标签，不同标签间用英文的逗号(,)隔开。

目的：

在被测试网页中，同时选择多个不同的页面元素。

CSS定位表达式：

```
div#div1,input,a
```

Python定位语句：

```
tagList = driver.find_elements_by_css_selector('div#div1,input,a')
```

代码解释：

上述CSS表达式表示同时查找所有的ID属性值为div1的div元素，所有的input元素，所有的a元素。

8.9.3 XPath定位与CSS定位的比较

XPath定位方式与CSS定位方式很相似，XPath定位功能相对更强大一些，但CSS定位方式执行速度更快。鉴于某些浏览器并不支持CSS定位方式，并且一般在自动化测试实施过程中使用XPath定位方式要比使用CSS定位方式更普遍，所以建议读者应首先掌握XPath定位方式。

XPath和CSS 3常用表达式语法比较如表8-7所示。

表 8-7

定位元素目标	XPath	CSS 3
所有元素	//*	*
所有的div元素	//div	div
所有的div元素的子元素	//div/*	div> *
根据ID属性获取元素	//*[@id='div1']	#div1
根据class属性获取元素	//*[contains(@class,'spread')]	.spread
拥有某个属性的元素	//*[@herf]	*[href]
所有div元素的第一个子元素	//div/*[1]	div> * :first-child
所有拥有子元素a的div元素	//div[a]	无法实现
input的下一个兄弟元素	//input/following-sibling::*[1]	input+ *

8.10 表格的定位

浏览器网页常常会包含各类表格，自动化测试工程师可能会经常操作表格中的行、列以及一些特定的单元格，因此熟练掌握表格定位方法是自动化测试实施过程中必要的技能。

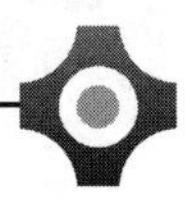

8.10.1　遍历表格所有的单元格

被测试网页的 HTML 代码：

```
<html>
<body>
    <table width="400" border="1" id="table">
        <tr>
            <td align="left">消费项目....</th>
            <td align="right">一月</th>
            <td align="right">二月</th>
        </tr>
        <tr>
            <td align="left">衣服</td>
            <td align="right">1000 元</td>
            <td align="right">500 元</td>
        </tr>
        <tr>
            <td align="left">化妆品</td>
            <td align="right">3000 元</td>
            <td align="right">500 元</td>
        </tr>
        <tr>
            <td align="left">食物</td>
            <td align="right">3000 元</td>
            <td align="right">650.00 元</td>
        </tr>
        <tr>
            <td align="left">总计</th>
            <td align="right">7000 元</th>
            <td align="right">1150 元</th>
        </tr>
    </table>
</body>
</html>
```

被测试页面内容展现如图 8-2 所示。

消费项目....	一月	二月
衣服	1000元	500元
化妆品	3000元	500元
食物	3000元	650.00元
总计	7000元	1150元

图　8-2

Python 实例代码：

```
#encoding=utf-8
```

```
from selenium import webdriver

driver = webdriver.Firefox(executable_path = "c:\\geckodriver")
driver.get(r"D:\table.html")
# 获取 id 定位方式获取整个表格对象
table = driver.find_element_by_id("table")
# 通过标签名获取表格中的所有行对象
trList = table.find_elements_by_tag_name("tr")
assert len(trList) == 5, "表格行数不符!"
for row in trList:
    # 遍历行对象,并获取每一行中所有列对象
    tdList = row.find_elements_by_tag_name("td")
    for col in tdList:
        # 遍历表格中的列,并打印单元格内容
        print(col.text + "\t",)
    print()
driver.quit()
```

输出结果：

```
消费项目....            一月              二月
衣服                   1000 元           500 元
化妆品                 3000 元           500 元
食物                   3000 元           650.00 元
总计                   7000 元           1150 元
```

实例代码逻辑：

(1) 先获取整个表格的页面对象。

```
table = driver.find_element_by_id("table")
```

(2) 在表格页面元素对象中,获取所有的 tr 元素对象,并存储在 trList 对象中。

```
trList = table.find_elements_by_tag_name("tr")
```

(3) 循环遍历存储表格行对象的 trList 对象,每获取一行中所有的单元格对象(并存储在 tdList 对象中),就循环遍历一次,并将每个单元格的文本内容输出。

```
for row in trList:
    # 遍历行对象,并获取每一行中所有列对象
    tdList = row.find_elements_by_tag_name("td")
    for col in tdList:
        # 遍历表格中的列,并打印单元格内容
        print col.text + "\t",
    print
```

代码 col.text 用于获取单元格的文本内容。

以上步骤完成表格中所有单元格的遍历输出,通过遍历可以实现读取任意单元格内容的操作。

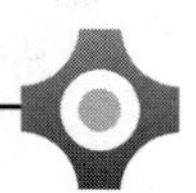

8.10.2　定位表格中的某个元素

目的：

在被测试网页中，定位显示表格的第二行第二列单元格。

XPath 表达式：

```
//*[@id='table']/tbody/tr[2]/td[2]
```

Python 定位语句：

```
cell = driver.find_element_by_xpath("//*[@id='table']/tbody/tr[2]/td[2]")
```

代码解释：

表达式中的 tr[2]表示第二行，td[2]表示第二列，组合起来就是第二行第二列的单元格。

CSS 表达式：

```
table#table>tbody>tr:nth-child(2)>td:nth-child(2)
```

Python 定位语句：

```
cell = driver.find_element_by_css_selector("table#table>tbody>tr:nth-child(2)>
td:nth-child(2)")
```

代码解释：

tr:nth-child(2)表示第二行，td:nth-child(2)表示第二列，组合起来就是第二行第二列的单元格。

8.10.3　定位表格中的子元素

被测试网页的 HTML 代码：

```
<html>
<body>
    <table width="400" border="1" id="table">
        <tr>
            <td align="left">消费项目....</th>
            <td align="right">一月</th>
            <td align="right">二月</th>
        </tr>
        <tr>
            <td align="left">衣服:
                <input type='checkbox'>外套</input>
                <input type='checkbox'>内衣</input>
            </td>
            <td align="right">1000 元</></td>
            <td align="right">500 元</td>
        </tr>
        <tr>
            <td align="left">化妆品:
```

```
                <input type='checkbox'>面霜</input>
                <input type='checkbox'>沐浴露</input>
            </td>
            <td align="right">3000 元</td>
            <td align="right">500 元</td>
        </tr>
        <tr>
            <td align="left">食物:
                <input type='checkbox'>主食</input>
                <input type='checkbox'>蔬菜</input>
            </td>
            <td align="right">3000 元</td>
            <td align="right">650.00 元</td>
        </tr>
        <tr>
            <td align="left">总计</th>
            <td align="right">7000 元</th>
            <td align="right">1150 元</th>
        </tr>
    </table>
</body>
</html>
```

页面内容展现如图 8-3 所示。

消费项目....	一月	二月
衣服 ☐外套 ☐内衣	1000元	500元
化妆品 ☐面霜 ☐沐浴露	3000元	500元
食物 ☐主食 ☐蔬菜	3000元	650.00元
总计	7000元	1150元

图 8-3

目的：

在被测试网页中，定位表格中第三行中“面霜”文字前的复选框。

XPath 表达式：

```
//td[contains(., "化妆")]/input[1]
```

Python 定位语句：

```
checkbox = driver.find_element_by_xpath('//td[contains(., "化妆")]/input[1]')
```

代码解释：

先找到包含子元素的单元格，在此单元格中再寻找子元素即可。表达式//td[contains(.,"化妆")]表示模糊匹配文本内容包含“化妆”关键字的单元格 td 元素，/input[1]表示找到的单元格 td 下的第一个 input 子元素。也可以通过 XPath 轴方式来查找该子元素，比如//td[contains(text(),'化妆')]/ descendant::input[1]。

第二篇　实战应用篇

第9章　WebDriver的多浏览器测试

本章主要讲 WebDriver 编写的脚本实例在不同浏览器上执行，在 Selenium WebDriver 官方支持的浏览器中，这里只针对 IE、Chrome 和 Firefox 这三个浏览器进行讲解，其他浏览器（比如：Opera、Safari）原理基本一样，请读者自行练习。

9.1 使用 IE 浏览器进行测试

环境准备：

（1）在使用 IE 浏览器进行 WebDriver 自动化测试之前，需要从 http://docs.seleniumhq.org/download/网站上下载一个 WebDriver 连接 IE 浏览器的驱动程序，文件名为 IEDriverServer.exe。下载页面的下载链接如图 9-1 所示。

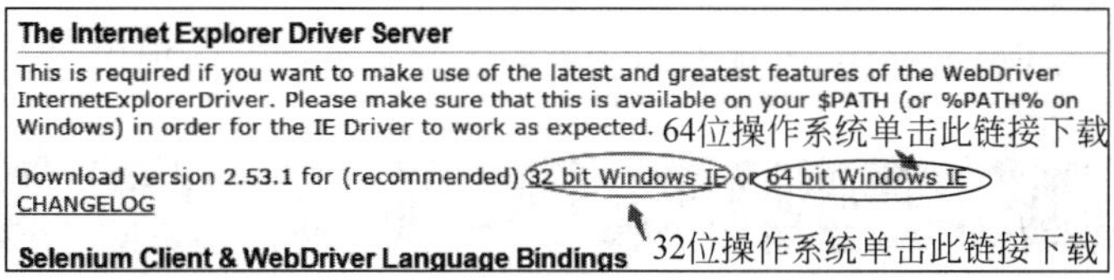

The Internet Explorer Driver Server

This is required if you want to make use of the latest and greatest features of the WebDriver InternetExplorerDriver. Please make sure that this is available on your $PATH (or %PATH% on Windows) in order for the IE Driver to work as expected. 64位操作系统单击此链接下载

Download version 2.53.1 for (recommended) 32 bit Windows IE or 64 bit Windows IE
CHANGELOG

Selenium Client & WebDriver Language Bindings 32位操作系统单击此链接下载

图　9-1

（2）解压下载后的压缩文件，并将其里面的 IEDriverServer.exe 文件保存本地磁盘的任意位置，比如 C:\下。

基于 unittest 框架的 Python 实例代码：

```
#encoding = utf - 8
from selenium import webdriver
import unittest

class VisitSogouByIE(unittest.TestCase):

    def setUp(self):
        # 启动 IE 浏览器
        self.driver = webdriver.Ie(executable_path = "c:\\IEDriverServer")

    def test_visitSogou(self):
        # 访问搜索首页
```

```
        self.driver.get("http://www.sogou.com")
        # 打印当前网页的网址
        print(self.driver.current_url)

    def tearDown(self):
        # 退出 IE 浏览器
        self.driver.quit()

if __name__ == '__main__':
    unittest.main()
```

执行输出结果：

```
https://www.sogou.com/
.
----------------------------------------------------------------------
Ran 1 test in 18.027s

OK
```

代码解释：

(1) 在 setUp()函数中通过 webdriver.Ie()方法获取 IE 浏览器的对话实例，函数所传的参数 executable_path = "c:\\IEDriverServer"指明了 WebDriver 连接 IE 浏览器所用驱动程序的存放路径。

(2) 在测试方法 test_visitSogou()中实现的是访问搜狗首页，并打印当前网页访问的网址。

(3) 在 tearDown()方法中，实现关闭浏览器实例等后期的清理工作。

9.2 使用 Firefox 浏览器进行测试

环境准备：

(1) 需要本地操作系统安装 Firefox 浏览器。

(2) 如果用的 Selenium 3 的版本，需要下载 WebDriver 连接 Firefox 浏览器的驱动程序文件，详细操作步骤见 5.2 小节。但如果用的是 Selenium 2 版本，则不需要准备驱动程序。

基于 unittest 的 Python 实例代码：

```
#encoding = utf-8
from selenium import webdriver
import unittest

class VisitSogouByFirefox(unittest.TestCase):

    def setUp(self):
        # 启动 Firefox 浏览器
        self.driver = webdriver.Firefox(executable_path = "c:\\geckodriver")
```

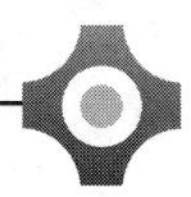

```
    def test_visitSogou(self):
        # 访问搜索首页
        self.driver.get("http://www.sogou.com")
        # 打印当前网页的网址
        print(self.driver.current_url)

    def tearDown(self):
        # 退出 Firefox 浏览器
        self.driver.quit()

if __name__ == '__main__':
    unittest.main()
```

更多说明：

使用 Firefox 浏览器进行自动化测试时，都默认 Firefox 浏览器安装在默认路径下，后续章节也一样，如果未安装到默认路径下，解决办法详见 5.2 小节。

9.3 使用 Chrome 浏览器进行测试

环境准备：

(1) 使用 WebDriver 在 Chrome 浏览器上进行测试时，需要从 http://chromedriver.storage.googleapis.com/index.html 网址下载 WebDriver 操作 Chrome 浏览器的驱动程序，需要下载的程序文件名为 chromedriver.exe，笔者这里选择的是 2.43 版(注意，2.43 版本的驱动要求 Chrome 浏览器版本必须是 70 的版本，Chrome 浏览器历史版本下载地址 https://www.chromedownloads.net/chrome64win/)。下载页面如图 9-2 所示，读者可以根据自己的操作系统类型选择相应的版本下载。

Index of /2.43/

Name	Last modified	Size
Parent Directory		-
chromedriver_linux64.zip	2018-10-17 02:46:13	3.89MB
chromedriver_mac64.zip	2018-10-17 04:09:49	5.71MB
chromedriver_win32.zip	2018-10-17 03:01:50	3.45MB
notes.txt	2018-10-17 05:00:45	0.02MB

图　9-2

(2) 解压下载后的文件，将 chromedriver.exe 文件保存在本地硬盘的任意位置，比如 C:\下。

基于 unittest 的 Python 实例代码：

```
#encoding = utf-8
from selenium import webdriver
import unittest

class VisitSogouByChrome(unittest.TestCase):
```

```
    def setUp(self):
        # 启动 Chrome 浏览器
        self.driver = webdriver.Chrome(executable_path = "c:\\chromedriver")

    def test_visitSogou(self):
        # 访问搜索首页
        self.driver.get("http://www.sogou.com")
        # 打印当前网页的网址
        print(self.driver.current_url)

    def tearDown(self):
        # 退出 Chrome 浏览器
        self.driver.quit()

if __name__ == '__main__':
    unittest.main()
```

说明：

在实施自动化测试过程中，经常会遇到正确的程序执行报错，而且有些错报还不清不楚。此时我们应该首先根据错误信息检查测试代码，如果在确定代码没有问题的情况下，考虑更换浏览器版本、驱动版本等方法来解决，因为 Selenium 3 版本开始，浏览器驱动均由各浏览器官方提供支持，并且浏览器的更新速度远超驱动更新速度，由此出现驱动不兼容最新版浏览器的情况也很正常，根据笔者的经验，一般通过降低浏览器版本或者更新驱动，再或者更换浏览器等都能解决。

第10章 WebDriver API详解

本章主要讲解 WebDriver 常用 API 的使用方法，所有实例都会给出被测试网页的 HTML 代码或者被测试网页的网址，方便读者在本地实践。API 调用代码会包含在以 test 命名开始的方法中，本章不再赘述 WebDriver 使用前的初始准备工作以及用于测试的浏览器准备工作。读者只需要将本章节提供的调用 API 的测试用例方法添加到完整的 unittest 单元测试框架代码中即可执行，完整的 unittest 框架代码参见第 8 章，单元测试框架代码替换如图 10-1 所示。

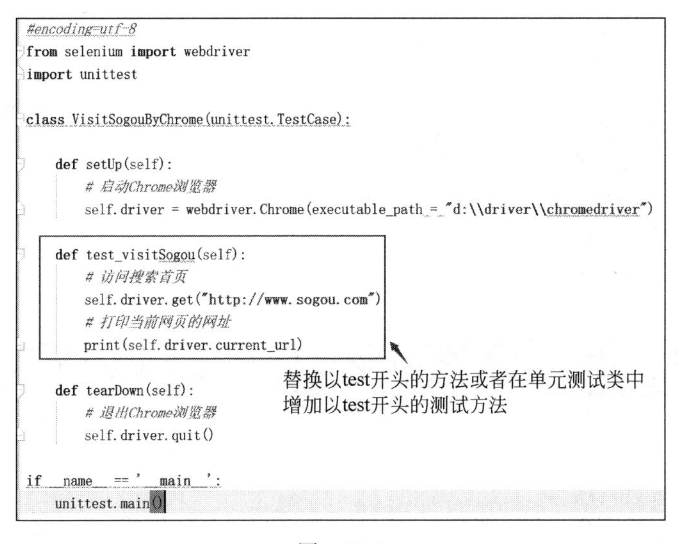

```
#encoding=utf-8
from selenium import webdriver
import unittest

class VisitSogouByChrome(unittest.TestCase):

    def setUp(self):
        # 启动Chrome浏览器
        self.driver = webdriver.Chrome(executable_path = "d:\\driver\\chromedriver")

    def test_visitSogou(self):
        # 访问搜索首页
        self.driver.get("http://www.sogou.com")
        # 打印当前网页的网址
        print(self.driver.current_url)

    def tearDown(self):
        # 退出Chrome浏览器
        self.driver.quit()

if __name__ == '__main__':
    unittest.main()
```


图 10-1

10.1 访问某个网址

目的：

打开浏览器访问指定的网址。

用于测试的网址：

http://www.sogou.com

调用 API 的实例代码：

```
def test_visitURL(self):
    visitURL = "http://www.sogou.com"
    # 通过 driver 对象的 get 方法,访问指定的网址
    self.driver.get(visitURL)
    assert self.driver.title.find("搜狗搜索引擎") >= 0, "assert error"
```

10.2 网页的前进和后退

目的：

本小节所讲的网页的前进和后退，模拟的是浏览器上的前进和后退功能。

用于测试的网址：

http://www.sogou.com

http://www.baidu.com

调用 API 的实例代码：

```
def test_visitRecentURL(self):
    firstVisitURL = "http://www.sogou.com"
    secondVisitURL = "http://www.baidu.com"
    # 首先访问搜狗首页
    self.driver.get(firstVisitURL)
    # 然后访问百度首页
    self.driver.get(secondVisitURL)
    # 返回上一次访问过的搜狗首页
    self.driver.back()
    # 再次回到百度首页
    self.driver.forward()
```

10.3 刷新当前网页

目的：

通过程序实现刷新网页。

用于测试的网址：

http://www.sogou.com

调用 API 的实例代码：

```
def test_refreshCurrentPage(self):
    url = "http://www.sogou.com"
    self.driver.get(url)
    # 刷新当前页面
    self.driver.refresh()
```

10.4 浏览器窗口最大化

目的：

将浏览器窗口全屏展示。

用于测试的网址：

http://www.baidu.com

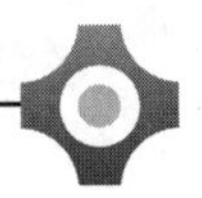

调用 API 的实例代码：

```
def test_maximizeWindow(self):
    url = "http://www.baidu.com"
    self.driver.get(url)
    # 最大化浏览器窗口,以便占满整个计算机屏幕
    self.driver.maximize_window()
```

10.5 获取并设置当前窗口的位置

目的：

改变浏览器窗口的位置。

用于测试的网址：

http://www.baidu.com

调用 API 的实例代码：

```
def test_window_position(self):
    url = "http://www.baidu.com"
    self.driver.get(url)
    # 获取当前浏览器在屏幕上的位置,返回的是字典对象
    position = self.driver.get_window_position()
    print("当前浏览器所在位置的横坐标:", position['x'])
    print("当前浏览器所在位置的纵坐标:", position['y'])
    # 设置当前浏览器在屏幕上的位置
    self.driver.set_window_position(y = 200, x = 400)
    # 设置浏览器的位置后,再次获取浏览器的位置信息
    print(self.driver.get_window_position())
```

更多说明：

(1) 获取的浏览器位置是指浏览器左上角所在的屏幕上的位置，返回的是 x，y 坐标值，即横纵坐标。

(2) get_window_position()和 set_window_position()方法在部分浏览器的部分版本上失效。

10.6 获取并设置当前窗口的大小

用于测试的网址：

http://www.baidu.com

调用 API 的实例代码：

```
def test_window_size(self):
    url = "http://www.baidu.com"
    self.driver.get(url)
    # 获取浏览器窗口的大小,返回字典类型
```

```
        sizeDict = self.driver.get_window_size()
        print("当前浏览器窗口的宽:", sizeDict['width'])
        print("当前浏览器窗口的高:", sizeDict['height'])
        # 设置浏览器窗口的大小
        self.driver.set_window_size(width = 200, height = 400, windowHandle = 'current')
        # 设置浏览器窗口大小以后,再次获取浏览器窗口大小信息
        print(self.driver.get_window_size(windowHandle = 'current'))
```

10.7 获取页面的 Title 属性值

用于测试的网址:

http://www.baidu.com

调用 API 的实例代码:

```
def test_getTitle(self):
    url = "http://www.baidu.com"
    self.driver.get(url)
    # 调用 driver 的 title 属性获取页面的 title 属性值
    title = self.driver.title
    print("当前网页的 title 属性值为:", title)
    # 断言页面的 title 属性值是否是"百度一下,你就知道"
    self.assertEqual(title, "百度一下,你就知道", "页面 title 属性值错误!")
```

更多说明:

获取页面的 title 属性值,在自动化测试中一般用于断言是否成功打开了某个网址,来证明测试过程正确性。

10.8 获取页面 HTML 源代码

用于测试网址:

http://www.sogou.com

调用 API 的实例代码:

```
def test_getPageSource(self):
    url = "http://www.sogou.com"
    self.driver.get(url)
    # 调用 driver 的 page_source 属性获取页面源码
    pageSource = self.driver.page_source
    # 打印页面源码
    print(pageSource)
    # 断言页面源码中是否包含"新闻"两个关键字,以此判断页面内容是否正确
    self.assertTrue("新闻" in pageSource, "页面源码中未找到'新闻'关键字")
```

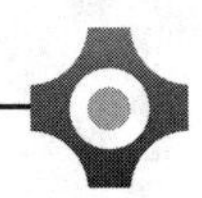

10.9　获取当前页面的 URL 地址

用于测试的网址：

http://www.sogou.com

调用 API 的实例代码：

```
def test_getCurrentPageUrl(self):
    url = "http://www.sogou.com"
    self.driver.get(url)
    # 获取当前页面的 URL
    currentPageUrl = self.driver.current_url
    # 打印当前 URL
    print(currentPageUrl)
    # 断言当前网页的网址是否为 https://www.sogou.com/
    self.assertEqual(currentPageUrl,
        "https://www.sogou.com/", "当前网页网址非预期!")
```

10.10　获取与切换浏览器窗口句柄

目的：

获取所有打开窗口的句柄,并在这些句柄间互相切换。

用于测试网址：

http://www.baidu.com

调用 API 的实例代码：

```
def test_operateWindowHandle(self):
    url = "http://www.baidu.com"
    self.driver.get(url)
    # 获取当前窗口句柄
    now_handle = self.driver.current_window_handle
    # 打印当前获取的窗口句柄
    print (now_handle)
    # 在百度搜索输入框中输入"selenium"
    self.driver.find_element_by_id("kw").send_keys("w3cschool")
    # 单击搜索按钮
    self.driver.find_element_by_id("su").click()
    # 导入 time 包
    import time
    # 等待 3 秒,以便网页加载完成
    time.sleep(3)
    # 单击 w3school 在线教育链接
    self.driver.find_element_by_xpath('//a[text() = "w3"]').click()
    time.sleep(5)
    # 获取所有窗口句柄
```

```
        all_handles = self.driver.window_handles
        print("++++", self.driver.window_handles[-1])
        # 循环遍历所有新打开的窗口句柄,也就是说不包括主窗口
        for handle in all_handles:
            if handle != now_handle:
                # 输出待选择的窗口句柄
                '''
                切换窗口,也可以用下面的方法,
                但此种方法在 selenium3.x 以后官方已经不推荐使用了
                self.driver.switch_to_window(handle)
                '''
                # 切换窗口
                self.driver.switch_to.window(handle)
                # 单击 HTML5 链接
                self.driver.find_element_by_link_text('HTML5').click()
                time.sleep(3)
                # 关闭当前的窗口
                self.driver.close()
        time.sleep(3)
        # 打印主窗口句柄
        print(now_handle)
        # 返回主窗口
        self.driver.switch_to.window(now_handle)
        time.sleep(2)
        self.driver.find_element_by_id("kw").clear()
        self.driver.find_element_by_id("kw").send_keys("光荣之路自动化测试培训")
        self.driver.find_element_by_id("su").click()
        time.sleep(5)
```

更多说明:

(1) driver.switch_to_window()方法在 selenium3.x 版本以后,官方开始推荐使用 driver.switch_to.window()方法代替。

(2) driver.window_handles 以列表对象形式返回所有打开窗口的句柄,包括主窗口,可以通过 driver.window_handles[-1]来获取当前打开窗口的句柄。

10.11 获取页面元素的基本信息

用于测试的网址:

http://www.baidu.com

调用 API 的实例代码:

```
def test_getBasicInfo(self):
        url = "http://www.baidu.com"
        # 访问百度首页
        self.driver.get(url)
        # 查找百度首页上的"新闻"链接元素
```

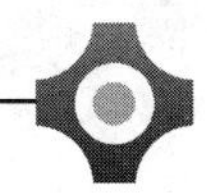

```
    newsElement = self.driver.find_element_by_xpath("//a[text() = '新闻']")
    # 获取查找到的"新闻"链接元素的基本信息
    print("元素的标签名:", newsElement.tag_name)
    print("元素的 size:", newsElement.size)
```

输出结果：

```
元素的标签名: a
元素的 size: {'width': 26, 'height': 24}
```

更多说明：

通过调用已定位元素内建的一些属性和方法，可以查看元素的基本信息，能够更好地支持自动化的实施。

10.12 获取页面元素的文本内容

用于测试的网址：

http://www.baidu.com

调用 API 的实例代码：

```
def test_getWebElementText(self):
    url = "http://www.baidu.com"
    # 访问百度首页
    self.driver.get(url)
    import time
    time.sleep(3)
    # 通过 xpath 定位方式找到 class 属性值为"mnav"的 div 元素下的第一个链接元素
    aElement = self.driver.find_element_by_xpath("//*[@class = 'mnav'][1]")
    # 通过找到的链接元素对象的 text 属性获取链接元素的文本内容
    a_text = aElement.text
    self.assertEqual(a_text, "新闻")
```

10.13 判断页面元素是否可见

用于测试的 HTML 代码：

```
<html>
<head>
<title>HTML 中显示与隐藏元素</title>
<meta http-equiv = "Content-Type" content = "text/html; charset = utf-8" />
<script type = "text/javascript">
    function showAndHidden1(){
        var div1 = document.getElementById("div1");
        var div2 = document.getElementById("div2");
        if(div1.style.display == 'block') div1.style.display = 'none';
        else div1.style.display = 'block';
        if(div2.style.display == 'block') div2.style.display = 'none';
        else div2.style.display = 'block';
```

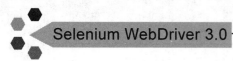

```
        }
        function showAndHidden2(){
            var div3 = document.getElementById("div3");
            var div4 = document.getElementById("div4");
            if(div3.style.visibility == 'visible') div3.style.visibility = 'hidden';
            else div3.style.visibility = 'visible';
            if(div4.style.visibility == 'visible') div4.style.visibility = 'hidden';
            else div4.style.visibility = 'visible';
        }
    </script>
    </head>
    <body>
        <div>display:元素不占用页面位置</div>
        <div id="div1" style="display:block;">DIV 1</div>
        <div id="div2" style="display:none;">DIV 2</div>
        <input id="button1" type="button" onclick="showAndHidden1();" value="DIV 切换" />
        <hr>
        <div>visibility:元素占用页面位置</div>
        <div id="div3" style="visibility:visible;">DIV 3</div>
        <div id="div4" style="visibility:hidden;">DIV 4</div>
        <input id="button2" type="button" onclick="showAndHidden2();" value="DIV 切换" />
    </body>
    </html>
```

调用 API 的实例代码：

```
def test_getWebElementIsDisplayed(self):
    url = "file:///" + "d:\\test.html"
    # 访问自定义的 html 网页
    self.driver.get(url)
    # 通过 id="div2"找到第二个 div 元素
    div2 = self.driver.find_element_by_id("div2")
    # 判断第二个 div 元素是否在页面上可见
    print(div2.is_displayed())
    # 单击第一个切换 div 按钮,将第二个 div 元素显示在页面上
    self.driver.find_element_by_id("button1").click()
    # 再次判断第二个 div 元素是否可见
    print(div2.is_displayed())
    # 通过 id="div4"找到第四个 div 元素
    div4 = self.driver.find_element_by_id("div4")
    # 判断第四个 div 元素是否在页面上可见
    print(div4.is_displayed())
    # 单击第二个切换 div 按钮,将第四个 div 元素显示在页面上
    self.driver.find_element_by_id("button2").click()
    # 再次判断第四个 div 元素是否可见
    print(div4.is_displayed())
```

更多说明：

页面不可见的元素虽不在页面上显示，但是存在于 DOM 树中，这些元素 WebDriver 也是可以找到的。

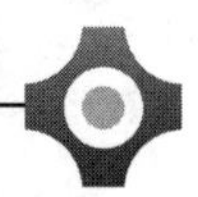

10.14 判断页面元素是否可操作

用于测试的 HTML 代码：

```
<html>
<head>
    <title>HTML中不可操作元素</title>
    <meta http-equiv="Content-Type" content="text/html; charset=utf-8" />
</head>
<body>
    <input id="input1" type="text" size="40" value="可操作">
    <br />
    <input id="input2" type="text" size="40" value="不可用" disabled>
    <br />
    <input id="input3" type="text" size="40" value="只读" readonly>
</body>
</html>
```

调用 API 的实例代码：

```
def test_getWebElementIsEnabled(self):
    url = "file:///" + "d:\\test.html"
    # 访问自定义的 html 网页
    self.driver.get(url)
    # 通过 id 找到第一个 input 元素
    input1 = self.driver.find_element_by_id("input1")
    # 判断第一个 input 元素是否可操作
    print(input1.is_enabled())
    # 通过 id 找到第二个 input 元素
    input2 = self.driver.find_element_by_id("input2")
    # 判断第二个 input 元素是否可操作
    print(input2.is_enabled())
    # 通过 id 找到第三个 input 元素
    input3 = self.driver.find_element_by_id("input3")
    # 判断第三个 input 元素是否可操作
    print(input3.is_enabled())
```

输出结果：

```
True
False
True
```

更多说明：

从执行结果可看出，对元素添加 disabled 属性以后，元素将会处于不可操作状态。

10.15 获取页面元素的属性

用于测试的网址：

http://www.sogou.com

调用 API 的实例代码：

```
def test_getWebElementAttribute(self):
    url = "http://www.sogou.com"
    # 访问搜狗首页
    self.driver.get(url)
    # 找到搜索输入框元素
    searchBox = self.driver.find_element_by_id("query")
    # 获取搜索输入框页面元素的 name 属性值
    print(searchBox.get_attribute("name"))
    # 向搜索输入框中输入"测试工程师指定的输入内容"内容
    searchBox.send_keys("测试工程师指定的输入内容")
    # 获取页面搜索框的 value 属性值(即搜索输入框的文字内容)
    print(searchBox.get_attribute("value"))
```

10.16 获取页面元素的 CSS 属性值

用于测试的网址：

http://www.baidu.com

调用 API 的实例代码：

```
def test_getWebElementCssValue(self):
    url = "http://www.baidu.com"
    # 访问百度首页
    self.driver.get(url)
    # 找到搜索输入框元素
    searchBox = self.driver.find_element_by_id("kw")
    # 使用页面元素对象的 value_of_css_property()方法获取元素的 CSS 属性值
    print("搜索输入框的高度是:", searchBox.value_of_css_property("height"))
    print("搜索输入框的宽度是:", searchBox.value_of_css_property("width"))
    font = searchBox.value_of_css_property("font-family")
    print("搜索输入框的字体是:", font)
    # 判断搜索输入框的字体是否是 arial 字体
    self.assertEqual(font, "arial")
```

10.17 清空输入框中的内容

用于测试的网址：

http://www.baidu.com

调用 API 的实例代码：

```
def test_clearInputBoxText(self):
    url = "http://www.baidu.com"
    # 访问百度网页
    self.driver.get(url)
    # 获取输入框页面对象
    input = self.driver.find_element_by_id("kw")
```

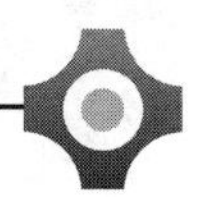

```
    input.send_keys("光荣之路自动化测试")
    import time
    time.sleep(3)
    # 清除输入框中默认内容
    input.clear()
    # 等待 3 秒,主要看清空输入框内容后的效果
    time.sleep(3)
```

更多说明：

在测试代码中直接导入需要的 time 包(import time)，是为了更清晰地说明测试脚本中的代码具体来自哪个 Python 包，以后的代码中也采用此种方法，读者可以将导包代码放到脚本文件的最顶端，以方便管理。

10.18　在输入框中输入指定内容

被测试网页的 HTML 代码：

```
<html>
    <body>
     <input type = "text" id = "text" value = "文本框默认内容">文本框</input>
    </body>
</html>
```

调用 API 的实例代码：

```
def test_sendTextToInputBoxText(self):
    url = "file:///" + "d:\\test.html"
    # 访问自定义的 html 网页
    self.driver.get(url)
    # 获取输入框页面对象
    input = self.driver.find_element_by_id("text")
    # 清除输入框中默认内容
    input.clear()
    input.send_keys("我是输入的文本内容")
    # 导入 time 包
    import time
    # 等待 3 秒,主要看清空输入框内容后的效果
    time.sleep(3)
```

10.19　单击按钮

目的：

模拟鼠标左键单击操作。

被测试网页的 HTML 代码：

```
<html>
<body>
    <input type = "text" id = "text" value = "文本框默认内容">文本框
```

```
    <input type="button" id ="button" value="改变文本框的文字"
    onClick=document.getElementById("text").value="改变了!">
</body>
</html>
```

调用 API 的实例代码：

```
def test_clickButton(self):
    url = "file:///" +"d:\\test.html"
    # 访问自定义的 html 网页
    self.driver.get(url)
    # 获取按钮页面对象
    button = self.driver.find_element_by_id("button")
    # 模拟鼠标左键单击操作
    button.click()
    import time
    time.sleep(3)
```

10.20 双击某个元素

目的：

模拟鼠标左键双击操作。

被测试网页的 HTML 代码：

```
<html>
<body>
    <input id='inputBox' type="text"
    ondblclick="javascript:this.style.background='red'">请双击
</body>
</html>
```

调用 API 的实例代码：

```
def test_doubleClick(self):
    url = "file:///" + "d:\\test.html"
    # 访问自定义的 html 网页
    self.driver.get(url)
    # 获取页面输入元素
    inputBox = self.driver.find_element_by_id("inputBox")
    # 导入支持双击操作的模块
    from selenium.webdriver import ActionChains
    # 开始模拟鼠标双击操作
    action_chains = ActionChains(self.driver)
    action_chains.double_click(inputBox).perform()
    import time
    time.sleep(3)
```

执行后双击 input 框，背景颜色将变为红色。

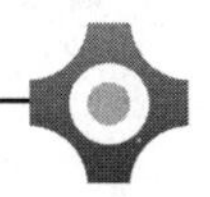

更多说明：

selenium.webdriver.ActionChains 包是 WebDriver 针对 Python 语言提供的专门用于模拟鼠标操作事件的包，比如双击、悬浮、拖曳等，这些在后面都会陆续介绍。

10.21 操作单选下拉列表

被测试网页的 HTML 代码：

```
<html>
<body>
    <select name='fruit' size=1>
        <option id='peach' value='taozi'>桃子</option>
        <option id='watermelon' value='xigua'>西瓜</option>
        <option id='orange' value='juzi' selected="selected">橘子</option>
        <option id='kiwifruit' value='mihoutao'>猕猴桃</option>
        <option id='maybush' value='shanzha'>山楂</option>
        <option id='litchi' value='lizhi'>荔枝</option>
    </select>
</body>
</html>
```

10.21.1 遍历所有选项并打印选项显示的文本和选项值

Python 语言实例代码：

```python
def test_printSelectText(self):
    url = "file:///" + "d:\\select.html"
    # 访问自定义的 html 网页
    self.driver.get(url)
    # 使用 name 属性找到页面上 name 属性为 fruit 的下拉列表元素
    select = self.driver.find_element_by_name("fruit")
    all_options = select.find_elements_by_tag_name("option")
    for option in all_options:
        print("选项显示的文本:", option.text)
        print("选项值为:", option.get_attribute("value"))
        option.click()
        import time
        time.sleep(1)
```

输出结果：

```
选项显示的文本: 桃子
选项值为: taozi
选项显示的文本: 西瓜
选项值为: xigua
选项显示的文本: 橘子
选项值为: juzi
选项显示的文本: 猕猴桃
选项值为: mihoutao
```

```
选项显示的文本: 山楂
选项值为: shanzha
选项显示的文本: 荔枝
选项值为: lizhi
```

10.21.2 选择下拉列表元素的三种方法

Python 语言实例代码：

```
def test_operateDropList(self):
    url = "file:///" + "d:\\select.html"
    # 访问自定义的 html 网页
    self.driver.get(url)
    # 导入 Select 模块
    from selenium.webdriver.support.ui import Select
    # 使用 Xpath 定位方式获取 select 页面元素对象
    select_element = Select(self.driver.find_element_by_xpath("//select"))
    # 打印默认选中项的文本
    print(select_element.first_selected_option.text)
    # 获取所有选择项的页面元素对象
    all_options = select_element.options
    # 打印选项总个数
    print(len(all_options))
    '''
    is_enabled():判断元素是否可操作
    is_selected():判断元素是否被选中
    '''
    if all_options[1].is_enabled() and not all_options[1].is_selected():
        # 方法 1:通过序号选择第二个元素,序号从 0 开始
        select_element.select_by_index(1)
        # 打印已选中项的文本
        print(select_element.all_selected_options[0].text)
        # 通过 assertEqual()方法判断当前选中的选项文本是否是"西瓜"
        self.assertEqual(select_element.all_selected_options[0].text, "西瓜")
    import time
    time.sleep(2)
    # 方法 2:通过选项的显示文本选择文本为"猕猴桃"选项
    select_element.select_by_visible_text("猕猴桃")
    # 判断已选中项的文本是否是"猕猴桃"
    self.assertEqual(select_element.all_selected_options[0].text, "猕猴桃")
    time.sleep(2)
    # 方法三:通过选项的 value 属性值选择 value = "shanzha"选项
    select_element.select_by_value("shanzha")
    print(select_element.all_selected_options[0].text)
    self.assertEqual(select_element.all_selected_options[0].text, "山楂")
```

更多说明：

select_element.all_selected_options 属性获取的是所有被选中项的对象组成的列表对象，由于本实例中是单选下拉列表，因此选中项只有一个，通过 select_element.all_selected_options[0].text 这句代码获取被选中项的文本内容。

10.22 断言单选列表选项值

目的：

判断单选列表内容是否与预期内容一致。

用于测试的 HTML 代码：

被测试的 HTML 代码同 10.21 节。

Python 语言实例代码：

```
def test_checkSelectText(self):
    url = "file:///" + "d:\\select.html"
    # 访问自定义的 html 网页
    self.driver.get(url)
    # 导入 Select 模块
    from selenium.webdriver.support.ui import Select
    # 使用 xpath 定位方式获取 select 页面元素对象
    select_element = Select(self.driver.find_element_by_xpath("//select"))
    # 获取所有选择项的页面元素对象
    actual_options = select_element.options
    # 声明一个 list 对象,存储下拉列表中所期望出现的文字内容
    expect_optionsList = ["桃子", "西瓜", "橘子", "猕猴桃", "山楂", "荔枝"]
    # 使用 Python 内置 map()函数获取页面中下拉列表展示的选项内容组成的列表对象
    actual_optionsList = list(map(lambda option: option.text, actual_options))
    # 判断期望列表对象和实际列表对象是否完全一致
    self.assertListEqual(expect_optionsList, actual_optionsList)
```

10.23 操作多选的选择列表

用于测试的 HTML 代码：

```
<html>
<body>
    <select name='fruit' size=6 multiple=true>
        <option id='peach' value='taozi'>桃子</option>
        <option id='watermelon' value='xigua'>西瓜</option>
        <option id='orange' value='juzi'>橘子</option>
        <option id='kiwifruit' value='mihoutao'>猕猴桃</option>
        <option id='maybush' value='shanzha'>山楂</option>
        <option id='litchi' value='lizhi'>荔枝</option>
    </select>
</body>
</html>
```

Python 语言实例代码：

```
def test_operateMultipleOptionSelectList(self):
    url = "file:///" + "d:\\select.html"
    # 访问自定义的 html 网页
```

```
self.driver.get(url)
# 导入 Select 模块
from selenium.webdriver.support.ui import Select
import time
# 使用 xpath 定位方式获取 select 页面元素对象
select_element = Select(self.driver.find_element_by_xpath("//select"))
# 通过序号选择第一个元素
select_element.select_by_index(0)
# 通过选项的文本选择"山楂"选项
select_element.select_by_visible_text("山楂")
# 通过选项的 value 属性值选择 value = "mihoutao"的选项
select_element.select_by_value("mihoutao")
# 打印所有的选中项文本
for option in select_element.all_selected_options:
    print(option.text)
# 取消所有已选中项
select_element.deselect_all()
time.sleep(2)
print("------------ 再次选中 3 个选项 --------------")
select_element.select_by_index(1)
select_element.select_by_visible_text("荔枝")
select_element.select_by_value("juzi")
# 通过选项文本取消已选中的文本为"荔枝"选项
select_element.deselect_by_visible_text("荔枝")
# 通过序号取消已选中的序号为 1 的选项
select_element.deselect_by_index(1)
# 通过选项的 value 属性值取消已选中的 value = "juzi"的选项
select_element.deselect_by_value("juzi")
```

10.24 操作可以输入的下拉列表(输入的同时模拟按键)

目的:

实现输入的同时模拟按键操作。

用于测试的 HTML 代码:

```
<html>
<body>
    <div style="position:relative;">
        <input list="pasta" id="select">
        <datalist id="pasta">
            <option>Bavette</option>
            <option>Rigatoni</option>
            <option>Fiorentine</option>
            <option>Gnocchi</option>
            <option>Tagliatelle</option>
            <option>Penne lisce</option>
            <option>Pici</option>
            <option>Pappardelle</option>
            <option>Spaghetti</option>
```

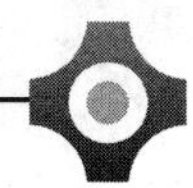

```
            <option>Cannelloni</option>
            <option>Cancl</option>
        </datalist>
    </div>
</body>
</html>
```

调用 API 的实例代码：

```
def test_operateMultipleOptionInputSelectList(self):
    url = "file:///" + "d:\\option_input_select_list.html"
    # 访问自定义的 html 网页
    self.driver.get(url)
    from selenium.webdriver.common.keys import Keys
    self.driver.find_element_by_id("select").clear()
    import time
    time.sleep(1)
    # 输入的同时按下箭头键
    self.driver.find_element_by_id("select").send_keys("c", Keys.ARROW_DOWN)
    self.driver.find_element_by_id("select").send_keys(Keys.ARROW_DOWN)
    self.driver.find_element_by_id("select").send_keys(Keys.ENTER)
    time.sleep(3)
```

更多说明：

运行这段测试代码可以看到输入字符 c 的同时看到筛选出的数据项中第一项被选中。但在某些浏览器的某些版本效果会不明显。除了下箭头，Keys 模块还提供了很多其他的模拟按键，后面会继续介绍一部分，读者也可以通过 dir()函数查看。

10.25　操作单选框

用于测试的 HTML 代码：

```
<html>
<body>
    <form>
        <input type="radio" name="fruit" value="berry" />草莓</input>
        <br />
        <input type="radio" name="fruit" value="watermelon" />西瓜</input>
        <br />
        <input type="radio" name="fruit" value="orange" />橙子</input>
    </form>
</body>
</html>
```

调用 API 实例代码：

```
def test_operateRadio(self):
    url = "file:///" + "d:\\radio.html"
    # 访问自定义的 html 网页
    self.driver.get(url)
```

```
# 使用 xpath 定位获取 value 属性值为'berry'的 input 元素对象,也就是"草莓"选项
berryRadio = self.driver.find_element_by_xpath("//input[@value = 'berry']")
# 单击选择"草莓"选项
berryRadio.click()
# 断言"草莓"单选框被成功选中
self.assertTrue(berryRadio.is_selected(), "草莓单选框未被选中!")
if berryRadio.is_selected():
    # 如果"草莓"单选框被成功选中,重新选择"西瓜"选项
    watermelonRadio = self.driver.find_element_by_xpath("//input[@value = 'watermelon']")
    watermelonRadio.click()
    # 选择"西瓜"选项以后,断言"草莓"选项处于未被选中状态
    self.assertFalse(berryRadio.is_selected())
# 查找所有 name 属性值为"fruit"的单选框元素对象,并存放在 radioList 列表中
radioList = self.driver.find_elements_by_xpath("//input[@name = 'fruit']")
'''
循环遍历 radioList 中的每个单选按钮,查找 value 属性值为"orange"的单选框,
如果找到此单选框以后,发现未处于选中状态,则调用 click 方法选中该选项。
'''
for radio in radioList:
    if radio.get_attribute("value") == "orange":
        if not radio.is_selected():
            radio.click()
            self.assertEqual(radio.get_attribute("value"), "orange")
```

10.26 操作复选框

用于测试的 HTML 代码：

```
<html>
<body>
    <form name = 'form1'>
        <input type = "checkbox" name = "fruit" value = "berry" />草莓</input>
        <br />
        <input type = "checkbox" name = "fruit" value = "watermelon" />西瓜</input>
        <br />
        <input type = "checkbox" name = "fruit" value = "orange" />橙子</input>
    </form>
</body>
</html>
```

调用 API 的实例代码：

```
def test_operateCheckBox(self):
    url = "file:///" + "d:\\CheckBox.html"
    # 访问自定义的 html 网页
    self.driver.get(url)
    # 使用 xpath 定位获取 value 属性值为'berry'的 input 元素对象,也就是"草莓"选项
    berryCheckBox = self.driver.find_element_by_xpath("//input[@value='berry']")
    # 单击选择"草莓"选项
    berryCheckBox.click()
```

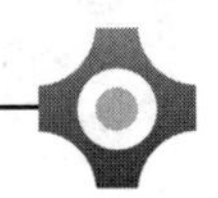

```
    # 断言"草莓"复选框被成功选中
    self.assertTrue(berryCheckBox.is_selected(), "草莓复选框未被选中!")
    if berryCheckBox.is_selected():
        # 如果"草莓"复选框被成功选中,再次单击取消选中
        berryCheckBox.click()
        # 断言"草莓"复选框处于未选中状态
        self.assertFalse(berryCheckBox.is_selected())
    # 查找所有 name 属性值为"fruit"的复选框元素对象,并存放在 checkBoxList 列表中
    checkBoxList = self.driver.find_elements_by_xpath("//input[@name='fruit']")
    # 遍历 checkBoxList 列表中的所有复选框元素,让全部复选框处于被选中状态
    for box in checkBoxList:
        if not box.is_selected():
            box.click()
```

10.27　断言页面源码中的关键字

目的：

确定所加载的页面是否出现了预期内容。

用于测试的网址：

http://www.baidu.com

Python 语言实例代码：

```
def test_assertKeyWord(self):
    url = "http://www.baidu.com"
    # 访问百度首页
    self.driver.get(url)
    self.driver.find_element_by_id("kw").send_keys("光荣之路自动化测试")
    self.driver.find_element_by_id("su").click()
    import time
    time.sleep(4)
    # 通过断言页面是否存在某些关键字来确定页面按照预期加载
    assert "首页 -- 光荣之路" in self.driver.page_source, "页面源码中不存在该关键字!"
```

有时会出现页面存在要断言的内容，但结果仍断言失败，这可能是由于页面还未加载完全就开始执行断言语句，导致要断言的内容在页面源码中找不到。

10.28　对当前浏览器窗口截屏

用于测试网址：

http://www.sogou.com

调用 API 的实例代码：

```
def test_captureScreenInCurrentWindow(self):
    url = "http://www.sogou.com"
    # 访问搜狗首页
```

```
    self.driver.get(url)
    try:
        '''
        调用 get_screenshot_as_file(filename)方法,对浏览器当前打开页面
        进行截图,并保存为 C 盘下的 screenPicture.png 文件.
        '''
        result = self.driver.get_screenshot_as_file("c:\\screenPicture.png")
        print (result)
    except IOError as err:
        print("error: {0}".format(err))
```

更多说明：

（1）调用截屏函数 get_screenshot_as_file()截图成功后会返回 True，如果发生了 IOError 异常，会返回 False。函数中传递的存放图片的路径可以是绝对路径，也可以是相对路径。

（2）当自动化测试过程中，未实现预期结果，可以将页面截图保存，方便更快速地定位问题。

10.29 拖曳页面元素

用于测试的网址：

http://jqueryui.com/resources/demos/draggable/scroll.html

调用 API 的实例代码：

```
def test_dragPageElement(self):
    url = "http://jqueryui.com/resources/demos/draggable/scroll.html"
    # 访问被测试网页
    self.driver.get(url)
    # 获取页面上第一个能拖曳的页面元素
    initialPosition = self.driver.find_element_by_id("draggable")
    # 获取页面上第二个能拖曳的页面元素
    targetPosition = self.driver.find_element_by_id("draggable2")
    # 获取页面上第三个能拖曳的页面元素
    dragElement = self.driver.find_element_by_id("draggable3")
    # 导入提供拖曳元素方法的模块 ActionChains
    from selenium.webdriver import ActionChains
    import time
    '''
    创建一个新的 ActionChains,将 WebDriver 实例对象 driver 作为参数值传入
    然后通过 WebDriver 实例执行用户动作。
    '''
    action_chains = ActionChains(self.driver)
    # 将页面上第一个能被拖曳的元素拖曳到第二个元素位置
    action_chains.drag_and_drop(initialPosition, targetPosition).perform()
    # 将页面上第三个能被拖曳的元素,向右下拖动 10 个像素,共拖动 5 次
    for i in range(5):
```

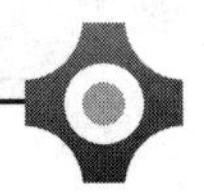

```
        action_chains.drag_and_drop_by_offset(dragElement, 10, 10).perform()
        time.sleep(2)
```

10.30 模拟键盘单个按键操作

用于测试的网址：

http://www.sogou.com

调用 API 的实例代码：

```
def test_simulateASingleKeys(self):
    url = "http://www.sogou.com"
    # 访问搜狗首页，焦点会自动定位到搜索输入框中
    self.driver.get(url)
    # 导入模拟按键模块 Keys
    from selenium.webdriver.common.keys import Keys
    import time
    # 通过 id 获取搜索输入框的页面元素
    query = self.driver.find_element_by_id("query")
    # 通过 WebDriver 实例发送一个 F12 键
    query.send_keys(Keys.F12)
    time.sleep(3)
    # 再次通过 WebDriver 实例模拟发送一个 F12 键
    query.send_keys(Keys.F12)
    # 在搜索输入框中输入 selenium
    query.send_keys("selenium")
    # 通过 WebDriver 实例模拟发送一个回车键，
    # 或者使用 query.send_keys(Keys.RETURN)
    query.send_keys(Keys.ENTER)
    time.sleep(3)
```

说明：

本实例推荐在 IE 浏览器上测试，其他浏览器部分版本可能看不到效果。部分计算机的功能键可能需要同时按下 Fn 键才能生效，比如 Fn+F12 组合键。

更多说明：

在本章的 10.24 小节中，讲解的是在输入操作的同时模拟一个键盘按键，而本小节讲的是，在其他页面动作完成后再进行键盘按键模拟操作。这两种情况下的按键都只能针对单个按键，并且是按下马上松开按键。Keys 模块还提供了一部分其他键可以模拟，请读者根据自己需要进行练习使用。

10.31 模拟组合按键操作

环境准备：

访问 http://sourceforge.net/projects/pywin32/files/pywin32/Build%20219/网站，下载支持 Python3.x 模拟 Windows 组合按键的 pywin 安装包。读者必须根据自己 Python3.x 的位数(64 位或 32 位)选择相应位数的安装包，否则安装会报错，安装方法同 Windows 应用程序

一致，安装过程中无须做任何更改。

安装完成后，在 Python 交换模式下执行如下两句代码：

```
import win32clipboard as w
import win32con
```

如果没报错，说明安装成功。

10.31.1 通过 WebDriver 内建的模块模拟组合键

目的：

通过模拟按键实现全选、剪切以及粘贴操作。

用于测试的网址：

http://www.baidu.com

Python 语言实例代码：

```
# 导入模拟组合按键需要的包
from selenium.webdriver import ActionChains
from selenium.webdriver.common.keys import Keys
import time
```

以上代码放到单元测试类外边区域。测试用例方法如下：

```
def test_simulationCombinationKeys(self):
    url = "http://www.baidu.com"
    # 访问百度首页
    self.driver.get(url)
    # 将焦点切换到搜索输入框中
    input = self.driver.find_element_by_id("kw")
    input.click()
    input.send_keys("光荣之路")
    time.sleep(2)
    ActionChains(self.driver).key_down(Keys.CONTROL).send_keys('a').\
    key_up(Keys.CONTROL).perform()
    time.sleep(2)
    ActionChains(self.driver).key_down(Keys.CONTROL).send_keys('x').\
    key_up(Keys.CONTROL).perform()
    self.driver.get(url)
    self.driver.find_element_by_id("kw").click()
    # 模拟 Ctrl + V 组合键,将从剪切板中获取的内容粘贴到搜索输入框中
    ActionChains(self.driver).key_down(Keys.CONTROL).send_keys('v').\
    key_up(Keys.CONTROL).perform()
    # 单击"百度一下"搜索按钮
    self.driver.find_element_by_id('su').click()
    time.sleep(3)
```

代码解释：

ActionChains(self.driver).key_down(Keys.CONTROL).send_keys('v').key_up(Keys.CONTROL).perform()这行代码中，key_down(Keys.CONTROL)表示按下 Ctrl 键，send_keys('v')类似模拟了 V 键，组合起来就是 Ctrl+V 组合键，而 key_up(Keys.CONTROL)表示释放 Ctrl 键。

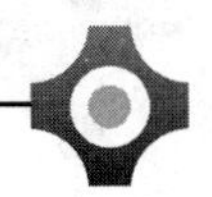

更多说明：

Selenium 3 的 ActionChains 模块在某些类型的浏览器的某些版本中失效或不稳定，比如 Firefox 49 版本，这可能是 Mozilla/geckodriver 驱动的一个 bug，如果读者确实想在 Firefox 浏览器上完成按键模拟，请考虑使用 Selenium 2。

本实例代码笔者在 Chrome 70.0.3538.102(64 位)版本的浏览器上实验成功，并能稳定运行。通过上面实例的方法可以模拟更多的组合按键，请读者自行模拟练习。

后续所有涉及 ActionChains 模块的实例代码默认均使用 Chrome 浏览器。

10.31.2　通过第三方模块模拟组合按键

目的：

通过模拟按键实现全选、剪切、粘贴以及回车键操作。

用于测试网址：

http://www.baidu.com

http://www.sogou.com

Python 语言实例代码：

```
# 导入模拟组合按键需要的包
import win32api
import win32con
import time

VK_CODE = {
    'enter':0x0D,
    'ctrl':0x11,
    'a':0x41,
    'v':0x56,
    'x':0x58
    }
#键盘键按下
def keyDown(keyName):
    win32api.keybd_event(VK_CODE[keyName], 0, 0, 0)
#键盘键抬起
def keyUp(keyName):
    win32api.keybd_event(VK_CODE[keyName], 0, win32con.KEYEVENTF_KEYUP, 0)
```

上面这段代码中，提供了模拟按键所需要导入的包，键盘键对应的命令以及完成按下键与抬起键动作方法，放到测试类外面。测试用例方法如下：

```
def test_simulationCombinationKeys(self):
    url = "http://www.sogou.com"
    # 访问搜狗首页
    self.driver.get(url)
    # 找到搜索输入框元素
    searchBox = self.driver.find_element_by_id("query")
```

```
# 将焦点切换到搜索输入框中
searchBox.click()
# 搜索输入框中输入"光荣之路自动化测试"
searchBox.send_keys("光荣之路自动化测试")
# 稍微等待几秒,防止太快串命令
time.sleep(3)
# 模拟 Ctrl + A,选中输入框中所有的内容
keyDown('ctrl')
keyDown('a')
# 释放 Ctrl + A 组合键
keyUp('a')
keyUp('ctrl')
# 模拟 Ctrl + X 剪切所选中的内容
keyDown('ctrl')
keyDown('x')
keyUp('x')
keyUp('ctrl')
# 访问百度首页
self.driver.get("http://www.baidu.com")
# 将焦点切换到搜索输入框中
self.driver.find_element_by_id("kw").click()
# 模拟 Ctrl + V 组合键,进行粘贴
keyDown("ctrl")
keyDown("v")
keyUp('v')
keyUp('ctrl')
# 模拟回车键
keyDown('enter')
keyUp('enter')
time.sleep(5)
```

更多说明:

上面实例代码中模拟键盘按键是通过将 Windows 键名与 Windows 命令一一映射来实现的,更多的按键命令映射如下:

```
#键盘上所有键的映射
VK_CODE = {
    'backspace':0x08,
    'tab':0x09,
    'clear':0x0C,
    'enter':0x0D,
    'shift':0x10,
    'ctrl':0x11,
    'alt':0x12,
    'pause':0x13,
    'caps_lock':0x14,
    'esc':0x1B,
    'spacebar':0x20,
    'page_up':0x21,
    'page_down':0x22,
    'end':0x23,
```

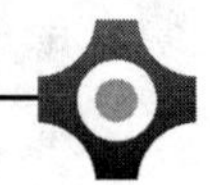

```
'home':0x24,
'left_arrow':0x25,
'up_arrow':0x26,
'right_arrow':0x27,
'down_arrow':0x28,
'select':0x29,
'print':0x2A,
'execute':0x2B,
'print_screen':0x2C,
'ins':0x2D,
'del':0x2E,
'help':0x2F,
'0':0x30,
'1':0x31,
'2':0x32,
'3':0x33,
'4':0x34,
'5':0x35,
'6':0x36,
'7':0x37,
'8':0x38,
'9':0x39,
'a':0x41,
'b':0x42,
'c':0x43,
'd':0x44,
'e':0x45,
'f':0x46,
'g':0x47,
'h':0x48,
'i':0x49,
'j':0x4A,
'k':0x4B,
'l':0x4C,
'm':0x4D,
'n':0x4E,
'o':0x4F,
'p':0x50,
'q':0x51,
'r':0x52,
's':0x53,
't':0x54,
'u':0x55,
'v':0x56,
'w':0x57,
'x':0x58,
'y':0x59,
'z':0x5A,
'numpad_0':0x60,
'numpad_1':0x61,
'numpad_2':0x62,
```

```
'numpad_3':0x63,
'numpad_4':0x64,
'numpad_5':0x65,
'numpad_6':0x66,
'numpad_7':0x67,
'numpad_8':0x68,
'numpad_9':0x69,
'multiply_key':0x6A,
'add_key':0x6B,
'separator_key':0x6C,
'subtract_key':0x6D,
'decimal_key':0x6E,
'divide_key':0x6F,
'F1':0x70,
'F2':0x71,
'F3':0x72,
'F4':0x73,
'F5':0x74,
'F6':0x75,
'F7':0x76,
'F8':0x77,
'F9':0x78,
'F10':0x79,
'F11':0x7A,
'F12':0x7B,
'F13':0x7C,
'F14':0x7D,
'F15':0x7E,
'F16':0x7F,
'F17':0x80,
'F18':0x81,
'F19':0x82,
'F20':0x83,
'F21':0x84,
'F22':0x85,
'F23':0x86,
'F24':0x87,
'num_lock':0x90,
'scroll_lock':0x91,
'left_shift':0xA0,
'right_shift ':0xA1,
'left_control':0xA2,
'right_control':0xA3,
'left_menu':0xA4,
'right_menu':0xA5,
'browser_back':0xA6,
'browser_forward':0xA7,
'browser_refresh':0xA8,
'browser_stop':0xA9,
'browser_search':0xAA,
'browser_favorites':0xAB,
```

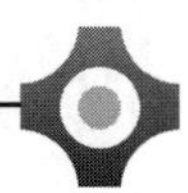

```
    'browser_start_and_home':0xAC,
    'volume_mute':0xAD,
    'volume_Down':0xAE,
    'volume_up':0xAF,
    'next_track':0xB0,
    'previous_track':0xB1,
    'stop_media':0xB2,
    'play/pause_media':0xB3,
    'start_mail':0xB4,
    'select_media':0xB5,
    'start_application_1':0xB6,
    'start_application_2':0xB7,
    'attn_key':0xF6,
    'crsel_key':0xF7,
    'exsel_key':0xF8,
    'play_key':0xFA,
    'zoom_key':0xFB,
    'clear_key':0xFE,
    '+':0xBB,
    ',':0xBC,
    '-':0xBD,
    '.':0xBE,
    '/':0xBF,
    '`':0xC0,
    ';':0xBA,
    '[':0xDB,
    '\\':0xDC,
    ']':0xDD,
    "'":0xDE,
    '`':0xC0
}
```

10.31.3　通过设置剪贴板实现复制和粘贴

目的：

通过按键实现复制粘贴操作。

用于测试的网址：

http://www.baidu.com

Python 语言实例代码：

```
# 导入模拟组合按键需要的包
from selenium.webdriver import ActionChains
from selenium.webdriver.common.keys import Keys
import win32clipboard as w
import win32con
import time

# 读取剪切板
def getText():
```

```
        w.OpenClipboard()
        d = w.GetClipboardData(win32con.CF_TEXT)
        w.CloseClipboard()
        return d

    # 设置剪切板内容
    def setText(aString):
        w.OpenClipboard()
        w.EmptyClipboard()
        w.SetClipboardData(win32con.CF_UNICODETEXT, aString)
        w.CloseClipboard()
```

以上代码放到单元测试类外边区域。测试用例方法如下：

```
    def test_copyAndPaste(self):
        url = "http://www.baidu.com"
        # 访问百度首页
        self.driver.get(url)
        # 定义即将要被设置到剪切板中的内容
        content = '光荣之路'
        # 将 content 变量中的内容设置到剪切板中
        setText(content)
        # 从剪切板中获取刚设置到剪切板中的内容
        getContent = getText()
        print(getContent)
        # 将焦点切换到搜索输入框中
        self.driver.find_element_by_id("kw").click()
        # 模拟 Ctrl + V 组合键,将从剪切板中获取到的内容粘贴到搜索输入框中
        ActionChains(self.driver).key_down(Keys.CONTROL).send_keys('v').\
            key_up(Keys.CONTROL).perform()
        # 单击"百度一下"搜索按钮
        self.driver.find_element_by_id('su').click()
        time.sleep(3)
```

代码解释：

通过 win32clipboard 和 win32con 包实现将数据设置到剪切板中；通过 WebDriver 内建的 ActionChains 和 Keys 模块共同实现组合按键的操作。

ActionChains(self.driver).key_down(Keys.CONTROL).send_keys('v').key_up(Keys.CONTROL).perform()这行代码中，key_down(Keys.CONTROL)表示按下 Ctrl 键，send_keys('v')类似模拟了 V 键，组合起来就是 Ctrl+V 组合键，而 key_up(Keys.CONTROL)表示释放 Ctrl 键。

更多说明：

上面实例代码其实是将 WebDriver 内建的模拟按键的方法与第三方提供的模拟按键的方法结合起来了，通过这种方法可以模拟更多的组合按键功能，读者可以根据自己的需要进行组合。

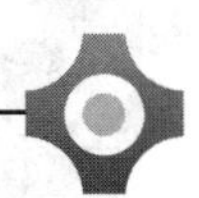

10.32 模拟鼠标右击

目的：

模拟右键菜单实现粘贴效果。

用于测试网址：

http://www.sogou.com

调用 API 的实例代码：

```
# 导入模拟组合按键需要的包
from selenium.webdriver import ActionChains
import win32clipboard as w
import win32con
import time

# 设置剪切板内容
def setText(aString):
    w.OpenClipboard()
    w.EmptyClipboard()
    w.SetClipboardData(win32con.CF_UNICODETEXT, aString)
    w.CloseClipboard()
```

将上面这段代码放到单元测试类外边的区域，测试用例方法代码如下：

```
def test_rigthClickMouse(self):
    url = "http://www.sogou.com"
    # 访问搜狗首页
    self.driver.get(url)
    # 找到搜索输入框
    searchBox = self.driver.find_element_by_id("query")
    # 将焦点切换到搜索输入框
    searchBox.click()
    time.sleep(2)
    # 在搜索输入框上执行一个鼠标右击操作
    ActionChains(self.driver).context_click(searchBox).perform()
    # 将 gloryroad 数据设置到剪切板中,相当于执行了复制操作
    setText(u'gloryroad')
    # 发送一个粘贴命令,字符 P 指代粘贴操作
    ActionChains(self.driver).send_keys('P').perform()
    # 单击搜索按钮
    self.driver.find_element_by_id('stb').click()
    time.sleep(2)
```

10.33 模拟鼠标左键按下与释放

目的：

在灰色的 div 区域，模拟鼠标左键单击，并在鼠标左键按下和释放过程中显示一些特定的文字。

用于测试 HTML 代码：

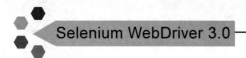

```
<html>
<head>
<meta http-equiv="Content-Type" content="text/html; charset=utf-8" />
<script type="text/javascript">
    function mouseDownFun()
    {
        document.getElementById('div1').innerHTML +=
        '鼠标左键被按下<br/>';
    }
    function mouseUpFun()
    {
        document.getElementById('div1').innerHTML +=
        '已经被按下的鼠标左键被释放抬起<br/>';
    }
    function clickFun()
    {
        document.getElementById('div1').innerHTML +=
        '单击动作发生<br/>';
    }
</script>
</head>
<body>
    <div id="div1" onmousedown="mouseDownFun();" onmouseup="mouseUpFun();"
    onclick="clickFun();" style="background:#CCC; border:3px solid #999;
    width:200px; height:200px; padding:10px">
    </div>
    <input style="margin-top:10px" type="button"
    onclick="document.getElementById('div1').innerHTML='';" value="清除信息" />
</body>
</html>
```

调用 API 的实例代码：

```
def test_simulationLeftClickMouseOfProcess(self):
    url = "d:\\test.html"
    # 访问自定义的 html 网页
    self.driver.get(url)
    div = self.driver.find_element_by_id("div1")
    from selenium.webdriver import ActionChains
    import time
    # 在 id 属性值为"div1"的元素上执行按下鼠标左键,并保持
    ActionChains(self.driver).click_and_hold(div).perform()
    time.sleep(2)
    # 在 id 属性值为"div1"的元素上释放一直按下的鼠标左键
    ActionChains(self.driver).release(div).perform()
    time.sleep(2)
    ActionChains(self.driver).click_and_hold(div).perform()
    time.sleep(2)
    ActionChains(self.driver).release(div).perform()
```

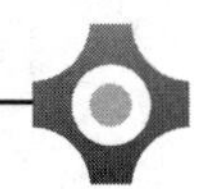

10.34　保持鼠标指针悬停在某个元素上

目的：

在网页上的链接上方悬浮鼠标指针，在页面上可以显示出一个蓝色的长方形图案，鼠标指针离开链接上方后，蓝色的长方形图案消失。

用于测试的 HTML 网页代码：

```
<html>
<head>
    <meta http-equiv="Content-Type" content="text/html; charset=utf-8" />
    <script language="javascript">
        function showNone()
        {
            document.getElementById('div1').style.display = "none";
        }
        function showBlock(){
        document.getElementById('div1').style.display = "block";
    }
    </script>
    <style type="text/css">
        #div1 {
            position:absolute;
            width:200PX;
            height:115px;
            z-index:1;
            left: 28px;
            top: 34px;
            background-color:#0033CC;}
    </style>
</head>
<body onload="showNone()">
    <div id="div1"></div>
    <a onmouseover="showBlock()" onmouseout="showNone()" id="link1">鼠标指过来 1</a>
    <a onmouseover="showBlock()" onmouseout="showNone()" id="link2">鼠标指过来 2</a>
    <p>鼠标指针悬浮这里的时候,蓝色的图形框就消失了<p>
</body>
</html>
```

调用 API 的实例代码：

```
def test_roverOnElement(self):
    url = "d:\\test.html"
    # 访问自定义的 html 网页
    self.driver.get(url)
    # 找到页面上第一个链接元素
    link1 = self.driver.find_element_by_partial_link_text("指过来 1")
    # 找到页面上第二个链接元素
    link2 = self.driver.find_element_by_partial_link_text("指过来 2")
```

```
    # 找到页面上的 p 元素
    p = self.driver.find_element_by_xpath("//p")
    print(link1.text, link2.text)
    # 导入需要的 Python 包
    from selenium.webdriver import ActionChains
    import time
    # 将鼠标指针悬浮到第一个链接元素上
    ActionChains(self.driver).move_to_element(link1).perform()
    time.sleep(2)
    # 将鼠标指针从第一个链接元素移动到 p 元素上
    ActionChains(self.driver).move_to_element(p).perform()
    time.sleep(2)
    # 将鼠标指针悬浮到第二个链接元素上
    ActionChains(self.driver).move_to_element(link2).perform()
    time.sleep(2)
    # 将鼠标指针从第二个链接元素移动到 p 元素上
    ActionChains(self.driver).move_to_element(p).perform()
    time.sleep(2)
```

结果说明：

当鼠标指针悬浮到页面上两个链接上时，页面会出现一个蓝色长方形图案，当将鼠标指针从链接元素上移动到 p 元素上时，蓝色长方形图案消失。

更多说明：

使用鼠标指针悬浮在某个元素的操作可以完成对某些网页上需要用鼠标指针悬浮后才能出现的页面元素的操作。

10.35 判断页面元素是否存在

用于测试的网址：

http://www.sogou.com

调用 API 的实例代码：

```
def isElementPresent(self, by, value):
    # 从 selenium.common.exceptions 模块导入 NoSuchElementException 异常类
    from selenium.common.exceptions import NoSuchElementException
    try:
        element = self.driver.find_element(by = by, value = value)
    except NoSuchElementException as err:
        # 打印异常信息
        print("erorr:{0}".format(err))
        # 发生了 NoSuchElementException 异常，说明页面中未找到该元素，返回 False
        return False
    else:
        # 没有发生异常，表示在页面中找到了该元素，返回 True
        return True

def test_isElementPresent(self):
    url = "http://www.sogou.com"
```

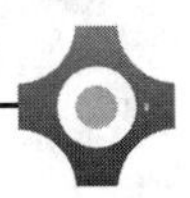

```
    # 访问 sogou 首页
    self.driver.get(url)
    # 判断页面元素 id 属性值为"query"的页面元素是否存在
    res = self.isElementPresent("id", "query")
    if res is True:
        print("所查找的元素存在于页面上!")
    else:
        print("页面中未找到所需要的页面元素!")
```

代码解释：

from selenium.common.exceptions import NoSuchElementException 表示从 selenium.common.exceptions 这个异常模块导入 NoSuchElementException 异常类，在使用 WebDriver 实施自动化的过程中抛出的所有异常都是从这个模块导入的，比如 TimeoutException 异常等。

10.36　隐式等待

隐式等待表示在自动化实施过程中，为查找页面元素或者执行命令设置一个最长等待时间，如果在规定时间内页面元素被找到或者命令被执行完成，则执行下一步，否则继续等待直到设置的最长等待时间截止。

用于测试网址：

http://www.sogou.com

调用 API 的实例代码：

```
def test_implictWait(self):
    # 导入异常类
    from selenium.common.exceptions import NoSuchElementException, TimeoutException
    # 导入堆栈类
    import traceback
    url = "http://www.sogou.com"
    # 访问搜狗首页
    self.driver.get(url)
    # 通过 driver 对象 implicitly_wait()方法来设置隐式等待时间,最长等待 10 秒
    self.driver.implicitly_wait(10)
    try:
        # 查找搜狗首页的搜索输入框页面元素
        searchBox = self.driver.find_element_by_id("query")
        # 在搜索输入框中输入"光荣之路自动化测试"
        searchBox.send_keys("光荣之路自动化测试")
        # 查找搜狗首页搜索按钮页面元素
        click = self.driver.find_element_by_id("stb")
        # 单击搜索按钮
        click.click()
    except (NoSuchElementException, TimeoutException):
        # 打印异常的堆栈信息
        print(traceback.print_exc())
```

更多说明：

隐式等待的好处是不用像强制等待(time. sleep(n))方法一样死等固定时间 n 秒，可以在一定程度上提升测试用例的执行效率。不过这种方法也存在一个弊端，那就是程序会一直等待整个页面加载完成，也就是说浏览器窗口标签栏中不再出现转动的小圆圈，才会继续执行下一步，比如某些时候想要的页面元素早就加载完成了，但由于个别 JS 等资源加载稍慢，此时程序仍然会等待页面全部加载完成才会继续执行下一步，这无形中加长了测试用例执行时间。

隐式等待时间只需要被设置一次，然后它将在 driver 的整个生命周期都起作用。

10.37 显式等待

上面我们介绍了强制等待和隐式等待，这里再介绍一种更智能的等待方式——显式等待。通过 selenium. webdriver. support. ui 模块提供的 WebDriverWait 类，再结合该类的 until()和 until_not()方法，并自定义好显式等待的条件，然后根据判断条件而进行灵活的等待。显式等待比隐式等待更节约测试脚本执行时间，推荐尽量使用显式等待方式来判断页面元素是否存在。

显式等待工作原理：

程序会每隔一段时间(该时间一般都很短，默认为 0.5 秒，也可以自定义)执行一下自定义的判定条件，如果条件成立，就执行下一步，否则继续等待，直到超过设定的最长等待时间，然后抛出 TimeoutException 异常。

WebDriverWait 类解析：

WebDriverWait 类的构造方法：

```
__init__(self, driver, timeout, poll_frequency = 0.5, ignored_exceptions = None)
```

参数解释：

driver：WebDriver 实例对象(IE，Firefox，Chrome 或 Remote)。

timeout：最长的显式等待时间，单位秒。

poll_frequency：调用频率，也就是在 timeout 时间段内，每隔 poll_frequency 时间执行一次判断条件，默认为 0.5 秒。

ignored_exceptions：执行过程中忽略的异常类型，默认只忽略 NoSuchElementException 异常类。

WebDriverWait 类提供的方法：

(1) until(method，message='')

在规定等待时间内，每隔一段时间调用一下 method 方法，直到其返回值不为 False，如果超时抛出带有 message 异常信息的 TimeoutException 异常。

(2) until_not(method，message='')

与 until()方法相反，表示在规定时间内，每隔一段时间调用一下 method 方法，直到其

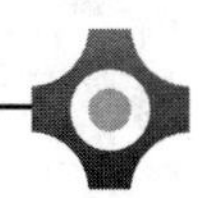

返回值为 False,如果超时,则抛出带有 message 异常信息的 TimeoutException 异常。

被测试网页的 HTML 代码:

```
<html>
    <head>
        <meta http-equiv='Content-Type' content='text/html; charset=utf-8'>
        <title>你喜欢的水果</title>
        <script type="text/javascript">
        function display_alert()
         {
         alert("I am an alert box!!")
         }
        </script>
    </head>
<body>
    <p>请选择你爱吃的水果</p>
    <br>
    <select name='fruit'>
        <option id='peach'          value='taozi'>桃子</option>
        <option id='watermelon'value='xigua'>西瓜</option>
    </select>
    <br>
    <input type="button" onclick="display_alert()" value="Display alert box" />
    <input id="check" type='checkbox'>是否喜欢吃水果?</input>
    <br><br>
    <input type="text" id="text" value="今年夏天西瓜相当甜!">文本框</input>
</body>
</html>
```

调用 API 的实例代码:

```
def test_explicitWait(self):
    # 导入堆栈类
    import traceback
    # 导入 By 类
    from selenium.webdriver.common.by import By
    # 导入显示等待类
    from selenium.webdriver.support.ui import WebDriverWait
    # 导入期望场景类
    from selenium.webdriver.support import expected_conditions as EC
    from selenium.common.exceptions import TimeoutException, NoSuchElementException
    url = "d:\\test.html"
    # 访问自动以测试网页
    self.driver.get(url)
    try:
        wait = WebDriverWait(self.driver, 10, 0.2)
        wait.until(EC.title_is("你喜欢的水果"))
        print("网页标题是"你喜欢的水果"")
        # 等待 10 秒,直到要找的按钮出现
        element = WebDriverWait(self.driver, 10).until\
            (lambda x: x.find_element_by_xpath\
```

```
            ("//input[@value = 'Display alert box']"))
        element.click()
        # 等待 alert 框出现
        alert = wait.until(EC.alert_is_present())
        # 打印 alert 框体消息
        print(alert.text)
        # 确认警告信息
        alert.accept()
        # 获取 id 属性值为 peach 的页面元素
        peach = self.driver.find_element_by_id("peach")
        # 判断 id 属性值为 peach 的页面元素是否能被选中
        peachElement = wait.until(EC.element_to_be_selected(peach))
        print("下拉列表的选项"桃子"目前处于选中状态")
        # 判断复选框是否可见并且能被单击
        wait.until(EC.element_to_be_clickable((By.ID, 'check')))
        print("复选框可见并且能被单击")
    except TimeoutException as err:
        # 捕获 TimeoutException 异常
        print(traceback.print_exc())
    except NoSuchElementException as err:
        # 捕获 NoSuchElementException 异常
        print(traceback.print_exc())
    except Exception as err:
        # 捕获其他异常
        print(traceback.print_exc())
```

10.38 显式等待中期望的场景

在自动化实施过程中，常常会有在执行某步操作或者某个命令之前，先看看要操作的元素是否处于显式状态、是否可操作等需求，也就是看看我们期望的场景是否存在。

WebDriver 中支持的场景方法均来自 selenium.webdriver.support.expected_conditions 模块，所有的场景方法如下。被测试网页 HTML 代码同本章 10.37 小节。

- **alert_is_present()**：判断页面是否出现 alert 框。显式等待中使用方法：

```
wait = WebDriverWait(driver, 10)
# 打印一下 alert 框消息
wait.until(EC.alert_is_present()).text
```

- **element_located_selection_state_to_be(locator, state)**：判断一个元素的状态是否是给定的选择状态，第一个传入的参数是一个定位器，定位器是一个元组(by, path)，第二个参数表示期望的元素状态，True 表选中状态，False 表未选中状态。相等返回 True，否则返回 False。显式等待中使用方法：

```
# True
wait.until(EC.element_located_selection_state_to_be((By.ID, "peach"), True))
# False
wait.until_not(EC.element_located_selection_state_to_be((By.ID, "watermelon"), True))
# 原型
```

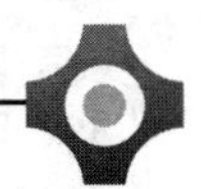

```
EC.element_located_selection_state_to_be((By.ID, "peach"), True).is_selected。
```

- **element_selection_state_to_be(driverObject, state)**：判断给定的元素是否被选中，第一个参数是一个 WebDriver 对象，第二个是期望的元素的状态，相等返回 True，否则返回 False。显式等待中使用方法：

```
# True
wait.until(EC.element_selection_state_to_be(driver.find_element_by_id("peach"), True))
# False
wait.until_not(EC.element_selection_state_to_be(driver.find_element_by_id("peach"),
False))
# 原型
EC.element_selection_state_to_be(driver.find_element_by_id("peach"), True).is_
selected。
```

- **element_located_to_be_selected(locator)**：期望某个元素处于选中状态，参数为一个定位器。显式等待中使用方法：

```
wait.until(EC.element_located_to_be_selected((By.ID, "peach")))
```

- **element_to_be_selected(driverObject)**：期望某个元素处于选中状态，参数为一个 WebDriver 实例对象。显式等待中使用方法：

```
wait.until(EC.element_to_be_selected(driver.find_element_by_id("peach")))
```

- **element_to_be_clickable(locator)**：判断某元素是否可见并且能被单击，条件满足返回该页面元素对象，否则返回 False。显式等待中使用方法：

```
# 存在并可见
wait.until(EC.element_to_be_clickable((By.XPATH, '//input[@value="Display alert box"]')))
# 不存在
wait.until_not(EC.element_to_be_clickable((By.XPATH, '//input/a"]')))
```

- **frame_to_be_available_and_switch_to_it(parm)**：判断 frame 是否可用，如果可用返回 True 并切入到该 frame，参数 parm 可以是定位器 locator 也就是(by, xpath)组成的元组，或者定位方式：id、name、index(该 frame 在页面上索引号)，或者 WebElement 对象。显式等待中使用方法：

```
测试网址:http://www.126.com 首页
wait = WebDriverWait(driver, 10, 0.2)
# 传入 ID 值"x-URS-iframe"
wait.until(EC.frame_to_be_available_and_switch_to_it((By.ID, "x-URS-iframe")))
# 传入 frame 的 WebElement 对象
wait.until(EC.frame_to_be_available_and_switch_to_it(driver.find_element_by_id("x-URS
-iframe")))
# 传入 frame 在页面中的索引号
wait.until(EC.frame_to_be_available_and_switch_to_it(1))
```

- **invisibility_of_element_located(locator)**：希望某个元素不可见或者不存在于 DOM 中，满足条件返回 True，否则返回定位到的元素对象。显式等待中使用方法：

```
wait.until(EC.invisibility_of_element_located((By.ID, "watermelon2")))
```

- **visibility_of_element_located(locator)**：希望某个元素出现在页面的 DOM 中，并且可见，如果满足条件返回该元素的页面元素对象。显式等待中使用方法：

```
element = wait.until(EC.visibility_of_element_located((By.ID, "peach")))
```

- **visibility_of(WebElement)**：希望某个元素出现在页面的 DOM 中，并且可见，如果满足条件返回该元素的页面元素对象。显式等待中使用方法：

```
element = wait.until(EC.visibility_of(driver.find_element_by_id("peach")))
```

- **visibility_of_any_elements_located(locator)**：判断页面上至少一个元素可见，返回满足条件的所有页面元素对象。显式等待中使用方法：

```
inputs = wait.until(EC.visibility_of_any_elements_located((By.TAG_NAME, "input")))
```

- **presence_of_all_elements_located(locator)**：判断页面上至少有一个元素出现，如果满足条件返回所有满足定位表达式的页面元素。显式等待中使用方法：

```
elements = wait.until(EC.presence_of_all_elements_located((By.ID, "text")))
```

- **presence_of_element_located(locator)**：判断某个元素是否出现在 DOM 中，不一定可见，存在返回该页面元素对象。显式等待中使用方法：

```
# 存在
wait.until(EC.presence_of_element_located((By.ID, "check")))
# 不存在
wait.until_not(EC.presence_of_element_located((By.ID, "div2")))
```

- **staleness_of(WebElement)**：判断一个元素是否仍在 DOM 中，如果在规定时间内已经移除返回 True，否则返回 False。显式等待中使用方法：

```
wait.until(EC.staleness_of(driver.find_element_by_id("check")))
```

- **text_to_be_present_in_element(locator, text)**：判断文本内容 text 是否出现在某个元素中，判断的是元素的 text。显式等待中使用方法：

```
wait.until(EC.text_to_be_present_in_element((By.TAG_NAME, "P"), "请选择你爱吃的水果"))
```

- **text_to_be_present_in_element_value(locator, text)**：判断 text 是否出现在元素的 value 属性值中。显式等待中使用方法：

```
wait.until(EC.text_to_be_present_in_element_value((By.ID, "peach"), "taozi"))
```

- **title_contains(partial_title)**：判断页面 title 标签内容是否包含 partial_title，只需要部分匹配即可，包含返回 True，否则返回 False。显式等待中使用方法：

```
wait.until(EC.title_contains("欢的水果"))
```

- **title_is(title_text)**：判断页面 title 内容是与传入的 title_text 内容完全匹配，匹配返

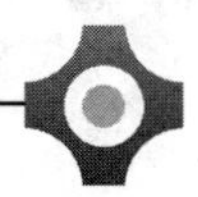

回 True，否则返回 False。显式等待中使用方法：

```
# 完全匹配
wait.until(EC.title_is("你喜欢的水果"))
# 不完全匹配
wait.until_not(EC.title_is("你喜欢的水果 3"))
```

10.39 使用 Title 属性识别和操作新弹出的浏览器窗口

用于测试的网址：

http://www.baidu.com 和 http://www.sogou.com

调用 API 的实例代码：

```
def test_identifyPopUpWindowByTitle(self):
    # 导入多个异常类型
    from selenium.common.exceptions import NoSuchWindowException, \
        TimeoutException
    # 导入期望场景类
    from selenium.webdriver.support import expected_conditions as EC
    # 导入 By 类
    from selenium.webdriver.common.by import By
    # 导入 WebDriverWait 类
    from selenium.webdriver.support.ui import WebDriverWait
    # 导入堆栈类
    import traceback
    # 导入时间模块
    import time
    url = "http://www.baidu.com"
    # 访问百度首页
    self.driver.get(url)
    # 通过执行 JavaScript 来新开一个窗口,访问搜狗
    js = 'window.open("http://www.sogou.com");'
    self.driver.execute_script(js)
    # 获取当前所有打开的浏览器窗口句柄
    all_handles = self.driver.window_handles
    # 打印当前浏览器窗口句柄
    print(self.driver.current_window_handle)
    # 打印打开的浏览器窗口的个数
    print(len(all_handles))
    # 等待 2 秒,以便更好查看效果
    time.sleep(2)
    # 如果存储浏览器窗口句柄的容器不为空,再遍历 all_handles 中所有的浏览器句柄
    if len(all_handles) > 0:
        try:
            for windowHandle in all_handles:
                # 切换窗口
                self.driver.switch_to.window(windowHandle)
                print(" --- ", self.driver.title)
                # 判断当前浏览器窗口的 title 属性是否等于
```

```
                # "搜狗搜索引擎 - 上网从搜狗开始"
                if self.driver.title == "搜狗搜索引擎 - 上网从搜狗开始":
                    # 显示等待页面搜索输入框加载完成,
                    # 然后输入"sogou 首页的浏览器窗口被找到"
                    WebDriverWait(self.driver, 10, 0.2).until(lambda x: \
                        x.find_element_by_id("query")).\
                         send_keys("sogou 首页的浏览器窗口被找到")
                    time.sleep(2)
        except NoSuchWindowException as err:
            # 捕获 NoSuchWindowException 异常
            print(traceback.print_exc())
        except TimeoutException as err:
            # 捕获 TimeoutException 异常
            print(traceback.print_exc())
    # 将浏览器窗口切换回默认窗口
    self.driver.switch_to.window(all_handles[0])
    print(self.driver.title)
    # 断言当前浏览器窗口的 title 属性是"百度一下,你就知道"
    self.assertEqual(self.driver.title, "百度一下,你就知道")
```

10.40 通过页面的关键内容识别和操作新浏览器窗口

用于测试的网址:

http://www.baidu.com 和 http://www.sogou.com

调用 API 的实例代码:

```
def test_identifyPopUpWindowByPageSource(self):
    # 导入多个异常类型
    from selenium.common.exceptions import NoSuchWindowException, \
        TimeoutException
    # 导入期望场景类
    from selenium.webdriver.support import expected_conditions as EC
    # 导入 By 类
    from selenium.webdriver.common.by import By
    # 导入 WebDriverWait 类
    from selenium.webdriver.support.ui import WebDriverWait
    # 导入堆栈类
    import traceback
    # 导入时间模块
    import time
    url = "http://www.baidu.com"
    # 访问百度首页
    self.driver.get(url)
    # 通过执行 JavaScript 来新开一个窗口,访问搜狗
    js = 'window.open("http://www.sogou.com");'
    self.driver.execute_script(js)
    # 获取当前所有打开的浏览器窗口句柄
    all_handles = self.driver.window_handles
    print(all_handles)
```

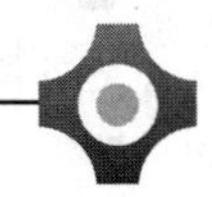

```
# 打印当前浏览器窗口句柄
print("当前浏览器窗口句柄:", self.driver.current_window_handle)
# 打印打开的浏览器窗口的个数
print("已打开窗口的个数:", len(all_handles))
# 等待 2 秒,以便更好查看效果
time.sleep(2)
# 如果存储浏览器窗口句柄的容器不为空,再遍历 all_handles 中所有的浏览器句柄
if len(all_handles) > 0:
    try:
        for windowHandle in all_handles:
            # 切换窗口
            # self.driver.switch_to.window(windowHandle)
            self.driver.switch_to.window(windowHandle)
            print("当前浏览器窗口句柄:", self.driver.current_window_handle)
            # 获取当前浏览器窗口的页面源代码
            pageSource = self.driver.page_source
            if "搜狗搜索" in pageSource:
                # 显示等待页面搜索输入框加载完成,
                # 然后输入"sogou 首页的浏览器窗口被找到"
                WebDriverWait(self.driver, 10, 0.2).until \
                    (lambda x: x.find_element_by_id("query")). \
                    send_keys("sogou 首页的浏览器窗口被找到")
                time.sleep(2)
    except NoSuchWindowException as err:
        # 如果没找到浏览器的句柄,会抛出 NoSuchWindowException 异常,
        # 打印异常的堆栈信息
        print(traceback.print_exc())
    except TimeoutException as err:
        # 显示等待超过规定时间后抛出 TimeoutException 异常
        # 打印异常的堆栈信息
        print(traceback.print_exc())
# 将浏览器窗口切换回默认窗口
self.driver.switch_to.window(all_handles[0])
# 断言当前浏览器窗口页面源代码中是否包含"百度一下"关键内容
self.assertTrue("百度一下" in self.driver.page_source)
```

10.41 操作 Frame 中的页面元素

有时网页中会嵌套一个或多个 Frame,此时如果我们直接去找嵌套在 Frame 里的页面元素就会抛出 NoSuchElementException 异常,所以在操作嵌套在 Frame 里面的页面元素之前,需要将页面焦点切换到 Frame 里。

目的:

将当前焦点切换到 Frmae 中,并在不同的 Frame 间互相切换,同时操作 frame 中的页面元素。

用于测试网页的 HTML 代码:

frameset.html 页面代码:

```
<html>
```

```
<head>
    <title>frameset 页面</title>
    <meta http-equiv="Content-Type" content="text/html; charset=utf-8" />
</head>
<frameset cols="25%,50%,25%">
    <frame id="leftframe" src="frame_left.html" />
    <frame id="middleframe" src="frame_middle.html" />
    <frame id="rightframe" src="frame_right.html" />
</frameset>
    </html>
```

frame_left.html 页面代码：

```
<html>
<head>
    <title>左侧 frame</title>
    <meta http-equiv="Content-Type" content="text/html; charset=utf-8" />
    <script type="text/javascript">
        function display_alert()
          {
          alert("I am an alert box!!")
          }
    </script>
</head>
<body>
    <p>这是左侧 frame 页面上的文字</p>
    <input type="button" onclick="display_alert()" value="Display alert box" />
</body>
</html>
```

frame_middle.html 页面代码：

```
<html>
<head>
    <title>中间 frame</title>
    <meta http-equiv="Content-Type" content="text/html; charset=utf-8" />
</head>
<body>
    <p>这是中间 frame 页面上的文字</p>
    <input type="text" id="text" value="">文本框</input>
</body>
</html>
```

frame_right.html 页面代码：

```
<html>
<head>
    <title>右侧 frame</title>
    <meta http-equiv="Content-Type" content="text/html; charset=utf-8" />
</head>
<body>
    <p>这是右侧 frame 页面上的文字</p>
```

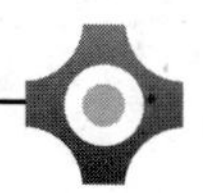

```
        <input id="python" type='radio' name="book" checked>python selenium</input>
        <br />
        <input id="java" type='radio' name="book">java selenium</input>
    </body>
    </html>
```

将这四个 HTML 文件放到同一目录下。

将 frameset.html 文件直接拖曳到浏览器窗口，展示的页面效果如图 10-2 所示。

图　10-2

调用 API 的实例代码：

```
def test_HandleFrame(self):
    from selenium.webdriver.support import expected_conditions as EC
    from selenium.webdriver.support.ui import WebDriverWait
    from selenium.common.exceptions import TimeoutException
    import traceback
    url = "file:///" + "d:\\frameset.html"
    # 访问自定义测试网页
    self.driver.get(url)
    # 使用索引方式进入指定的 frame 页面，索引号从 0 开始
    # 所以想进入中间的 frame，需要使用索引号 1
    # 如果没有使用此行代码，则无法找到页面中左侧 frame 中的任何页面元素
    self.driver.switch_to.frame(0)
    # 找到左侧 frame 中的 p 标签元素
    leftFrameText = self.driver.find_element_by_xpath("//p")
    # 断言左侧 frame 中的文字是否和"这是左侧 frame 页面上的文字"几个关键字相一致
    self.assertAlmostEqual(leftFrameText.text, "这是左侧 frame 页面上的文字")
    # 找到左侧 frame 中的按钮元素，并单击该元素
    self.driver.find_element_by_tag_name("input").click()
    try:
        # 动态等待 alert 窗体出现
        alertWindow = WebDriverWait(self.driver, 10).until(EC.alert_is_present())
        # 打印 alert 消息
        print(alertWindow.text)
        alertWindow.accept()
    except TimeoutException as err:
        print(traceback.print_exc())
    # 使用 driver.switchTo.default_content 方法，从左侧 frame 中返回到 frameset 页面
    # 如果不调用此行代码，则无法从左侧 frame 页面中直接进入其他 frame 页面
    self.driver.switch_to.default_content()

    # 通过标签名找到页面中所有的 frame 元素，然后通过索引进入该 frame
```

```
self.driver.switch_to.frame(self.driver.find_elements_by_tag_name("frame")[1])
# 断言页面源码中是否存在"这是中间 frame 页面上的文字"关键字串
assert "这是中间 frame 页面上的文字" in self.driver.page_source
# 再输入框中输入"我在中间 frame"
self.driver.find_element_by_tag_name("input").send_keys("我在中间 frame")
self.driver.switch_to.default_content()
self.driver.switch_to.frame(self.driver.find_element_by_id("rightframe"))
assert "这是右侧 frame 页面上的文字" in self.driver.page_source
self.driver.switch_to.default_content()
```

更多说明：

切进 frame 和切出 frame 的方法，在较低版本的 Selenium 中，提供的是 driver.switch_to_frame()和 driver.switch_to_default_content()方法，而本书使用的 Selenium 版本已经推荐用户使用 driver.switch_to.frame()和 driver.switch_to.default_content()方法进行代替，但同时也是兼容老版本的。

10.42 使用 Frame 中的 HTML 源码内容操作 Frame(x)

目的：

能够使用 Frame 页面的 HTML 源码定位指定的 Frame 页面并进行操作。

用于测试的 HTML 代码：

同 10.41 节被测试页面的 HTML 代码。

调用 API 实例代码：

```
def test_HandleFrameByPageSource(self):
    url = "d:\\frameset.html"
    # 访问自定义测试网页
    self.driver.get(url)
    # 找到页面上所有的 frame 页面对象,并存储到名为 framesList 列表中
    framesList = self.driver.find_elements_by_tag_name("frame")
    # 通过 for 循环遍历 framesList 中所有的 frame 页面,查找页面源码中含有
    # "中间 frame"的 frame 页面
    for frame in framesList:
        # 进入到 frame 页面
        self.driver.switch_to.frame(frame)
        # 判断每个 frame 的 HTML 源码中是否包含"中间 frame"几个关键词
        if "中间 frame" in self.driver.page_source:
            # 如果包含需要查找的关键字,则查找到页面上的 p 标签元素
            p = self.driver.find_element_by_xpath("//p")
            # 断言页面上 p 元素文本内容是否是"这是中间 frame 页面上的文字"
            self.assertAlmostEqual("这是中间 frame 页面上的文字", p.text)
            # 退出 frame
            self.driver.switch_to.default_content()
            # 找到指定的 frame 页面,并做相应的操作后退出循环
            break
        else:
            # 如果没找到指定的 frame,则调用此行代码,返回到 frameset 页面中
            # 以便下次 for 循环中能继续调用 driver.switch_to.frame 方法,否则会报错
            self.driver.switch_to.default_content()
```

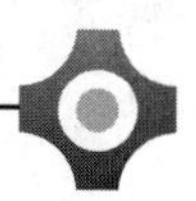

10.43 操作 IFrame 中的页面元素

用于测试的 HTML 代码：

同 10.41 节被测试网页的 HTML 代码，其中需要更新一下如下页面的 HTML 代码。

修改 frame_left.html 页面代码如下：

```
<html>
<head>
    <title>左侧 frame</title>
    <meta http-equiv="Content-Type" content="text/html; charset=utf-8" />
    <script type="text/javascript">
        function display_alert()
          {
          alert("I am an alert box!!")
          }
    </script>
</head>
<body>
    <p>这是左侧 frame 页面上的文字</p>
    <input type="button" onclick="display_alert()" value="Display alert box" />
    <iframe id="showIfame" src='iframe.html' style="width:200px";height:50px></iframe>
</body>
</html>
```

在 frame_left.html 同级目录下新增 iframe.html 文件：

```
<html>
    <head>
        <title>iframe</title>
        <meta http-equiv="Content-Type" content="text/html; charset=utf-8" />
    </head>
<body>
    <p>这是 iframe 页面上的文字</p>
</body>
</html>
```

页面展示效果如图 10-3 所示。

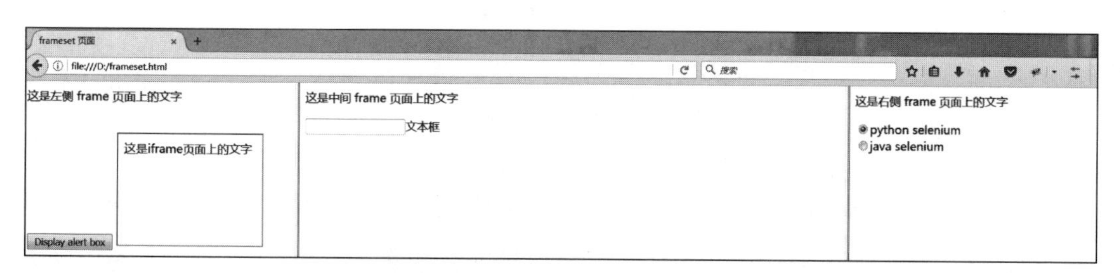

图　10-3

调用 API 的实例代码：

```
def test_HandleIFrame(self):
    url = "d:\\frameset.html"
    # 访问自定义测试网页
    self.driver.get(url)
    # 改变操作区域,切换进入页面上第一个 frame,也就是左边的 frame
    self.driver.switch_to.frame(0)
    # 断言页面是否存在"这是左侧 frame 页面上的文字"关键字串,
    # 以判断是否成功切换进 frame 页面
    assert "这是左侧 frame 页面上的文字" in self.driver.page_source

    # 改变操作区域,切换进入 id 为"showIfame"的 iframe 页面
    self.driver.switch_to.frame(self.driver.find_element_by_id("showIfame"))
    # 断言页面是否存在"这是 iframe 页面上的文字"这样的关键字串,
    # 以便判断是否成功切换进 iframe 页面
    assert "这是 iframe 页面上的文字" in self.driver.page_source

    # 将操作区域切换到 frameset 页面,以便能成功进入其他 frame
    self.driver.switch_to.default_content()
    # 断言页面的 title 值是否为"frameset 页面"
    assert "frameset 页面" == self.driver.title
    # 改变操作区域,切换进入中间 frame 页面
    self.driver.switch_to.frame(1)
    # 断言页面上是否存在"这是中间 frame 页面上的文字"这样的关键字串
    assert "这是中间 frame 页面上的文字" in self.driver.page_source
```

更多说明：

（1）如果在一个 frame 中又内嵌了 iframe，再想进入这个内嵌的 iframe 时，必须先进入 frame，然后才能进入 iframe 页面。

（2）在一个 frame 下无论依次进入多少层内嵌的 frame 或 iframe，调用一次 driver.switch_to.default_content()函数都会直接从所有的 frame 中切换出来回到默认页面，比如本例的 frameset 页面。

10.44 操作 JavaScript 的 Alert 弹窗

目标：

能够模拟鼠标单击弹出的 Alert 窗口上的“确定”按钮。

用于测试网页的 HTML 代码：

```
<html>
    <head>
        <title>你喜欢的水果</title>
        <meta http-equiv="Content-Type" content="text/html; charset=utf-8" />
    </head>
<body>
    <input id='button' type='button' onclick="alert('这是一个 alert 弹出框');"
    value='单击此按钮,弹出 alert 弹窗'/>
```

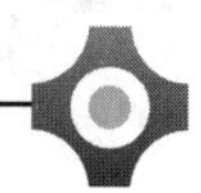

```
</body>
</html>
```

该 HTML 代码在浏览器中展示弹窗的效果图如图 10-4 所示。

图　10-4

调用 API 的实例代码：

```
def test_HandleAlert(self):
    from selenium.common.exceptions import NoAlertPresent-
Exception
    import time
    url = "d:\\alert.html"
    # 访问自定义测试网页
    self.driver.get(url)
    # 通过 id 属性值查找页面上的按钮元素
    button = self.driver.find_element_by_id("button")
    # 单击按钮元素,则会弹出一个 Alert 消息框,
    # 上面显示"这是一个 alert 弹出框"和"确定"按钮
    button.click()
    try:
        # 使用 driver.switch_to_alert()方法获取 alert 对象
        alert = self.driver.switch_to_alert
        time.sleep(2)
        # 使用 alert.text 属性获取 alert 框中的内容,
        # 并断言文字内容是否是"这是一个 alert 弹出框"
        self.assertEqual(alert.text, "这是一个 alert 弹出框")
        # 调用 alert 对象的 accept()方法,模拟鼠标单击 alert 弹窗上的"确定"按钮
        # 以便关闭 alert 窗
        alert.accept()
    except NoAlertPresentException as err:
        # 如果 Alert 框未弹出显示在页面上,则会抛出 NoAlertPresentException 的异常
        print("尝试操作的 alert 框未被找到")
        print("error: {0}".format(err))
```

更多说明：

在本书使用的 Selenium 版本中,已经推荐用户使用 driver.switch_to_alert 来代替 driver.switch_to.alert()方法获取 Alert 对象。

10.45　操作 JavaScript 的 confirm 弹窗

目标：

能够模拟鼠标单击 JavaScript 弹出的 confirm 框中的“确定”和“取消”按钮。

用于测试网页的 HTML 代码：

```
<html>
    <head>
        <title>你喜欢的水果</title>
```

```
        <meta http-equiv="Content-Type" content="text/html; charset=utf-8" />
    </head>
<body>
    <input id='button' type='button' onclick="confirm('这是一个 confirm 弹出框');"
    value='单击此按钮,弹出 confirm 弹出窗'/>
</body>
</html>
```

该 HTML 代码在浏览器中展示弹窗效果图如图 10-5 所示。

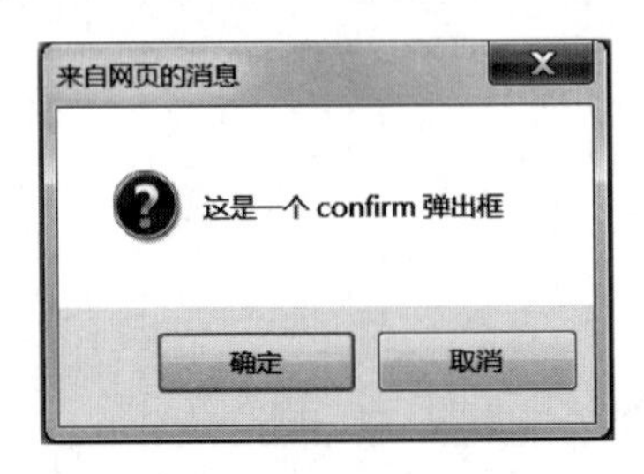

图 10-5

调用 API 的实例代码：

```
def test_Handleconfirm(self):
    from selenium.common.exceptions import NoAlertPresentException
    import time
    url = "d:\\confirm.html"
    # 访问自定义测试网页
    self.driver.get(url)
    # 通过 id 属性值查找页面上的按钮元素
    button = self.driver.find_element_by_id("button")
    # 单击按钮元素,则会弹出一个 confirm 提示框,
    # 上面显示"这是一个 confirm 弹出框""确定"和"取消"按钮
    button.click()
    try:
        # 较高版本的 Selenium 推荐使用 driver.switch_to.alert 方法代替
        # driver.switch_to.alert 方法来获取 alert 对象
        alert = self.driver.switch_to.alert
        time.sleep(2)
        # 使用 alert.text 属性获取 confirm 框中的内容,
        # 并断言文字内容是否是"这是一个 confirm 弹出框"
        self.assertEqual(alert.text, "这是一个 confirm 弹出框")
        # 调用 alert 对象的 accept()方法,模拟鼠标单击 confirm 弹窗上的"确定"按钮
        # 以便关闭 confirm 窗
        alert.accept()
        # 取消下面一行代码的注释,就会模拟单击 confirm 框上的"取消"按钮
        # alert.dismiss()
    except NoAlertPresentException as err:
        # 如果 confirm 框未弹出显示在页面上,则会抛出 NoAlertPresentException 的异常
        print("尝试操作的 confirm 框未被找到")
        print("error: {0}".format(err))
```

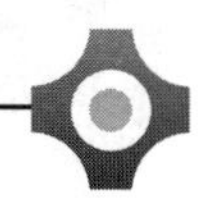

10.46　操作 JavaScript 的 prompt 弹窗

目标：

能够在 JavaScript 的 prompt 弹窗中输入自定义的内容，并单击“确定”按钮或“取消”按钮。

用于测试网页的 HTML 代码：

```
<html>
    <head>
        <title>你喜欢的水果</title>
        <meta http-equiv = "Content-Type" content = "text/html; charset = utf-8" />
    </head>
<body>
    <input id = 'button' type = 'button' onclick = "prompt('这是一个 prompt 弹出框');"
    value = '单击此按钮,弹出 prompt 弹出框'/>
</body>
</html>
```

该 HTML 代码在浏览器中展示弹窗效果图如图 10-6 所示。

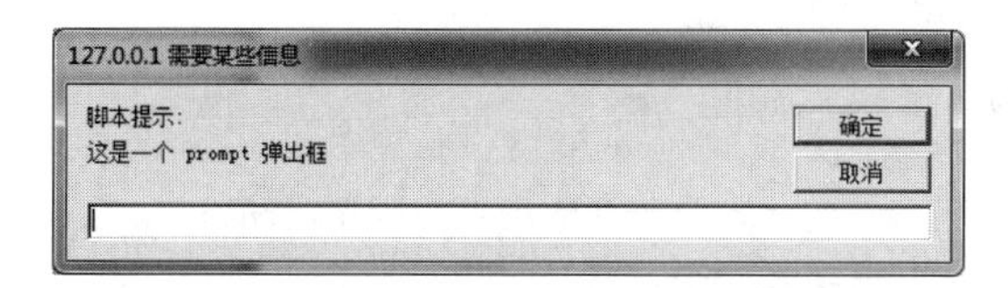

图　10-6

调用 API 的实例代码：

```
def testHandlePrompt(self):
    url = "file:///d:\\prompt.html"
    # 访问自定义网页
    self.driver.get(url)
    # 使用 id 定位方式,找到被测试网页上唯一按钮元素
    element = self.driver.find_element_by_id("button")
    element.click()
    import time
    time.sleep(1)
    # 单击按钮元素,弹出一个 prompt 提示框,
    # 上面将显示"这是一个 prompt 弹出框"、输入框、
    # "确定"按钮和"取消"按钮
    # 使用 driver.switch_to.alert 方法获取 Alert 对象
    alert = self.driver.switch_to.alert
    # 使用 alert.text 方法获取 prompt 框上面的文字,
    # 并断言文字内容是否和"这是一个 prompt 弹出框"一致
    self.assertEqual("这是一个 prompt 弹出框", alert.text)
    time.sleep(1)
    # 调用 alert.send_keys()方法,在 prompt 窗体的输入框中输入
    # "光荣之路:要想改变命运,必须每天学习 2 小时!"
```

```
    alert.send_keys("光荣之路:要想改变命运,必须每天学习2小时!")
    time.sleep(1)
    # 使用alert对象的accept方法,
    # 单击prompt框的"确定"按钮,关闭prompt框
    alert.accept()
    # 使用alert对象的dismiss方法,单击prompt框上的"取消"按钮,关闭prompt框
    # 取消下面一行代码的注释,就会模拟单击prompt框上的"取消"按钮
    # alert.dismiss()
```

10.47 操作浏览器的 Cookie

目标：

能够遍历输出 Cookie 信息中所有的 key 和 value；能够删除指定的 Cookie 对象；能够删除所有的 Cookie 对象。

用于测试的网址：

http://www.sogou.com

调用 API 的实例代码：

```
def test_Cookie(self):
    url = "http://www.sogou.com"
    # 访问搜狗首页
    self.driver.get(url)
    # 得到当前页面下所有的Cookies,并输出它们所在域、name、value、有效期和路径
    cookies = self.driver.get_cookies()
    for cookie in cookies:
        print("%s -> %s -> %s -> %s -> %s"
              % (cookie['domain'], cookie["name"], cookie["value"],
                 cookie["expiry"], cookie["path"]))

    # 根据Cookie的name值获取该条Cookie信息,获取name值为'SUV'的Cookie信息
    ck = self.driver.get_cookie("SUV")
    print("%s -> %s -> %s -> %s -> %s"
          % (ck['domain'], ck["name"], ck["value"],
             ck["expiry"], ck["path"]))

    # 删除cookie有2种方法
    # 第一种:通过Cookie的name属性,删除name值为"ABTEST"的Cookie信息
    print(self.driver.delete_cookie("ABTEST"))

    # 第二种:一次性删除全部Cookie信息
    self.driver.delete_all_cookies()
    # 删除全部Cookie后,再次查看Cookies,确认是否已被全部删除
    cookies = self.driver.get_cookies()
    print(cookies)

    # 添加自定义Cookie信息
    self.driver.add_cookie({"name":"gloryroadTrain", 'value': '1479697159269020'})
    # 查看添加的Cookie信息
```

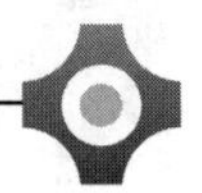

```
        cookie = self.driver.get_cookie("gloryroadTrain")
        print(cookie)
```

10.48 指定页面加载时间

在实施自动化测试过程中，经常会遇到加载某一个页面需要等待很长时间，其实页面基本元素都已经加载完成，可以进行后续操作，而 Selenium WebDriver 在执行 get 方法时会一直等待页面完全加载完毕以后才会执行后续操作，这无形中增加了自动化测试的时间，针对此种情况，就需要指定一下页面加载超时时间，到达等待时间点不再继续等待加载，而是继续执行后续操作。

访问 http://phantomjs.org/download.html 网址下载基于 Webkit 的 JavaScript API，它使用 Webkit 来编译解释执行 JavaScript 代码。任何可以在基于 Webkit 浏览器做的事情，它都能做到。这里我们选择 Windows 版本的(phantomjs-2.1.1-windows.zip)，然后解压到某个目录下等待使用即可。

用于测试网址：

http://mail.126.com

实例代码：

```
#encoding = utf - 8
from selenium import webdriver
from selenium.common.exceptions import TimeoutException
from selenium.webdriver.common.keys import Keys
import time
import unittest

class setPageLoadTime(unittest.TestCase):
    def setUp(self):
        # 启动火狐浏览器
        self.driver = webdriver.Firefox(executable_path = "c:\\geckodriver")

    def test_PageLoadTime(self):
        dr = webdriver.PhantomJS(executable_path =
            'C:\\phantomjs - 2.1.1 - windows\\bin\\phantomjs.exe')
        # 设定页面加载限制时间为 4 秒，下面方法同时使用生效
        dr.set_page_load_timeout(4)
        dr.set_script_timeout(4)
        self.driver.maximize_window()
        startTime = time.time()
        try:
            self.driver.get("http://mail.126.com")
        except TimeoutException:
            print('页面加载超过设定时间，超时')
            # 当页面加载时间超过设定时间，
            # 通过执行 JavaScript 来停止加载，然后继续执行后续动作
            self.driver.execute_script('window.stop()')
        end = time.time() - startTime
```

```
        print(end)
        time.sleep(2)
        # 切换进 frame 控件
        self.driver.switch_to.frame("x-URS-iframe")
        # 获取用户名输入框
        userName = self.driver.find_element_by_xpath('//input[@name="email"]')
        # 输入用户名
        userName.send_keys("xxx")
        # 获取密码输入框
        pwd = self.driver.find_element_by_xpath("//input[@name='password']")
        # 输入密码
        pwd.send_keys("xxx")
        # 发送一个回车键
        pwd.send_keys(Keys.RETURN)

    def tearDown(self):
        self.driver.quit()

if __name__ == '__main__':
    unittest.main()
```

第11章 WebDriver高级应用

上一章讲解了 WebDriver 常用 API 的使用方法，本章将作为 WebDriver 进阶部分，讲解 WebDriver 高级应用。读者如果想向中级水平的自动化测试工程师靠近，请务必掌握本章中的全部应用实例。

11.1 使用 JavaScript 操作页面元素

目的：

在 WebDriver 脚本代码中执行 JavaScript 代码，来实现对页面元素的操作。此种方式主要用于解决在某些情况下，页面元素的.clik()方法无法生效等问题。

用于测试的网址：

http://www.baidu.com

实例代码：

```
#encoding = utf-8
from selenium import webdriver
from selenium.common.exceptions import WebDriverException
import unittest
import traceback
import time

class TestDemo(unittest.TestCase):

    def setUp(self):
        # 启动 Chrome 浏览器
        self.driver = webdriver.Chrome(executable_path = "c:\\chromedriver")

    def test_executeScript(self):
        url = "http://www.baidu.com"
        # 访问百度首页
        self.driver.get(url)
        # 构造 JavaScript 查找百度首页的搜索输入框的代码字符串
        searchInputBoxJS = "document.getElementById('kw').value = '光荣之路';"
        # 构造 JavaScript 查找百度首页的搜索按钮的代码字符串
        searchButtonJS = "document.getElementById('su').click()"
        try:
            # 通过 JavaScript 代码在百度首页搜索输入框中输入"光荣之路"
            self.driver.execute_script(searchInputBoxJS)
            time.sleep(2)
            # 通过 JavaScript 代码单击百度首页上的搜索按钮
            self.driver.execute_script(searchButtonJS)
```

```
            time.sleep(2)
            self.assertTrue("百度百科" in self.driver.page_source)
        except WebDriverException as err:
            # 当定位失败时,会抛出 WebDriverException 异常
            print("在页面中没有找到要操作的页面元素 ",traceback.print_exc())
        except AssertionError as err:
            print("页面不存在断言的关键字串")
        except Exception as err:
            # 发生其他异常时,打印异常堆栈信息
            print(traceback.print_exc())

    def tearDown(self):
        # 退出 Chrome 浏览器
        self.driver.quit()

if __name__ == '__main__':
    unittest.main()
```

11.2 操作 Web 页面的滚动条

目的：

(1) 滚动页面的滚动条到页面最下面。

(2) 滚动页面的滚动条到页面的某个元素。

(3) 滚动页面的滚动条向下移动某个数量的像素。

用于测试的网址：

http://www.seleniumhq.org/

实例代码：

```
#encoding = utf - 8
from selenium import webdriver
import unittest
import traceback
import time

class TestDemo(unittest.TestCase):

    def setUp(self):
        # 启动 Chrome 浏览器
        self.driver = webdriver.Chrome(executable_path = "c:\\chromedriver")

    def test_scroll(self):
        url = "http://www.seleniumhq.org/"
        # 访问 Selenium 官网首页
        try:
            self.driver.get(url)
            # 使用 JavaScript 的 scrollTo 函数和 document.body.scrollHeight 参数
            # 将页面的滚动条滑动到页面的最下方
```

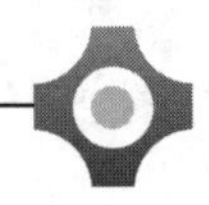

```
            self.driver.execute_script\
                ("window.scrollTo(100, document.body.scrollHeight);")
            # 停顿 3 秒,用于人工验证滚动条是否滑动到指定的位置
            # 根据测试需要,可注释下面的停顿代码
            time.sleep(3)

            # 使用 JavaScript 的 scrollIntoView 函数将被遮挡的元素滚动到可见屏幕上
            # scrollIntoView(true)表示将元素滚到屏幕中间
            # scrollIntoView(false)表示将元素滚动屏幕底部
            self.driver.execute_script\
                ("document.getElementById('choice').scrollIntoView(true);")
            # 停顿 3 秒,用于人工验证滚动条是否滑动到指定的位置
            # 根据测试需要,可注释下面的停顿代码
            time.sleep(3)

            # 使用 JavaScript 的 scrollBy 方法,使用 0 和 400 横纵坐标参数,
            # 将页面纵向向下滚动 400 像素
            self.driver.execute_script("window.scrollBy(0,400);")
            # 停顿 3 秒,用于人工验证滚动条是否滑动到指定的位置
            # 根据测试需要,可注释下面的停顿代码
            time.sleep(3)
        except Exception as err:
            # 打印异常堆栈信息
            print(traceback.print_exc())

    def tearDown(self):
        # 退出 Chrome 浏览器
        self.driver.quit()

if __name__ == '__main__':
    unittest.main()
```

更多说明：

这里我们操作页面滚动条的方法其实是调用 JavaScript 方法，通过这种方法就可以随意操作页面的滚动条了，无论是纵向滚动，还是横向滚动。

11.3 在 Ajax 方式产生的浮动框中，单击选择包含某个关键字的选项

目的：

有些被测试页面包含 Ajax 的局部刷新机制，并且会产生显示多条数据的浮动框，需要单击选择浮动框中包含某个关键字的选项。

用于测试的网址：

http://www.sogou.com

单击一下搜狗首页的搜索框，将焦点切换到搜索输入框中后，会看到弹出浮动框的效果，如图 11-1 所示。

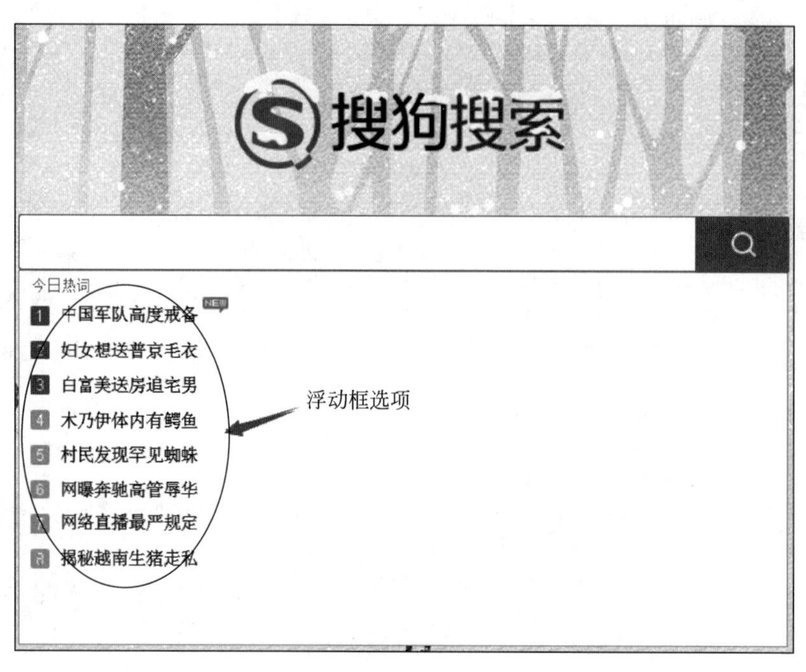

图 11-1

第一种方法，通过模拟键盘下箭头进行选择悬浮框选项

实例代码：

```
#encoding = utf-8
from selenium import webdriver
from selenium.webdriver.common.keys import Keys
import unittest
import time

class TestDemo(unittest.TestCase):

    def setUp(self):
        # 启动 Chrome 浏览器
        self.driver = webdriver.Chrome(executable_path = "c:\\chromedriver")

    def test_AjaxDivOptionByKeys(self):
        url = "http://www.sogou.com/"
        # 访问搜狗的首页
        self.driver.get(url)
        # 找到搜狗首页中的搜索输入框页面元素
        searchBox = self.driver.find_element_by_id("query")
        # 在搜索输入框中输入"光荣之路"
        searchBox.send_keys("光荣之路")
        # 等待 2 秒,以便悬浮框加载完成
        time.sleep(2)
        for i in range(3):
            # 选择悬浮框中第几个联想关键词选项就循环几次
            # 模拟键盘单击下箭头
```

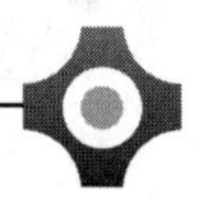

```
            searchBox.send_keys(Keys.DOWN)
            time.sleep(0.5)
        # 当按下箭头到想要选择的选项后,再模拟键盘单击回车键,选中该选项
        searchBox.send_keys(Keys.ENTER)
        time.sleep(3)

    def tearDown(self):
        # 退出 Chrome 浏览器
        self.driver.quit()

if __name__ == '__main__':
    unittest.main()
```

第二种方法,通过匹配模糊内容选择悬浮框中选项

实例代码:

```
#encoding=utf-8
from selenium import webdriver
from selenium.common.exceptions import NoSuchElementException
import traceback
import unittest
import time

class TestDemo(unittest.TestCase):

    def setUp(self):
        # 启动 Chrome 浏览器
        self.driver = webdriver.Chrome(executable_path = "c:\\chromedriver")

    def test_AjaxDivOptionByWords(self):
        url = "http://www.sogou.com/"
        # 访问搜狗的首页
        self.driver.get(url)
        try:
            # 找到搜狗首页中的搜索输入框页面元素
            searchBox = self.driver.find_element_by_id("query")
            # 在搜索输入框中输入"光荣之路"
            searchBox.send_keys("光荣之路")
            # 等待 2 秒,以便悬浮框加载完成
            time.sleep(2)
            # 查找内容包含"光荣之路电影"的悬浮选项
            suggetion_option = self.driver.\
                find_element_by_xpath("//ul/li[contains(., '光荣之路电影')]")
            # 单击找到的选项
            suggetion_option.click()
            time.sleep(3)
        except NoSuchElementException as err:
            # 打印异常堆栈信息
            print(traceback.print_exc())

    def tearDown(self):
```

```
        # 退出 Chrome 浏览器
        self.driver.quit()

if __name__ == '__main__':
    unittest.main()
```

更多说明：

因为浮动框的内容可能会时常发生变化，如果只想固定选择浮动框中的某一项，比如第三项，可以参考如下代码。

```
def test_AjaxDivOptionByIndex(self):
    url = "http://www.sogou.com/"
    # 访问搜狗的首页
    self.driver.get(url)
    try:
        # 找到搜狗首页中的搜索输入框页面元素
        searchBox = self.driver.find_element_by_id("query")
        # 在搜索输入框中输入"光荣之路"
        searchBox.send_keys("光荣之路")
        # 等待 2 秒，以便悬浮框加载完成
        time.sleep(2)
        # 查找浮动框中的第三选项，只要更改 li[3]中的索引数字，
        # 就可以实现任意单击选项浮动框中的选项.注意，索引从 1 开始
        suggetion_option = self.driver.\
            find_element_by_xpath("//*[@id = 'vl']/div[1]/ul/li[3]")
        # 单击找到的选项
        suggetion_option.click()
        time.sleep(3)
    except NoSuchElementException as err:
        # 打印异常堆栈信息
        print(traceback.print_exc())
```

11.4 结束 Windows 中浏览器的进程

目标：

通过代码关闭浏览器进程。

Python 语言实例代码：

```
#encoding = utf-8
from selenium import webdriver
import unittest

class TestDemo(unittest.TestCase):

    def test_killWindowsProcess(self):
        # 启动火狐浏览器
        firefoxDriver = webdriver.Firefox(executable_path = "c:\\geckodriver")
        # 启动 IE 浏览器
        ieDriver = webdriver.Ie(executable_path = "c:\\IEDriverServer")
```

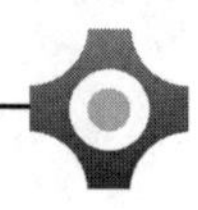

```
        # 启动 Chrome 浏览器
        chromeDriver = webdriver.Chrome(executable_path = "c:\\chromedriver")
        # 导入 Python 的 os 包
        import os
        # 结束 Firefox 浏览器进程
        returnCode = os.system("taskkill /F /iM firefox.exe")
        if returnCode == 0:
            print("成功结束 Firefox 浏览器进程!")
        else:
            print("结束 Firefox 浏览器进程失败!")
        # 结束 IE 浏览器进程
        returnCode = os.system("taskkill /F /iM iexplore.exe")
        if returnCode == 0:
            print("成功结束 IE 浏览器进程!")
        else:
            print("结束 IE 浏览器进程失败!")
        # 结束 Chrome 浏览器进程
        returnCode = os.system("taskkill /F /iM chrome.exe")
        if returnCode == 0:
            print("成功结束 Chrome 浏览器进程!")
        else:
            print("结束 Chrome 浏览器进程失败!")

if __name__ == '__main__':
    unittest.main()
```

代码说明：

将 Windows 上的 DOS 命令的字符串类型数据作为参数值传递给 Python 语言的 os.system(command)函数，然后该函数执行该 DOS 命令，并返回执行结果码(0 表示命令执行成功，非 0 表示命令执行失败)，以此来结束 Windows 中的浏览器进程。

11.5 更改一个页面对象的属性值

目的：

掌握设定页面对象的所有属性的方法，本节以设定文本框的可编辑状态和显示长度为目标。

用于测试的网页的 HTML 代码：

```
<html>
<head>
    <title>设置文本框属性</title>
    <meta http-equiv = "Content-Type" content = "text/html; charset = utf-8" />
</head>
<body>
    <input type = "text" id = "text" value = "今年夏天西瓜相当甜!" size = 100 >
    文本框
</body>
</html>
```

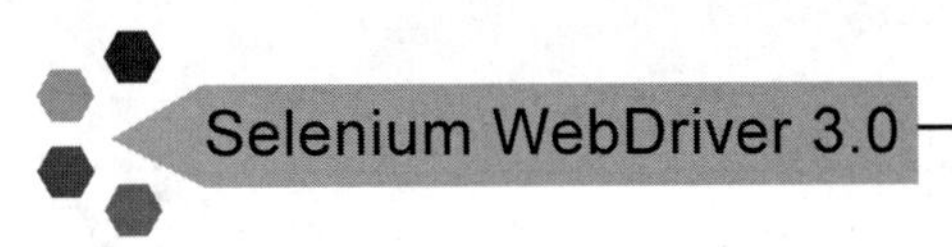

实例代码：

```
#encoding = utf - 8
from selenium import webdriver
import unittest

def addAttribute(driver, elementObj, attributeName, value):
    # 封装向页面标签中添加新属性方法
    # 调用 JavaScript 代码给页面标签添新属性,arguments[0] - [1]分别会用后面的
    # element、attributeName 和 value 参数值进行替换,并执行该 JavaScript 代码
    # 添加新属性的 JavaScript 代码语法为:element.attributeName = value
    # 比如 input.name = "test"
    driver.execute_script("arguments[0].%s = arguments[1]" %attributeName,
                          elementObj, value)

def setAttribute(driver, elementObj, attributeName, value):
    # 封装设置页面对象的属性值的方法
    # 调用 JavaScript 代码修改页面元素的属性值,arguments[0] - [2]分别会用后面的
    # element、attributeName 和 value 参数值进行替换,并执行该 JavaScript 代码
    driver.execute_script("arguments[0].setAttribute\
    (arguments[1],arguments[2])", elementObj, attributeName, value)

def getAttribute(elementObj, attributeName):
    # 封装获取页面对象的属性值的方法
    return elementObj.get_attribute(attributeName)

def removeAttribute(driver, elementObj, attributeName):
    # 封装删除页面元素属性的方法
    # 调用 JavaScript 代码删除页面元素的指定的属性,arguments[0] - [1]分别会用后面
    # 的 element、attributeName 参数值进行替换,并执行该 JavaScript 代码
    driver.execute_script("arguments[0].removeAttribute(arguments[1])",
                          elementObj, attributeName)

class TestDemo(unittest.TestCase):

    def setUp(self):
        # 启动 Chrome 浏览器
        self.driver = webdriver.Chrome(executable_path = "c:\\chromedriver")

    def test_dataPicker(self):
        url = "d:\\operateAttribute.html"
        # 访问自定义网页
        self.driver.get(url)
        # 找到页面上标签名为 input 的页面元素
        element = self.driver.find_element_by_xpath("//input")

        # 向页面文本框 input 标签中添加新属性 name = "search"
        addAttribute(self.driver, element, 'name', "search")
        # 添加新属性后,查看一下新添加的属性
        print('添加的新属性值%s = "%s"' % ("name", getAttribute(element, "name")))
```

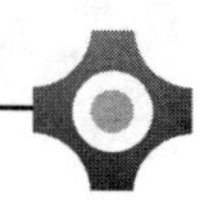

```
        # 查看修改前文本框 input 标签的 value 属性值
        print("更改文本框中的内容前的内容:", getAttribute(element, "value"))
        # 更改 input 页面元素的 value 属性值为"这是更改后的文字内容"
        setAttribute(self.driver, element, "value", "这是更改后的文字内容")
        # 更改 input 页面元素的 value 属性值后,再次查看其 value 属性值
        print("更改文本框中内容后的内容:", getAttribute(element, "value"))

        # 查看修改前文本框 input 页面元素中的 size 属性值
        print("更改前文本框标签中的 size 属性值:", getAttribute(element, "size"))
        # 更改 input 页面元素的 size 属性值为"20"
        setAttribute(self.driver, element, "size", 20)
        # 更改 input 页面元素的 size 属性值后,再次查看其 size 属性值
        print("更改后文本框标签中的 size 属性值:", getAttribute(element, "size"))

        # 查看删除 input 页面元素 value 属性前 value 属性值
        print("文本框 value 属性值:", getAttribute(element, "value"))
        # 删除文本框的 value 属性
        removeAttribute(self.driver, element, "value")
        # 删除文本框的 value 属性后,再次查看 value 属性值
        print("删除 value 属性值后 value 属性值:", getAttribute(element, "value"))

    def tearDown(self):
        # 退出 Chrome 浏览器
        self.driver.quit()

if __name__ == '__main__':
    unittest.main()
```

输出结果：

```
添加的新属性值 name = "search"
更改文本框中的内容前的内容: 今年夏天西瓜相当甜!
更改文本框中内容后的内容: 这是更改后的文字内容
更改前文本框标签中的 size 属性值: 100
更改后文本框标签中的 size 属性值: 20
文本框 value 属性值: 这是更改后的文字内容
删除 value 属性值后 value 属性值:
```

本节实例针对页面元素属性的新增、更改、查询以及删除都是临时的，只针对当前会话有效，页面源码并没有被真正修改。

本实例在 IE 浏览器上可能实验不成功，因为 IE 浏览器经常存在 JavaScript 兼容性问题。

11.6　无人工干预地自动下载某个文件

目的：

在网上下载文件时，通常需要人为设定下载文件并选择保存路径，这样就无法实现完全

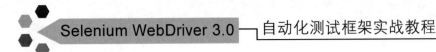

自动的下载过程。下面的例子实现了基于 Firefox 浏览器的全自动化文件下载操作，脚本执行后会将文件自动保存到指定目录的文件夹下。

用于测试的网址：

https://www.python.org/downloads/release/python－2712/

https://github.com/mozilla/geckodriver/releases

实例代码：

```
# encoding = utf - 8
from selenium import webdriver
import unittest, time

class TestDemo(unittest.TestCase):

    def setUp(self):
        # 创建一个 FirefoxProfile 实例,用于存放自定义配置
        profile = webdriver.FirefoxProfile()
        # 指定下载路径,默认只会自动创建一级目录,如果指定了
        # 多级不存在的目录,将会下载到默认路径
        profile.set_preference('browser.download.dir', 'd:\\iDownload')
        # 将 browser.download.folderList 设置为 2,表示将文件下载到指定路径
        # 设置成 2 表示使用自定义下载路径;
        # 设置成 0 表示下载到桌面;设置成 1 表示下载到默认路径
        profile.set_preference('browser.download.folderList', 2)
        # browser.helperApps.alwaysAsk.force 对于未知的 MIME 类型文件会弹出窗口
        # 让用户处理,默认值为 True,设定为 False 表示不会记录打开未知 MIME 类型
        # 文件的方式
        profile.set_preference("browser.helperApps.alwaysAsk.force", False)
        # 在开始下载时是否显示下载管理器
        profile.set_preference('browser.download.manager.showWhenStarting',
                               False)
        # 设定为 False 会把下载框隐藏
        profile.set_preference("browser.download.manager.useWindow", False)
        # 默认值为 True,设定为 False 表示不获取焦点
        profile.set_preference("browser.download.manager.focusWhenStarting",
                               False)
        # 下载.exe 文件弹出警告,默认值是 True,设定为 False 则不会弹出警告框
        profile.set_preference("browser.download.manager.alertOnEXEOpen",
                               False)
        # browser.helperApps.neverAsk.openFile 表示直接打开下载文件,不显示确认框
        # 默认值为空字符串,下行代码行设定了多种文件的 MIME 类型,
        # 例如 application/exe,表示.exe 类型的文件,
        # application/excel 表示 Excel 类型的文件
        profile.set_preference("browser.helperApps.neverAsk.openFile",
                               "application/pdf")
        # 对所给出文件类型不再弹出框进行询问,直接保存到本地磁盘
        profile.set_preference('browser.helperApps.neverAsk.saveToDisk',
                               'application/zip, application/octet - stream')
        # browser.download.manager.showAlertOnComplete 设定下载文件结束后是否显示下
        # 载完成提示框,默认为 True,设定为 False 表示下载完成后不显示下载完成提示框
```

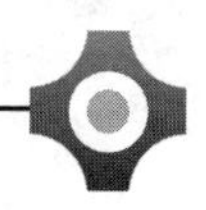

```
        profile.set_preference("browser.download.manager.showAlertOnComplete",
                               False);
        # browser.download.manager.closeWhenDone 设定下载结束后是否自动
        # 关闭下载框,默认值为 True,设定为 False 表示不关闭下载管理器
        profile.set_preference("browser.download.manager.closeWhenDone",
                               False)

        # 启动浏览器时,通过 firefox_profile 参数
        # 将自动将配置添加到 FirefoxProfile 对象中
        self.driver = webdriver.Firefox(executable_path = "c:\\geckodriver",
                                        firefox_profile = profile)

    def test_dataPicker(self):
        # 访问 WebDriver 驱动 Firefox 的驱动文件下载网址
        url1 = "https://github.com/mozilla/geckodriver/releases"
        self.driver.get(url1)
        # 选择下载 zip 类型文件,使用 application/zip 指代此类型文件
        self.driver.find_element_by_xpath\
            ('//strong[. = "geckodriver - v0.11.1 - win64.zip"]').click()
        # 等待加载下载文件
        time.sleep(10)

        # 访问 Python 2.7.12 文件下载页面,下载扩展名为 msi 的文件
        # 使用 application/octet - stream 来指明此类文件类型
        url = "https://www.python.org/downloads/release/python - 2712/"
        self.driver.get(url)
        # 找到 Python 2.7.12 下载页面中链接文字为"Windows x86 - 64 MSI installer"
        # 的链接页面元素,单击进行无人工干预的下载 Python 2.7.12 解释器文件
        self.driver.find_element_by_link_text\
            ("Windows x86 - 64 MSI installer").click()
        # 等待文件下载完成,根据各自的网络带宽情况设定等待相应的时间
        time.sleep(100)

    def tearDown(self):
        self.driver.quit()

if __name__ == '__main__':
    unittest.main()
```

代码解释：

通过 profile.set_preference('browser.helperApps.neverAsk.saveToDisk', 'xxxx')这种方式添加需要屏蔽下载询问弹出的文件类型,如果要同时添加多种文件类型,文件类型间用逗号隔开,如本例中的"application/zip, application/octet-stream"。

更多说明：

访问 http://www.w3school.com.cn/media/media_mimeref.asp 查看更多的文件类型及解释。

11.7 无人工干预地自动上传附件

本小节主要介绍通过程序代码无人工干预地上传文件附件,并进行提交操作。

11.7.1 使用 WebDriver 的 send_keys 方法上传文件

目的：

使用 send_keys 方法上传一个文件附件，并进行提价操作。

用于测试的网页的 HTML 代码：

```
<html>
<head>
    <title>上传文件</title>
    <meta http-equiv="Content-Type" content="text/html; charset=utf-8" />
</head>
<body>
    <form enctype="multipart/form-data" action="parse_file.jsp" method="post">
        <p>Browse for a file to upload: </p>
        <input id="file" name="file" type="file">
        <br/><br/>
        <input type="submit" id="filesubmit" value="SUBMIT">
    </form>
</body>
</html>
```

由于篇幅所限，这里不给出 parse_file.jsp 的源代码。上传文件成功后，会跳转到 parse_file.jsp 页面，此页面的 Title 显示为“文件上传成功”。

实例代码：

```
#encoding=utf-8
from selenium import webdriver
import unittest
import time
import traceback
from selenium.webdriver.support.ui import WebDriverWait
from selenium.webdriver.common.by import By
from selenium.webdriver.support import expected_conditions as EC
from selenium.common.exceptions import TimeoutException, NoSuchElementException

class TestDemo(unittest.TestCase):

    def setUp(self):
        # 启动 Chrome 浏览器
        self.driver = webdriver.Firefox(executable_path = "c:\\geckodriver")

    def test_uploadFileBySendKeys(self):
        url = "d:\\uploadFile.html"
        # 访问自定义网页
        self.driver.get(url)
        try:
            # 创建一个显示等待对象
            wait = WebDriverWait(self.driver, 10, 0.2)
            # 显示等待判断被测试页面上的上传文件按钮是否处于可被单击状态
```

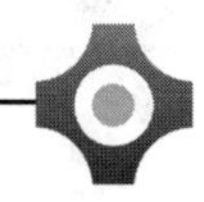

```
            wait.until(EC.element_to_be_clickable((By.ID, 'file')))
        except TimeoutException as err:
            # 捕获 TimeoutException 异常
            print(traceback.print_exc())
        except NoSuchElementException as err:
            # 捕获 NoSuchElementException 异常
            print(traceback.print_exc())
        except Exception as err:
            # 捕获其他异常
            print(traceback.print_exc())
        else:
            # 查找页面上 ID 属性值为 file 的文件上传框
            fileBox = self.driver.find_element_by_id("file")
            # 在文件上传框的路径框里输入要上传的文件路径"c:\\test.txt"
            fileBox.send_keys("c:\\test.txt")
            # 暂停查看上传的文件
            time.sleep(4)
            # 找到页面上 ID 属性值为 filesubmit 的文件提交按钮对象
            fileSubmitButton = self.driver.find_element_by_id("filesubmit")
            # 单击提交按钮,完成文件上传操作
            fileSubmitButton.click()
            # 因为文件上传需要时间,所以这里可以添加显示等待场景,
            # 判断文件上传成功后,页面是否跳转到文件上传成功的页面.
            # 通过 EC.title_is()方法判断跳转后的页面的 Title
            # 值是否符合期望,如果匹配将继续执行后续代码

            # 如果实现了 parse_file.jsp 页面,并且可以成功调转,
            # 可以将下面代码取消注释,断言文件上传成功
            # wait.until(EC.title_is("文件上传成功"))

    def tearDown(self):
        # 退出浏览器
        self.driver.quit()

if __name__ == '__main__':
    unittest.main()
```

11.7.2 模拟键盘操作,实现上传文件

目的:

通过模拟键盘按键操作,来完成文件上传功能。

用于测试网页的 HTML 代码:

被测试网页的 HTML 代码同 10.7.1 小节。

实例代码:

```
#encoding = utf-8
from selenium import webdriver
import unittest
import time
```

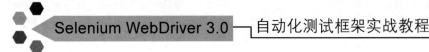

```
import traceback
import win32clipboard as w
import win32api
import win32con
from selenium.webdriver.support.ui import WebDriverWait
from selenium.webdriver.common.by import By
from selenium.webdriver.support import expected_conditions as EC
from selenium.common.exceptions import TimeoutException, NoSuchElementException

# 用于设置剪切板内容
def setText(aString):
    w.OpenClipboard()
    w.EmptyClipboard()
    w.SetClipboardData(win32con.CF_UNICODETEXT, aString)
    w.CloseClipboard()

# 键盘按键映射字典
VK_CODE = {
    'enter':0x0D,
    'ctrl':0x11,
    'v':0x56}

# 键盘键按下
def keyDown(keyName):
    win32api.keybd_event(VK_CODE[keyName], 0, 0, 0)
# 键盘键抬起
def keyUp(keyName):
    win32api.keybd_event(VK_CODE[keyName], 0, win32con.KEYEVENTF_KEYUP, 0)

class TestDemo(unittest.TestCase):

    def setUp(self):
        # 启动浏览器
        self.driver = webdriver.Firefox(executable_path = "c:\\ geckodriver")

    def test_uploadFileByKeyboard(self):
        url = "d:\\uploadFile.html"
        # 访问自定义网页
        self.driver.get(url)
        try:
            # 创建一个显示等待对象
            wait = WebDriverWait(self.driver, 10, 0.2)
            # 显示等待判断被测试页面上的上传文件按钮是否处于可被单击状态
            wait.until(EC.element_to_be_clickable((By.ID, 'file')))
        except TimeoutException as err:
            # 捕获 TimeoutException 异常
            print(traceback.print_exc())
        except NoSuchElementException as err:
            # 捕获 NoSuchElementException 异常
            print(traceback.print_exc())
        except Exception as err:
```

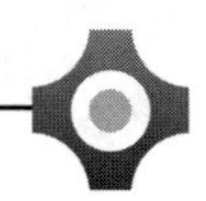

```
            # 捕获其他异常
            print(traceback.print_exc())
        else:
            # 将即将要上传的文件名及路径设置到剪切板中
            setText("c:\\test.txt")
            # 查找页面上 ID 属性值为 file 的文件上传框,
            # 并单击调出选择文件上传框
            self.driver.find_element_by_id("file").click()
            time.sleep(2)
            # 模拟键盘按下 Ctrl + V 组合键
            keyDown("ctrl")
            keyDown("v")
            # 模拟键盘释放 Ctrl + V 组合键
            keyUp("v")
            keyUp("ctrl")
            time.sleep(1)
            # 模拟键盘按下回车键
            keyDown("enter")
            # 模拟键盘释放回车键
            keyUp("enter")
            # 暂停查看上传的文件
            time.sleep(2)
            # 找到页面上 ID 属性值为 filesubmit 的文件提交按钮对象
            fileSubmitButton = self.driver.find_element_by_id("filesubmit")
            # 单击提交按钮,完成文件上传操作
            fileSubmitButton.click()
            # 因为文件上传需要时间,所以这里可以添加显示等待场景,
            # 判断文件上传成功后,页面是否跳转到文件上传成功的页面.
            # 通过 EC.title_is()方法判断跳转后的页面的 Title
            # 值是否符合期望,如果匹配将继续执行后续代码
            # wait.until(EC.title_is("文件上传成功"))

    def tearDown(self):
        # 退出浏览器
        self.driver.quit()

if __name__ == '__main__':
    unittest.main()
```

11.7.3 使用第三方工具 AutoIt 上传文件

目的：

能使用第三方工具 AutoIt 操作一些 WebDriver 无法操作的文件上传对象。

用于测试的网页的 HTML 代码：

被测试网页的 HTML 代码同 11.7.1 小节。

AutoIt 工具的安装方法：

（1）访问 https://www.autoitscript.com/site/autoit/downloads/网址，下载 AutoIt 工具，如图 11-2 所示。

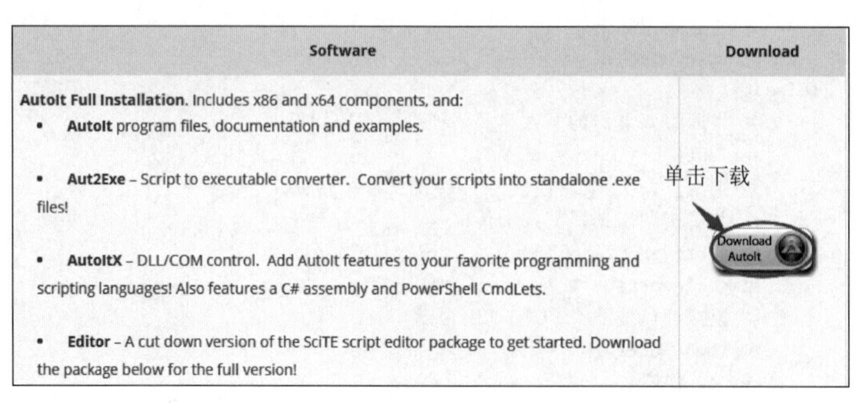

图 11-2

(2) AutoIt 软件下载成功后,会得到一个扩展名为.exe 的文件,双击该文件进行安装。

(3) 双击 exe 文件后,将显示如图 11-3 所示的界面,单击 Next 按钮。

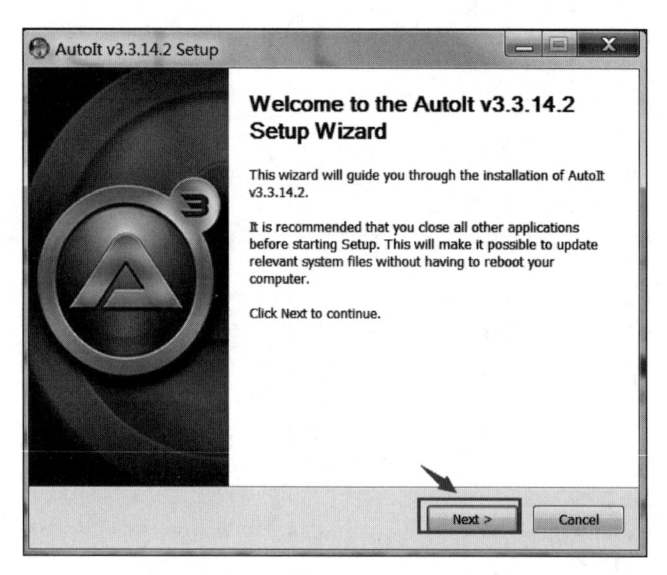

图 11-3

(4) 显示如图 11-4 所示的界面时,单击 I Agree 按钮。

(5) 然后一路单击 Next 按钮,直到出现如图 11-5 所示的界面,设置好文件即将安装的路径,然后单击 Install 按钮,开始安装。

(6) 等待安装的进度条走完,将出现如图 11-6 所示界面,单击 Finish 按钮,完成所有安装步骤。

(7) 从 https://www.autoitscript.com/site/autoit/downloads/网址上下载 AutoIt 的编辑器,如图 11-7 和图 11-8 所示。

(8) 下载完后将会得到一个 SciTE4AutoIt3.exe 文件,双击该文件进行安装,一路单击 Next 按钮,直到出现如图 11-9 所示界面,单击 I Agree 按钮进行正式安装。

(9) 待安装进度条走完,将出现如图 11-10 所示的界面,单击 Finish 按钮,完成 AutoIt 脚本编辑器的安装。

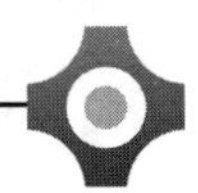

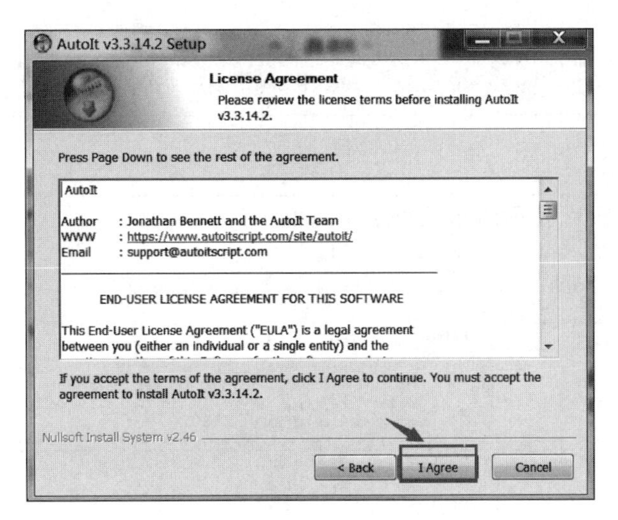

图　11-4

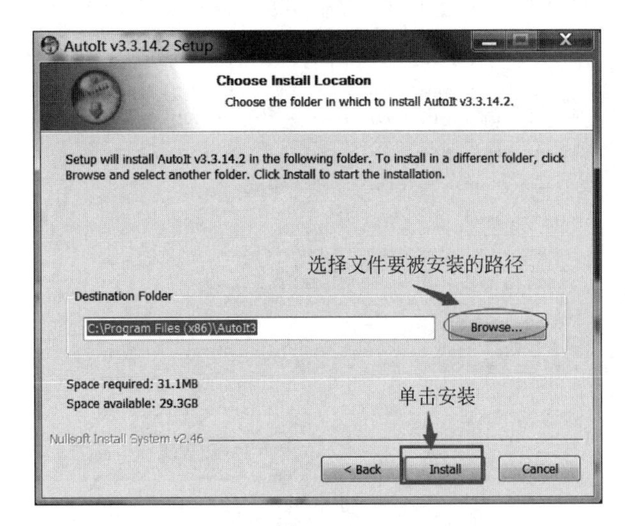

图　11-5

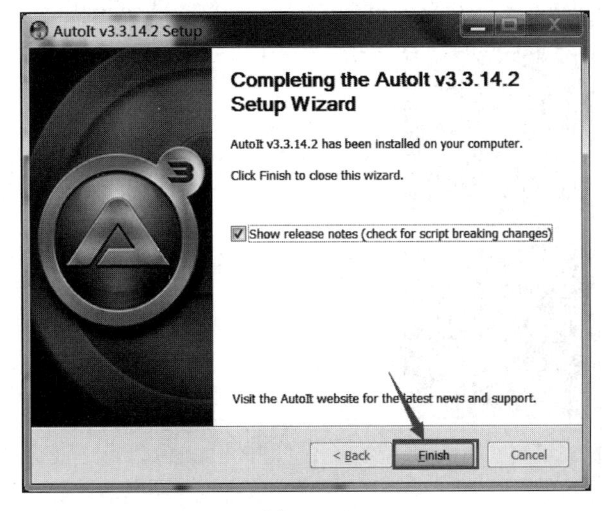

图　11-6

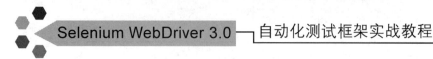

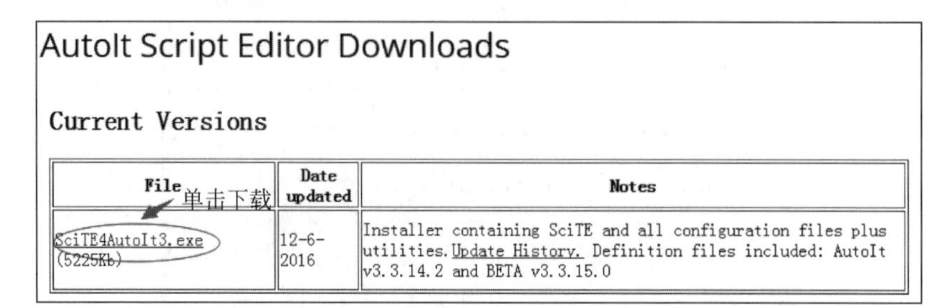

图 11-7

AutoIt Script Editor Downloads

Current Versions

File	Date updated	Notes
SciTE4AutoIt3.exe (5225Kb)	12-6-2016	Installer containing SciTE and all configuration files plus utilities. Update History. Definition files included: AutoIt v3.3.14.2 and BETA v3.3.15.0

单击下载

图 11-8

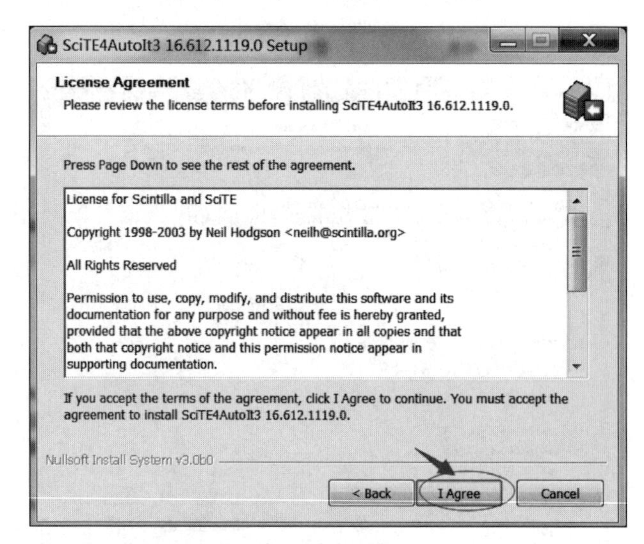

图 11-9

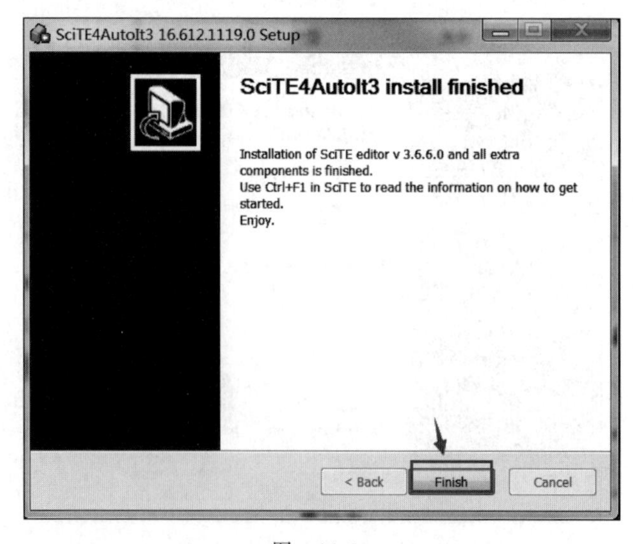

图 11-10

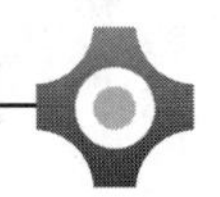

编辑操作文件上传框体的 AutoIt 脚本：

（1）单击"开始"→"所有程序"→AutoIt v3→SciTE→SciTE 命令，启动 AutoIt 的文本编辑器。

（2）在编辑器中输入如下脚本：

```
#include <Constants.au3>

Send("c:\test.txt")
Send("{ENTER}")
Send("{ENTER}")
```

脚本解释：

Send("c:\test.txt")表示使用键盘输入 c:\test.txt。

Send("{ENTER}")表示按回车键。

调用两次回车键，主要是解决某些操作系统默认的输入法是中文输入法，输入 c:\test.txt 以后，必须按一下回车键才能将输入的内容写入路径输入框中，再按一次回车键，就等价于单击文件打开窗体的"打开"按钮。

（3）将 AutoIt 脚本保存为文件名为 test.au3 的文件并存放在 D 盘驱动器中。

（4）单击"开始"→"所有程序"→AutoIt v3→Compile script to.exe(x64)（根据自己的操作系统位数选择正确的位数），调出将 AutoIt 脚本转换成 exe 文件的界面，如图 11-11 所示。

图 11-11

（5）在 Source 路径框中选择上面保存的 AutoIt 脚本 test.au3 文件，在 Destination 处选择.exe 单选项，并在接下来的输入框中设置好生成 exe 文件的保存路径，其他按默认即可。

（6）单击 Convert 按钮，将会把 AutoIt 脚本 test.au3 文件转换成 test.exe 可执行文件 test.exe 。

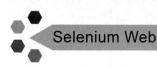

使用 AutoIt 脚本上传文件的实例代码：

```
#encoding = utf - 8
from selenium import webdriver
import unittest
import time, os
import traceback
from selenium.webdriver.support.ui import WebDriverWait
from selenium.webdriver.common.by import By
from selenium.webdriver.support import expected_conditions as EC
from selenium.common.exceptions import TimeoutException, NoSuchElementException

class TestDemo(unittest.TestCase):

    def setUp(self):
        # 启动 Chrome 浏览器
        self.driver = webdriver.Firefox(executable_path = "c:\\geckodriver")

    def test_uploadFileByAutoIt(self):
        url = "d:\\uploadFile.html"
        # 访问自定义网页
        self.driver.get(url)
        try:
            # 创建一个显示等待对象
            wait = WebDriverWait(self.driver, 10, 0.2)
            # 显示等待判断被测试页面上的上传文件按钮是否处于可被单击状态
            wait.until(EC.element_to_be_clickable((By.ID, 'file')))
        except TimeoutException as err:
            # 捕获 TimeoutException 异常
            print(traceback.print_exc())
        except NoSuchElementException as err:
            # 捕获 NoSuchElementException 异常
            print(traceback.print_exc())
        except Exception as err:
            # 捕获其他异常
            print(traceback.print_exc())
        else:
            # 查找页面上 ID 属性值为 file 的文件上传框,
            # 并单击调出选择文件上传框
            self.driver.find_element_by_id("file").click()
            # 通过 Python 提供的 os 模块的 system 方法执行生成的 test.exe 文件
            os.system("d:\\book\\test.exe")
            # 由于 AutoIt 脚本转换后的可执行文件 test.exe 可能执行速度比较慢,
            # 这里等待 5 秒,以确保 test.exe 脚本执行成功
            time.sleep(5)
            # 找到页面上 ID 属性值为 filesubmit 的文件提交按钮对象
            fileSubmitButton = self.driver.find_element_by_id("filesubmit")
            # 单击提交按钮,完成文件上传操作
            fileSubmitButton.click()
```

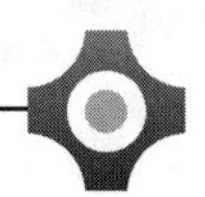

```
            # 因为文件上传需要时间,所以这里可以添加显示等待场景,
            # 判断文件上传成功后,页面是否跳转到文件上传成功的页面.
            # 通过 EC.title_is()方法判断跳转后的页面的 Title
            # 值是否符合期望,如果匹配将继续执行后续代码
            wait.until(EC.title_is("文件上传成功"))

    def tearDown(self):
        # 退出浏览器
        self.driver.quit()

if __name__ == '__main__':
    unittest.main()
```

11.8 右键另存为下载文件

目的：

能够模拟实现在下载链接上直接右击另存为下载某文件的功能，来解决 11.6 小节中对部分类型文件下载不能生效的问题，本小节以.exe 类型的文件下载为例。

用于测试的网址：

http://ftp.mozilla.org/pub/mozilla.org//firefox/releases/35.0b8/win32/zh-CN/

AutoItScript 文件准备：

新建一个名为 loadFile.au3 的 AutoItScript 编辑器，文件具体内容如下：

```
;ControlFocus("title","text",controlID)
;表示将焦点切换到标题为 title 窗体中的 controlID 上
;Edit1 表示第一个可以编辑的实例
;title 表示弹出的 Window 窗口标题,不同浏览器的标题可能不一样
ControlFocus("请输入要保存的文件名...","","Edit1")

;等待 10 秒以便 Window 窗口加载成功
WinWait("[CLASS:#32770]","",10)

;将焦点切换到 Edit1 输入框中
ControlFocus("另存为","","Edit1")

;等待 2 秒
Sleep(2000)

;将要下载的文件名及路径写入 Edit1 编辑框中
ControlSetText("另存为","", "Edit1", "d:\iDownload\Firefox Setup 35.0b8.exe")

Sleep(2000)

;单击窗体中的第一个按钮,也就是保存按钮
ControlClick("另存为","","Button1")
```

保存后将该文件编译成 exe 文件，并存放到本地磁盘。

实例代码：

```
# encoding = utf - 8
from selenium import webdriver
import unittest, time, os
from selenium.webdriver.common.keys import Keys
from selenium.webdriver import ActionChains
import win32api
import win32con

VK_CODE = {'enter':0x0D, 'down_arrow':0x28}

#键盘键按下
def keyDown(keyName):
    win32api.keybd_event(VK_CODE[keyName], 0, 0, 0)
#键盘键抬起
def keyUp(keyName):
    win32api.keybd_event(VK_CODE[keyName], 0, win32con.KEYEVENTF_KEYUP, 0)

class TestDemo(unittest.TestCase):
    def setUp(self):
        self.driver = webdriver.Firefox(executable_path = "c:\\ geckodriver")

    def test_dataPickerByRightKey(self):
        # 定义将要访问的网址
        url = "http://ftp.mozilla.org/pub/mozilla.org//firefox/releases/35.0b8/win32/zh - CN/"
        self.driver.get(url)
        # 将窗口最大化
        self.driver.maximize_window()
        # 暂停 5 秒，目的是防止页面有一些多余的弹窗抢占焦点
        time.sleep(5)
        # 找到文本内容为"Firefox Setup 35.0b8.exe"的超链接元素
        a = self.driver.find_element_by_link_text("Firefox Setup 35.0b8.exe")
        time.sleep(2)
        # 在找到的链接元素上模拟右击，
        # 以便调出选择"另存为"选项的菜单
        ActionChains(self.driver).context_click(a).perform()
        # 暂停 2 秒，防止命令执行太快
        time.sleep(2)
        for i in range(4):
            # 循环按 4 次下箭头，将焦点切换到"另存为"选项上
            # 不同浏览器此选项的位置可能不同
            a.send_keys(Keys.DOWN)
            keyDown("down_arrow")
            keyUp("down_arrow")
            print i
            time.sleep(2)
```

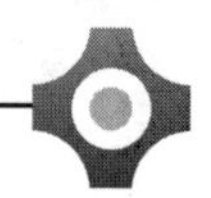

```
            time.sleep(2)
            # 当焦点切换到"另存为"选项上后,模拟按回车键
            # 调出保存下载文件路径的 Windows 窗体
            keyDown("enter")
            keyUp("enter")
            time.sleep(3)
            # 通过执行 AutoIt 编写的操作弹窗的 Windows 文件保存窗体
            # 完成文件保存路径的设置
            os.system("d:\\book\\loadFile.exe")
            # 等待文件下载完成,根据各自的网络带宽情况设定等待相应的时间
            time.sleep(100)

    def tearDown(self):
            self.driver.quit()

if __name__ == '__main__':
    unittest.main()
```

本例提供的 AutoItScript 脚本仅针对 Chrome 浏览器,如想使用其他浏览器,需要根据具体浏览器的情况修改脚本,因为不同浏览器调出的保存文件窗体标题不一样,同时保存下载文件的选项索引号也不一致。

11.9 操作日期控件

目的:

能够在日期控件上进行任意年、月、日的选择。

用于测试的网址:

http://jqueryui.com/resources/demos/datepicker/other-months.html

被测试网站如是国外的,有可能会出现访问不稳定的情况,如有条件请使用相关网络代理工具。被测试网页中的日期选择控件效果图如图 11-12 所示。

Date:

November 2016

Su	Mo	Tu	We	Th	Fr	Sa
30	31	1	2	3	4	5
6	7	8	9	10	11	12
13	14	15	16	17	18	19
20	21	22	23	24	25	26
27	28	29	30	1	2	3

图 11-12

操作日期选择控件的实例代码：

```
#encoding = utf - 8
from selenium import webdriver
import unittest, time, traceback
from selenium.webdriver.support.ui import WebDriverWait
from selenium.webdriver.common.by import By
from selenium.webdriver.support import expected_conditions as EC
from selenium.common.exceptions import TimeoutException, NoSuchElementException

class TestDemo(unittest.TestCase):

    def setUp(self):
        # 启动 Chrome 浏览器
        self.driver = webdriver.Firefox(executable_path = "c:\\geckodriver")

    def test_datePicker(self):
        url = "http://jqueryui.com/resources/demos/datepicker/other - months.html"
        # 访问指定的网址
        self.driver.get(url)
        try:
            # 创建一个显示等待对象
            wait = WebDriverWait(self.driver, 10, 0.2)
            # 显示等待判断被测试页面上的日期输入框是否可见并且能被单击
            wait.until(EC.element_to_be_clickable((By.ID, 'datepicker')))
        except TimeoutException as err:
            # 捕获 TimeoutException 异常
            print(traceback.print_exc())
        except NoSuchElementException as err:
            # 捕获 NoSuchElementException 异常
            print(traceback.print_exc())
        except Exception as err:
            # 捕获其他异常
            print(traceback.print_exc())
        else:
            # 查找被测试页面上的日期输入框页面元素
            dateInputBox = self.driver.find_element_by_id("datepicker")
            # 查找到日期输入框,直接输入指定格式的日期字符串
            # 就可以变相模拟在日期控件上进行选择了
            dateInputBox.send_keys("5/7/2016")
            time.sleep(3)

    def tearDown(self):
        # 退出浏览器
        self.driver.quit()

if __name__ == '__main__':
    unittest.main()
```

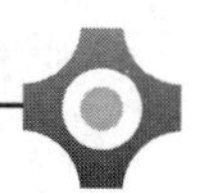

上面被测试网页中的日期控件支持输入，但有时候会遇到日期选择控件不允许用户输入的情况，此种情况下可以通过 JavaScript 语句改变页面元素属性值的方式将日期控件修改成可编辑状态，以便完成脚本直接输入日期来进行日期选择，具体代码请参阅 11.5 小节说明。

11.10 启动带有用户配置信息的 Firefox 浏览器窗口

目的：

由于 WebDriver 启动 Firefox 浏览器时会启用全新的 Firefox 浏览器窗口，导致当前机器的 Firefox 浏览器已经配置的信息在测试中均无法生效，例如已经安装的浏览器插件、个人收藏夹等。为了解决此问题，自动化测试脚本中需要使用指定的配置信息来启动 Firefox 浏览器窗口。

生成用户自定义的 Firefox 浏览器配置文件：

（1）单击桌面左下角的 Windows 图标，在“搜索程序和文件”输入框中输入“cmd”，并按回车键，以便调出 CMD 控制台。

（2）在 CMD 中使用 cd 命令进入 firefox. exe 文件所在目录（比如 C:\Program Files\Mozilla Firefox），并输入 firefox. exe -ProfileManager -no-remote 命令，然后按回车键，调出“Firefox- 选择用户配置文件”操作窗口，如图 11-13 所示。

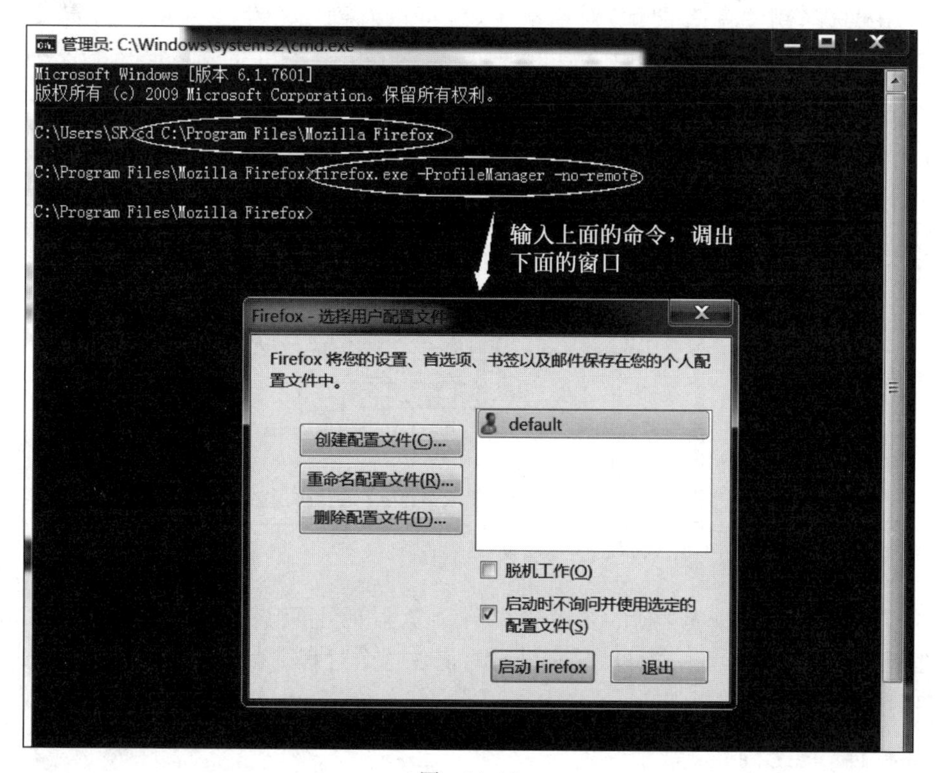

图 11-13

(3) 在弹出的 FireFox 的“选择用户配置文件”对话框中，单击“创建配置文件”按钮，在接下来弹出的窗口中直接单击“下一步”按钮，显示如图 11-14 所示的界面。

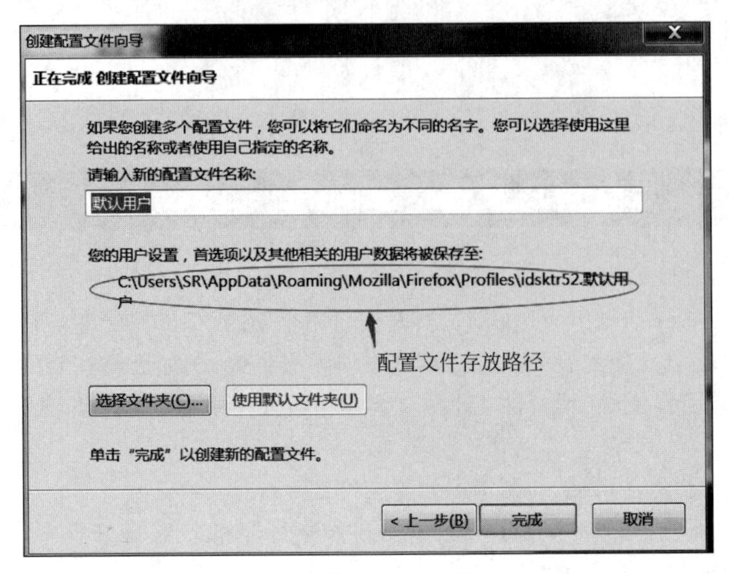

图 11-14

(4) 在弹出的“创建配置文件向导”窗体中的“请输入新的配置文件名称：”对应的输入框中输入自定义的配置文件名称(比如：WebDriver)，并单击“完成”按钮完成配置文件。

(5) 在“选择用户配置文件”对话框中，就可以看到已经生成的用户自定义的名为 Webdriver 的用户配置文件，选中它后单击“启动 Firefox”按钮完成自定义配置文件的生成，如图 11-15 所示。

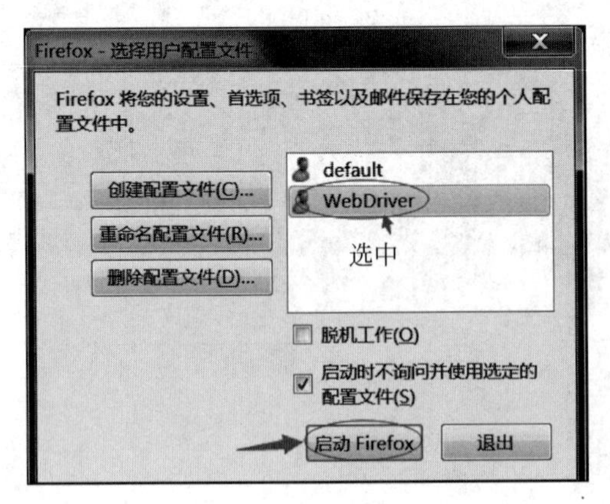

图 11-15

(6) 启动 Firefox 浏览器后，在设置中将浏览器的主页设置为 http://www.baidu.com，也就是让浏览器学习一遍设置主页操作，记住这个过程。

实例代码：

```
# encoding = utf - 8
from selenium import webdriver
from selenium.common.exceptions import NoSuchElementException
```

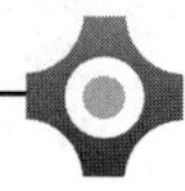

```
import unittest, time

class TestCustomConfigurationBrowser(unittest.TestCase):

    def setUp(self):
        # 创建存储自定义配置文件的路径变量
        proPath = "C:\\Users\\SR\AppData\\Roaming\\Mozilla\\Firefox\\Profiles\\6rfmnay8.
WebDriver"
        # 加载自定义配置文件到 FirefoxProfile 实例中,
        # 等价于 profile = webdriver.FirefoxProfile(proPath)
        profile = webdriver.FirefoxProfile(proPath)
        # 将添加了新配置文件的 Firefox 浏览器首页设为搜狗主页
        profile.set_preference("browser.startup.homepage",
                               "http://www.sogou.com")
        # 设置开始页面不是空白页,0 表示空白页,
        # 这一步必须做,否则设置的主页不会生效
        profile.set_preference("browser.startup.page", 1)
        # 启动带自定义配置文件的 Firefox 浏览器
        self.driver = webdriver.Firefox(executable_path = "c:\\geckodriver",
                                  firefox_profile = profile)

    def testSoGouSearch(self):
        # 等待 5 秒,以便浏览器启动完成
        time.sleep(5)
        try:
            # 找到搜狗主页搜索输入框页面元素
            searchBox = self.driver.find_element_by_id("query")
            # 在找到的搜索输入框中输入"光荣之路自动化测试"
            searchBox.send_keys("光荣之路自动化测试")
            # 找到搜索按钮,并单击
            self.driver.find_element_by_id("stb").click()
            time.sleep(10)
        except NoSuchElementException, e:
            print("修改带自定义配置文件的浏览器主页不成功!")

    def tearDown(self):
        # 退出 Firefox 浏览器
        self.driver.quit()

if __name__ == '__main__':
    unittest.main()
```

代码解释：

在创建 Firefox 浏览器自定义配置文件时，已经将浏览器的主页设定为 http://www.baidu.com，而在带着此自定义配置文件启动浏览器前，又通过 profile.set_preference("browser.startup.homepage", "http://www.sogou.com")语句，将浏览器此时的主页修改为搜狗首页，所以在启动浏览器时自动打开搜狗主页，并继续后续的搜索操作。

更多说明：

(1) 通过 driver = webdriver.Firefox(executable_path="c:\\geckodriver")这样的方

式启动的 Firefox 浏览器均是一个不带任何配置、不带任何插件等信息的全新的浏览器实例,通过本实例的介绍,读者可以在自动化实施过程中启动自定义配置信息的 Firefox 实例。如果读者想启动默认的 Firefox 浏览器,也就是平时我们手动单击 Firefox 快捷启动图标启动的 Firefox 浏览器,可以将本例中的自定义配置信息文件换成默认的配置文件(扩展名为 default 的文件夹,这是 Firefox 默认就创建好的),其存放路径和用户自定义配置文件存放在同一目录下。

(2) 如需获取更多的配置项,请读者在 Firefox 浏览器地址栏中访问 about:config 进行查询。

11.11 UI 对象库

目的:

能够使用配置文件存储被测试页面上页面元素的定位方式和定位表达式,做到定位数据和程序的分离。测试程序写好以后,可以方便不具备编码能力的测试人员进行自定义修改配置。此部分内容可以作为自定义的高级自动化框架的组成部分。

用于测试的网址:

http://www.sogou.com

实例代码:

新建一个名叫 SoGouTest 的工程,在工程下新建三个文件,分别为 SoGou.py、ObjectMap.py 以及 UiObjectMap.ini。

UiObjectMap.ini 页面元素定位表达式配置文件内容如下:

```
[sogou]
searchBox = id > query
searchButton = id > stb
```

ObjectMap.py 表示 ObjectMap 工具类文件,供测试程序调用,内容如下:

```
# encoding = utf - 8
from selenium.webdriver.support.ui import WebDriverWait
from configparser import ConfigParser
import os

class ObjectMap(object):
    def __init__(self):
        # 获取存放页面元素定位表达方式及定位表达式的配置文件所在绝对路径
        # os.path.abspath(__file__)表示获取当前文件所在路径目录
        self.uiObjMapPath = os.path.dirname(os.path.abspath(__file__))\
                            + "\\UiObjectMap.ini"
        print self.uiObjMapPath

    def getElementObject(self, driver, webSiteName, elementName):
        try:
            # 创建一个读取配置文件的实例
            cf = ConfigParser()
```

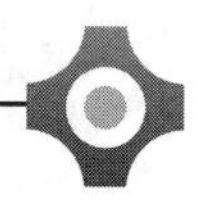

```
            # 将配置文件内容加载到内存
            cf.read(self.uiObjMapPath)
            # 根据 section 和 option 获取配置文件中页面元素的定位方式及
            # 定位表达式组成的字符串,并使用">"分割
            locators = cf.get(webSiteName, elementName).split(">")
            # 得到定位方式
            locatorMethod = locators[0]
            # 得到定位表达式
            locatorExpression = locators[1]
            print(locatorMethod, locatorExpression)
            # 通过显式等待方式获取页面元素
            element = WebDriverWait(driver, 10).until(lambda x: \
                    x.find_element(locatorMethod, locatorExpression))
        except Exception as err:
            raise err
        else:
            # 当页面元素被找到后,将该页面元素对象返回给调用者
            return element
```

SoGou.py 中调用 ObjectMap 工具类实现测试逻辑,该文件具体内容如下:

```
#encoding = utf - 8
from selenium import webdriver
import unittest
import time, traceback
from ObjectMap import ObjectMap

class TestSoGouByObjectMap(unittest.TestCase):

    def setUp(self):
        self.obj = ObjectMap()
        # 启动 Firefox 浏览器
        self.driver = webdriver.Firefox(executable_path = "c:\\geckodriver")

    def testSoGouSearch(self):
        url = "http://www.sogou.com"
        # 访问搜狗首页
        self.driver.get(url)
        try:
            # 查找页面搜索输入框
            searchBox = self.obj.getElementObject\
                (self.driver, "sogou", "searchBox")
            # 在找到的搜索输入框中输入"WebDriver 实战宝典"
            searchBox.send_keys("WebDriver 实战宝典")
            # 查找搜索按钮
            searchButton = self.obj.getElementObject\
                (self.driver, "sogou", "searchButton")
            # 单击找到的搜索按钮
            searchButton.click()
            # 等待 2 秒,以便页面加载完成
            time.sleep(2)
```

```
                # 断言关键字"吴晓华"是否按预期出现在页面源代码中
                self.assertTrue("吴晓华" in self.driver.page_source,
                                "assert error!")
            except Exception as err:
                # 打印异常堆栈信息
                print(traceback.print_exc())

    def tearDown(self):
        # 退出 Firefox 浏览器
        self.driver.quit()

if __name__ == '__main__':
    unittest.main()
```

更多说明：

本实例实现了程序与数据分离，首先从 UI 对象库文件 UiObjectMap.ini 中取得搜狗首页中需要操作的页面元素的定位方式和定位表达式，然后在 ObjectMap 类中的取得该页面元素的实例对象，最后返回给测试用例方法中进行后续处理。这样做的好处是可以在一定条件下满足一部分不会编码的测试人员实施自动化测试。

11.12 操作富文本框

富文本框的技术实现和普通的文本框的定位存在较大的区别，富文本框的常见技术用到了 Frame 标签，并且在 Frame 里面实现了一个完整的 HTML 网页结构，所以使用普通的定位模式将无法直接定位到富文本框对象，本节将会解决这个问题。

目的：

能够定位到页面中的富文本框对象并进入该富文本框，然后向富文本框中输入内容。

使用 JavaScript 代码实现向富文本框中输入 HTML 格式的内容。

用于测试的网址：

http://mail.sohu.com

实例代码 1：

```
#encoding = utf - 8
from selenium import webdriver
import unittest, time, traceback
from selenium.webdriver.support.ui import WebDriverWait
from selenium.webdriver.support import expected_conditions as EC
from selenium.common.exceptions import TimeoutException, NoSuchElementException
from selenium.webdriver.common.by import By

class TestDemo(unittest.TestCase):

    def setUp(self):
        # 启动 Firefox 浏览器
        self.driver = webdriver.Firefox(executable_path = "c:\\geckodriver")
```

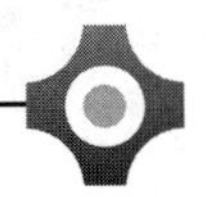

```
def test_SohuMailSendEMail(self):
    url = "http://mail.sohu.com"
    # 访问搜狐邮箱登录页
    self.driver.get(url)
    try:
        userName = self.driver.find_element_by_xpath\
            ('//input[@placeholder = "请输入您的邮箱"]')
        userName.clear()
        userName.send_keys("xxxx")
        passWord = self.driver.find_element_by_xpath\
            ('//input[@placeholder = "请输入您的密码"]')
        passWord.clear()
        passWord.send_keys("xxxx")
        login = self.driver.find_element_by_xpath('//input[@value = "登 录"]')
        login.click()
        wait = WebDriverWait(self.driver, 10)
        # 显示等待,确定页面是否成功登录并跳转到登录成功后的首页
        wait.until(EC.element_to_be_clickable\
                   ((By.XPATH, '//li[text() = "写邮件"]')))
        self.driver.find_element_by_xpath('//li[text() = "写邮件"]').click()
        time.sleep(2)
        receiver = self.driver.find_element_by_xpath\
            ('//div[@arr = "mail.to_render"]//input')
        # 输入收件人
        receiver.send_keys("xxxx")
        subject = self.driver.find_element_by_xpath\
            ('//input[@ng - model = "mail.subject"]')
        # 输入邮件标题
        subject.send_keys("一封测试邮件!")
        # 获取邮件正文编辑区域的 iframe 页面元素对象
        iframe = self.driver.find_element_by_xpath\
            ('//iframe[contains(@id, "ueditor_0")]')
        # 通过 switch_to.frame()方法切换进入富文本框中
        self.driver.switch_to.frame(iframe)
        # 获取富文本框中编辑页面元素对象
        editBox = self.driver.find_element_by_xpath("/html/body")
        # 输入邮件正文
        editBox.send_keys("邮件的正文内容")
        # 从富文本框中切换出,回到默认页面
        self.driver.switch_to.default_content()
        # 找到页面上的"发送"按钮,并单击它
        self.driver.find_element_by_xpath('//span[. = "发送"]').click()
        # 显示都等待含有关键字串"发送成功"的页面元素出现在页面中
        wait.until(EC.visibility_of_element_located\
                   ((By.XPATH, '//span[. = "发送成功"]')))
        print("邮件发送成功")
    except TimeoutException:
        print("显示等待页面元素超时")
    except NoSuchElementException:
        print("寻找的页面元素不存在", traceback.print_exc())
    except Exception:
```

```
                # 打印其他异常堆栈信息
                print(traceback.print_exc())

    def tearDown(self):
        # 退出 Firefox 浏览器
        self.driver.quit()

if __name__ == '__main__':
    unittest.main()
```

实例代码 2:

```
#encoding=utf-8
from selenium import webdriver
import unittest, time, traceback
from selenium.webdriver.support.ui import WebDriverWait
from selenium.webdriver.support import expected_conditions as EC
from selenium.common.exceptions import TimeoutException, NoSuchElementException
from selenium.webdriver.common.by import By

class TestDemo(unittest.TestCase):

    def setUp(self):
        # 启动 Firefox 浏览器
        self.driver = webdriver.Firefox(executable_path="c:\\geckodriver")

    def test_SohuMailSendEMail(self):
        url = "http://mail.sohu.com"
        # 访问搜狐邮箱登录页
        self.driver.get(url)
        try:
            userName = self.driver.find_element_by_xpath\
                ('//input[@placeholder="请输入您的邮箱"]')
            userName.clear()
            userName.send_keys("xxxx")
            passWord = self.driver.find_element_by_xpath\
                ('//input[@placeholder="请输入您的密码"]')
            passWord.clear()
            passWord.send_keys("xxxx")
            login = self.driver.find_element_by_xpath('//input[@value="登 录"]')
            login.click()
            wait = WebDriverWait(self.driver, 10)
            # 显示等待,确定页面是否成功登录并跳转到登录成功后的首页
            wait.until(EC.element_to_be_clickable\
                            ((By.XPATH, '//li[text()="写邮件"]')))
            self.driver.find_element_by_xpath('//li[text()="写邮件"]').click()
            time.sleep(2)
            receiver = self.driver.find_element_by_xpath\
                ('//div[@arr="mail.to_render"]//input')
            # 输入收件人
            receiver.send_keys("xxxx")
```

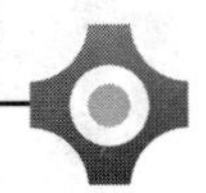

```
            subject = self.driver.find_element_by_xpath\
                ('//input[@ng-model="mail.subject"]')
            # 输入邮件标题
            subject.send_keys("一封测试邮件!")
            # 获取邮件正文编辑区域的 iframe 页面元素对象
            iframe = self.driver.find_element_by_xpath\
                ('//iframe[contains(@id, "ueditor_0")]')
            # 通过 switch_to.frame()方法切换进入富文本框中
            self.driver.switch_to.frame(iframe)
            # 通过 JavaScript 代码向邮件正文编辑框中输入正文
            self.driver.execute_script("document.getElementsByTagName('body')\
                [0].innerHTML = '<b>邮件的正文内容<b>;'")
            # 从富文本框中切换出,回到默认页面
            self.driver.switch_to.default_content()
            # 找到页面上的"发送"按钮,并单击它
            self.driver.find_element_by_xpath('//span[.="发送"]').click()
            # 显示都等待含有关键字串"发送成功"的页面元素出现在页面中
            wait.until(EC.visibility_of_element_located\
                        ((By.XPATH, '//span[.="发送成功"]')))
            print("邮件发送成功")
        except TimeoutException:
            print("显示等待页面元素超时")
        except NoSuchElementException:
            print("寻找的页面元素不存在", traceback.print_exc())
        except Exception:
            # 打印其他异常堆栈信息
            print(traceback.print_exc())

    def tearDown(self):
        # 退出 Firefox 浏览器
        self.driver.quit()

if __name__ == '__main__':
    unittest.main()
```

实例代码 3:

```
# encoding = utf-8
from selenium import webdriver
import unittest, time, traceback
from selenium.webdriver.support.ui import WebDriverWait
from selenium.webdriver.support import expected_conditions as EC
from selenium.common.exceptions import TimeoutException, NoSuchElementException
from selenium.webdriver.common.by import By
from selenium.webdriver.common.keys import Keys
import win32clipboard as w
import win32api, win32con

# 用于设置剪切板内容
def setText(aString):
    w.OpenClipboard()
```

```
    w.EmptyClipboard()
    w.SetClipboardData(win32con.CF_UNICODETEXT, aString)
    w.CloseClipboard()

# 键盘按键映射字典
VK_CODE = {'ctrl':0x11, 'v':0x56}

# 键盘键按下
def keyDown(keyName):
    win32api.keybd_event(VK_CODE[keyName], 0, 0, 0)
# 键盘键抬起
def keyUp(keyName):
    win32api.keybd_event(VK_CODE[keyName], 0, win32con.KEYEVENTF_KEYUP, 0)

class TestDemo(unittest.TestCase):

    def setUp(self):
        # 启动 Firefox 浏览器
        self.driver = webdriver.Firefox(executable_path="c:\\geckodriver")

    def test_SohuMailSendEMail(self):
        url = "http://mail.sohu.com"
        # 访问搜狐邮箱登录页
        self.driver.get(url)
        try:
            userName = self.driver.find_element_by_xpath\
                ('//input[@placeholder="请输入您的邮箱"]')
            userName.clear()
            userName.send_keys("xxxx")
            passWord = self.driver.find_element_by_xpath\
                ('//input[@placeholder="请输入您的密码"]')
            passWord.clear()
            passWord.send_keys("xxxx")
            login = self.driver.find_element_by_xpath('//input[@value="登 录"]')
            login.click()
            wait = WebDriverWait(self.driver, 10)
            # 显示等待,确定页面是否成功登录并跳转到登录成功后的首页
            wait.until(EC.element_to_be_clickable\
                        ((By.XPATH, '//li[text()="写邮件"]')))
            self.driver.find_element_by_xpath('//li[text()="写邮件"]').click()
            time.sleep(2)
            receiver = self.driver.find_element_by_xpath\
                ('//div[@arr="mail.to_render"]//input')
            # 输入收件人
            receiver.send_keys("xxxx")
            subject = self.driver.find_element_by_xpath\
                ('//input[@ng-model="mail.subject"]')
            # 输入邮件标题
            subject.send_keys("一封测试邮件!")
            # 输入完邮件标题后,按下 Tab 键可以将页面焦点切换到富文本框编辑区域
            subject.send_keys(Keys.TAB)
```

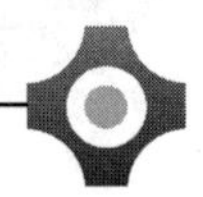

```
            # 设置剪切板内容,也就是将要输入的正文内容
            setText("邮件正文内容")
            # 模拟键盘的 Ctrl + V 组合键,将剪切板内容粘贴到富文本编辑区中
            keyDown("ctrl")
            keyDown("v")
            keyUp("v")
            keyUp("ctrl")
            # 找到页面上的"发送"按钮,并单击它
            self.driver.find_element_by_xpath('//span[. = "发送"]').click()
            # 显示都等待含有关键字串"发送成功"的页面元素出现在页面中
            wait.until(EC.visibility_of_element_located\
                       ((By.XPATH, '//span[. = "发送成功"]')))
            print("邮件发送成功")
        except TimeoutException:
            print("显示等待页面元素超时")
        except NoSuchElementException:
            print("寻找的页面元素不存在", traceback.print_exc())
        except Exception:
            # 打印其他异常堆栈信息
            print(traceback.print_exc())

    def tearDown(self):
        # 退出 Firefox 浏览器
        self.driver.quit()

if __name__ == '__main__':
    unittest.main()
```

在以上三个实例代码中,读者需要将代码中以“xxxx”字符串代替的搜狐邮箱用户名、密码以及收件人地址替换成有效的数据,否则程序会报错。

如果遇到登录邮箱时需要手机号验证,或者需要输入验证码,请读者手动完成。

更多说明:

以上三种方法,都能实现向搜狐邮箱写信页面的富文本框中输入内容,但它们之间又有什么区别呢？三种方法的比较如下。

方法 1:

优点:实现简单,只要调用 WebDriver 对页面元素对象提供的 send_keys()方法,即可实现内容输入。

缺点:必须能定位到要被操作元素,对脚本编写人员的定位能力要求比较高,同时不支持 HTML 格式的内容输入。

方法 2:

优点:可以支持 HTML 格式的文字内容作为富文本框的输入内容。

缺点:由于各种网页中富文本框实现的机制可能不同,有可能造成定位到富文本框的文本编辑区对象比较困难,此时就需要熟练了解 HMTL 代码含义以及 Frame 的进出方式,对脚本编写人员的能力要求就比较高。

方法 3：

优点：不管何种类型的富文本框，只要找到它上面的紧邻元素，然后通过模拟按 Tab 键的方式均可进入富文本框中，由此可以使用一种方法解决所有类型的富文本框定位问题。

缺点：不能在富文本框编辑器中进行 HTML 格式的内容输入。

以上三种方式各有利弊，只要能够相对稳定地完成对富文本框的操作，读者可以自行选择任意一种方法使用。

11.13 精确比较页面截图图片

目的：

在测试过程中，一般会对核心页面进行截图，并且使用测试过程中所截图和以前测试过程中的截图进行精确的比对。如果精确百分之百匹配，可以判断两张图片完全一致，如果页面中发生了任何微小的变化，则会认为图片不匹配。

环境准备：

（1）Windows 下按 Win + R 组合键，调出运行窗口，在“打开”输入框中输入 CMD 后回车，调出 CMD 窗口，然后在 CMD 窗口执行 pip install pillow 安装 Python 图像处理库，安装成功界面如图 11-16 所示。

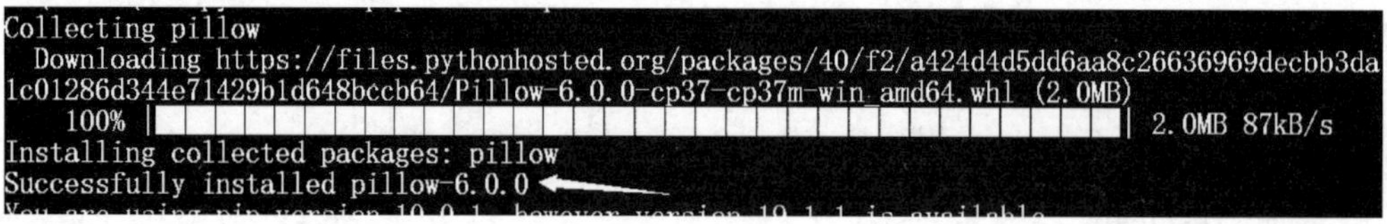

```
Collecting pillow
  Downloading https://files.pythonhosted.org/packages/40/f2/a424d4d5dd6aa8c26636969decbb3da
1c01286d344e71429b1d648bccb64/Pillow-6.0.0-cp37-cp37m-win_amd64.whl (2.0MB)
    100% |████████████████████████████████| 2.0MB 87kB/s
Installing collected packages: pillow
Successfully installed pillow-6.0.0
```

图 11-16

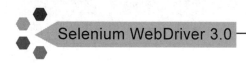

（2）CMD 下进入 Python 交互模式，执行 from PIL import Image，如果没有报错，说明 Python 图像处理库 Pillow 已经安装成功。

如果直接使用 pip 工具不能成功安装 Pillow 库，请访问 https://pypi.python.org/pypi/Pillow/4.0.0，根据安装的 Python 版本及位数选择相应的离线安装文件进行安装，注意下载的 Pillow 离线安装文件版本及位数必须与已安装的 Python 版本及位数保持一致，否则会安装失败。

用于测试的网址：

http://www.sogou.com

实例代码：

```
#encoding = utf - 8
from selenium import webdriver
import unittest, time
from PIL import Image

class ImageCompare(object):
    '''
    本类实现了对两张图片通过像素比对的算法，获取文件的像素个数大小
    然后使用循环的方式将两张图片的所有像素进行一一对比，
```

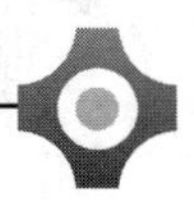

```
    并计算比对结果的相似度的百分比
    '''
    def make_regalur_image(self, img, size = (256, 256)):
        # 将图片尺寸强制重置为指定的 size 大小
        # 然后再将其转换成 RGB 值
        return img.resize(size).convert('RGB')

    def split_image(self, img, part_size = (64, 64)):
        # 按给定大小切分图片
        w, h = img.size
        pw, ph = part_size
        assert w % pw == h % ph == 0
        return [img.crop((i, j, i + pw, j + ph)).copy() \
                for i in xrange(0, w, pw) for j in xrange(0, h, ph)]

    def hist_similar(self, lh, rh):
        # 统计切分后每部分图片的相似度频率曲线
        assert len(lh) == len(rh)
        return sum(1 - (0 if l == r else float(abs(l - r)) / max(l, r)) \
                   for l, r in zip(lh, rh)) / len(lh)

    def calc_similar(self, li, ri):
        # 计算两张图片的相似度
        return sum(self.hist_similar(l.histogram(), r.histogram())\
            for l, r in zip(self.split_image(li), self.split_image(ri))) / 16.0

    def calc_similar_by_path(self, lf, rf):
        li, ri = self.make_regalur_image(Image.open(lf)), \
                 self.make_regalur_image(Image.open(rf))
        return self.calc_similar(li, ri)


class TestDemo(unittest.TestCase):

    def setUp(self):
        self.IC = ImageCompare()
        # 启动 Firefox 浏览器
        self.driver = webdriver.Firefox(executable_path = "c:\\geckodriver")

    def test_ImageComparison(self):
        url = "http://www.sogou.com"
        # 访问搜狗首页
        self.driver.get(url)
        time.sleep(3)
        # 截取第一次访问搜狗首页的图片,并保存在本地
        self.driver.save_screenshot("D:\\book\\sogou1.png")
        self.driver.get(url)
        time.sleep(3)
        # 截取第二次访问搜狗首页的图片,并保存在本地
        self.driver.save_screenshot("D:\\book\\sogou2.png")
        # 打印两张截图后比对相似度,100 表示完全匹配
```

```
        print(self.IC.calc_similar_by_path('D:\\book\\sogou1.png',
                'D:\\book\\sogou2.png') * 100)

    def tearDown(self):
        # 退出 Firefox 浏览器
        self.driver.quit()

if __name__ == '__main__':
    unittest.main()
```

11.14 高亮显示正在操作的页面元素

目的：

在测试过程中，经常会进行调试工作，使用高亮显示被操作页面元素的方式可以提高调试过程中的效率，以显眼的颜色高亮显示的方式提示测试人员目前正在操作的页面元素。

用于测试的网址：

http://www.sogou.com

实例代码：

```
#encoding = utf-8
import unittest
from selenium import webdriver
import time

def highLightElement(driver,element):
    # 封装好的高亮显示页面元素的方法
    # 使用 JavaScript 代码将传入的页面元素对象的背景颜色和边框颜色分别设置为
    # 绿色和红色
    driver.execute_script("arguments[0].setAttribute('style',\
    arguments[1]);", element,"background:green; border:2px solid red;")

class TestDemo(unittest.TestCase):
    def setUp(self):
        # 获取浏览器驱动实例
        self.driver = webdriver.Firefox(executable_path = "c:\\geckodriver")

    def test_HighLightWebElement(self):
        url = "http://www.sogou.com"
        # 访问搜狗首页
        self.driver.get(url)
        searchBox = self.driver.find_element_by_id("query")
        # 调用高亮显示元素的封装函数,将搜索输入框进行高亮显示
        highLightElement(self.driver, searchBox)
        # 等待 3 秒,以便查看高亮效果
        time.sleep(3)
        searchBox.send_keys("光荣之路自动化测试")
        submitButton = self.driver.find_element_by_id("stb")
        # 调用高亮显示元素的封装函数,将搜索按钮进行高亮显示
```

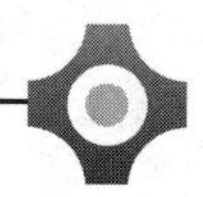

```
        highLightElement(self.driver, submitButton)
        time.sleep(3)
        submitButton.click()
        time.sleep(3)

    def tearDown(self):
        # 退出浏览器
        self.driver.quit()

if __name__ == '__main__':
    unittest.main()
```

11.15 浏览器中新开标签页

目的：

在同一浏览器中新开一个或多个新的标签页(Tab)，同时能在这些标签页中输入测试网址进行测试。

用于测试网址：

http://www.sogou.com

http://www.baidu.com

实例代码：

```
#encoding = utf-8
import unittest
from selenium import webdriver
import time
import win32api, win32con

VK_CODE = {'ctrl':0x11, 't':0x54, 'tab':0x09}

# 键盘键按下
def keyDown(keyName):
    win32api.keybd_event(VK_CODE[keyName], 0, 0, 0)
# 键盘键抬起
def keyUp(keyName):
    win32api.keybd_event(VK_CODE[keyName], 0, win32con.KEYEVENTF_KEYUP, 0)

# 封装的按键方法
def simulateKey(firstKey, secondKey):
    keyDown(firstKey)
    keyDown(secondKey)
    keyUp(secondKey)
    keyUp(firstKey)

class TestDemo(unittest.TestCase):
    def setUp(self):
        # 获取浏览器驱动实例
```

```
        self.driver = webdriver.Firefox(executable_path = "c:\\geckodriver")

    def test_newTab(self):
        # 等待 3 秒,等待浏览器启动完成
        time.sleep(3)
        # 使用 for 循环,再新开两个新的标签页
        for i in range(2):
            simulateKey("ctrl", "t")
        # 通过 Ctrl + Tab 组合键,将当前页面切换为默认页面,
        # 也就是最先打开的标签页
        simulateKey("ctrl", "tab")
        # 访问搜狗首页
        self.driver.get("http://www.sogou.com")
        self.driver.find_element_by_id("query").send_keys("光荣之路")
        self.driver.find_element_by_id("stb").click()
        time.sleep(3)
        self.assertTrue("乔什·卢卡斯" in self.driver.page_source)

        # 获取所有的打开的窗口句柄
        all_handles = self.driver.window_handles
        print(len(all_handles))
        # 将当前窗口句柄切换至第二个标签页
        self.driver.switch_to.window(all_handles[1])
        self.driver.get("http://www.baidu.com")
        self.driver.find_element_by_id("kw").send_keys("WebDriver 实战宝典")
        self.driver.find_element_by_id("su").click()
        time.sleep(3)
        self.assertTrue("吴晓华" in self.driver.page_source)

        # 将当前窗口的句柄切换至第三个标签页
        self.driver.switch_to.window(all_handles[2])
        self.driver.get("http://www.baidu.com")
        self.driver.find_element_by_id("kw").send_keys("selenium")
        self.driver.find_element_by_id("su").click()
        time.sleep(3)
        self.assertTrue("selenium " in self.driver.page_source)

    def tearDown(self):
        # 退出浏览器
        self.driver.quit()

if __name__ == '__main__':
    unittest.main()
```

说明：

IE 浏览器对本小节实例不支持。

11.16 测试过程中发生异常或断言失败时进行屏幕截图

目的：

在自动化测试执行过程中,有可能会抛异常或断言失败的情况,此时通过对发生异常或

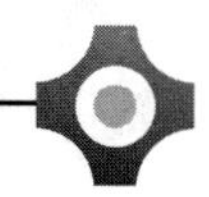

断言失败的页面截图操作，可以方便自动化测试人员进行 bug 定位及问题排查。

用于测试网址：

http://www.sogou.com

实例代码：

首先在 PyCharm 工具中新创建一个名为 GloryRoad 的 Python 工程，然后在该工程下面创建 DateUtil.py、FileUtil.py 以及 SoGou.py 这三个 Python 文件。

DateUtil.py 文件具体内容如下：

```
#encoding = utf-8
import time
from datetime import datetime

'''
本文件主要用于获取当前的日期以及时间，
用于生成保存截图文件目录名
'''

def currentDate():
    date = time.localtime()
    # 构造今天的日期字符串
    today = str(date.tm_year) + "-" + str(date.tm_mon) + "-" + str(date.tm_mday)
    return today

def currentTime():
    timeStr = datetime.now()
    # 构造当前时间字符串
    now = timeStr.strftime('%H-%M-%S')
    return now

if __name__ == "__main__":
    print(currentDate())
    print(currentTime())
```

FileUtil.py 文件的具体内容如下：

```
#encoding = utf-8
from DateUtil import currentDate, currentTime
import os

'''
本文件主要用于创建目录，用于存放异常截图，
创建目录的方法仅供大家参考，将来用于根据测试
需要创建测试人员需要的目录或文件等
'''

def createDir():
    # 获得当前文件所在目录绝对路径
    currentPath = os.path.dirname(os.path.abspath(__file__))
    # 获取今天的日期字符串
```

```
    today = currentDate()
    # 构造以今天日期命名的目录的绝对路径
    dateDir = os.path.join(currentPath, today)
    print(dateDir)
    if not os.path.exists(dateDir):
        # 如果以今天日期命名的目录不存在则创建
        os.mkdir(dateDir)
    # 获得当前的时间字符串
    now = currentTime()
    # 构造以当前时间命名的目录的绝对路径
    timeDir = os.path.join(dateDir, now)
    print(timeDir)
    if not os.path.exists(timeDir):
        # 如果以当前时间命名的目录不存在则创建
        os.mkdir(timeDir)
    return timeDir
```

SoGou.py 文件的具体内容如下：

```
#encoding = utf - 8
from selenium import webdriver
import unittest, time, os
from FileUtil import createDir
import traceback

# 创建存放异常截图的目录,并得到本次实例中存放图片目录的绝对路径,
# 并且作为全局变量,以供本次所有测试用例调用
picDir = createDir()

def takeScreenshot(driver, savePath, picName):
    # 封装截屏方法
    # 构造屏幕截图路径及图片名
    picPath = os.path.join(savePath, str(picName) + ".png")
    try:
        # 调用 WebDriver 提供的 get_screenshot_as_file()方法,
        # 将截取的屏幕图片保存为本地文件
        driver.get_screenshot_as_file(picPath)
    except Exception as err:
        # 打印异常堆栈信息
        print(traceback.print_exc())

class TestFailCaptureScreen(unittest.TestCase):

    def setUp(self):
        # 启动 Firefox 浏览器
        self.driver = webdriver.Firefox(executable_path = "c:\\geckodriver")

    def testSoGouSearch(self):
        url = "http://www.sogou.com"
        # 访问搜狗首页
        self.driver.get(url)
```

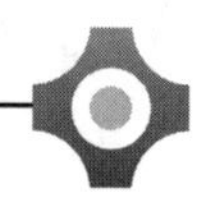

```
            try:
                self.driver.find_element_by_id("query").\
                    send_keys("光荣之路自动化测试")
                self.driver.find_element_by_id("stb").click()
                time.sleep(3)
                # 断言页面的代码中是否存在"事在人为"这 4 个关键字,
                # 因为页面中没有这 4 个字,所以会触发 except 语句的执行,并触发截图操作
                self.assertTrue("事在人为" in self.driver.page_source, \
                                事在人为"关键字串在页面源代码中未找到")
            except AssertionError as err:
                # 调用封装好的截图方法,进行截图并保存在本地磁盘
                takeScreenshot(self.driver, picDir, "error")
            except Exception as err:
                print(traceback.print_exc())
                takeScreenshot(self.driver, picDir, "error")

        def tearDown(self):
            # 退出 Firefox 浏览器
            self.driver.quit()

if __name__ == '__main__':
    unittest.main()
```

更多说明:

此实例借助了两个工具文件 DateUtil.py 和 FileUtil.py 来实现本次测试的目的,当然读者也可以将这两文件中的方法封装到类中形成工具类,此种方法将常用的代码进行封装,便于提高代码的复用度,提高测试脚本编写的效率。

11.17 使用日志模块记录测试过程中的信息

目的:

在自动化测试脚本的执行过程中,使用 Python 的日志模块(logging)记录在测试用例执行过程中一些重要信息或者错误日志等,用于监控和调试后续的脚本。

用于测试网址:

http://www.sogou.com

实例代码:

在 Pycharm 中新建名为 SoGouSearch 的 Python 工程,并在工程下创建 Logger.conf、Log.py 以及 SoGou.py 三个文件。

Logger.conf 文件具体内容如下:

```
###############################################################
[loggers]
keys = root,example01,example02
[logger_root]
level = DEBUG
handlers = hand01,hand02
```

```
[logger_example01]
handlers = hand01,hand02
qualname = example01
propagate = 0
[logger_example02]
handlers = hand01,hand03
qualname = example02
propagate = 0
###############################################
[handlers]
keys = hand01,hand02,hand03
[handler_hand01]
class = StreamHandler
level = DEBUG
formatter = form01
args = (sys.stderr,)
[handler_hand02]
class = FileHandler
level = DEBUG
formatter = form01
args = ('d:\\AutoTestLog.log', 'a')
[handler_hand03]
class = handlers.RotatingFileHandler
level = INFO
formatter = form01
args = ('d:\\AutoTestLog.log', 'a', 10 * 1024 * 1024, 5)
###############################################
[formatters]
keys = form01,form02
[formatter_form01]
format = %(asctime)s %(filename)s[line: %(lineno)d] %(levelname)s %(message)s
datefmt = %Y-%m-%d %H:%M:%S
[formatter_form02]
format = %(name)-12s: %(levelname)-8s %(message)s
datefmt = %Y-%m-%d %H:%M:%S
```

Log.py 文件具体内容如下：

```
# encoding = utf-8
import logging.config
import os

BASE_DIR = os.path.dirname(os.path.abspath(__file__))
conf_path = BASE_DIR + "\Logger.conf"
logging.config.fileConfig(conf_path)

def debug(message):
    # 打印 debug 级别的日志方法
    logging.debug(message)

def warning(message):
```

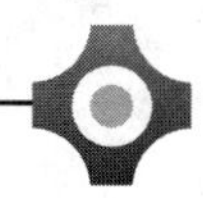

```
    # 打印 warning 级别的日志方法
    logging.warning(message)

def info(message):
    # 打印 info 级别的日志方法
    logging.info(message)
```

SoGou.py 文件具体内容如下：

```
#encoding=utf-8
from selenium import webdriver
import unittest
# 从当前文件所在目录中导入 Log.py 文件中所有内容
from Log import *

class TestSoGouSearch(unittest.TestCase):

    def setUp(self):
        # 启动 Firefox 浏览器
        self.driver = webdriver.Firefox(executable_path="c:\\geckodriver")

    def testSoGouSearch(self):
        info("============== 搜索 ==============")
        url = "http://www.sogou.com"
        # 访问搜狗首页
        self.driver.get(url)
        info("访问 sogou 首页")
        self.driver.find_element_by_id("query").send_keys("光荣之路自动化测试")
        info("在输入框中输入搜索关键字串"光荣之路自动化测试"")
        self.driver.find_element_by_id("stb").click()
        info("单击搜索按钮")
        info("========== 测试用例执行结束 ==========")

    def tearDown(self):
        # 退出 Firefox 浏览器
        self.driver.quit()

if __name__ == '__main__':
    unittest.main()
```

执行结果：

执行 SoGou.py 文件后，会在本地磁盘 D 盘中生成一个日志文件 AutoTestLog.log，日志文件中的内容如下：

```
2016-11-28 16:15:52 SoGou.py[line:15] INFO ============== 搜索 ==============
2016-11-28 16:15:57 SoGou.py[line:19] INFO 访问搜狗首页
2016-11-28 16:15:58 SoGou.py[line:21] INFO 在输入框中输入搜索关键字串"光荣之路自动化测试"
2016-11-28 16:15:58 SoGou.py[line:23] INFO 单击搜索按钮
2016-11-28 16:15:58 SoGou.py[line:24] INFO ========== 测试用例执行结束 ==========
```

由此可实现在测试过程中打印日志的目的，并用于后期分析哪些测试语句被正确执行了，以及哪些测试语句执行失败了。

更多说明：

本实例中日志的级别只配置到 DEBUG 级别，如果读者有需要打印其他级别的日志，请自行修改 Logger.conf 文件中的相关内容。上述针对打印日志的配置可以满足日常测试需要，如果对上面日志配置信息有不明白或需要更多自定义的日志需求，请访问 https://docs.python.org/2/howto/logging.html 进一步学习。

11.18 封装操作表格的公用类

目的：

能够使用自己编写操作表格的公用类，并基于公用类进行表格中元素的各类操作。

用于测试网页的 HTML 代码：

```
<html>
<head>
    <title>设置文本框属性</title>
    <meta http-equiv="Content-Type" content="text/html; charset=utf-8" />
</head>
<body>
    <table width="400" border="1" id="table">
        <tr>
            <td align="left">
                <p>第一行第一列</p>
                <input type="text">
            </td>
            <td align="left">
                <p>第一行第二列</p>
                <input type="text">
            </td>
            <td align="left">
                <p>第一行第三列</p>
                <input type="text">
            </td>
        </tr>
        <tr>
            <td align="left">
                <p>第二行第一列</p>
                <input type="text">
            </td>
            <td align="left">
                <p>第二行第二列</p>
                <input type="text">
            </td>
            <td align="left">
                <p>第二行第三列</p>
                <input type="text">
```

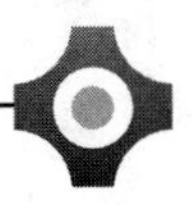

```
            </td>
        </tr>
        <tr>
            <td align = "left">
                <p>第三行第一列</p>
                <input type = "text">
            </td>
            <td align = "left">
                <p>第三行第二列</p>
                <input type = "text">
            </td>
            <td align = "left">
                <p>第三行第三列</p>
                <input type = "text">
            </td>
        </tr>
    </table>
</body>
</html>
```

实例代码：

Table.py 文件具体内容如下：

```
#encoding = utf - 8

class Table(object):
    # 定义一个私有属性__table,用于存放 table 对象
    __table = ''

    def __init__(self, table):
        # Table 类的构造方法
        self.setTable(table)

    def setTable(self, table):
        # 对私有属性__table 进行赋值操作
        self.__table = table

    def getTable(self):
        # 获取私有属性__table 的值
        return self.__table

    def getRowCount(self):
        # 返回 table 对象中所有的行 tr 标签元素对象
        return len(self.__table.find_elements_by_tag_name("tr"))

    def getColumnCount(self):
        # 获取表格对象中的列数
        return len(self.__table.find_elements_by_tag_name("tr")[0].\
                    find_elements_by_tag_name("td"))

    def getCell(self, rowNo, colNo):
```

```
        # 获取表格中某行某列的单元格对象
        try:
            # 找到表格中的某一行,因为行号从 0 开始,
            # 例如要找第三行,则需要进行 3 - 1 = 2 来获取第三行 tr 元素对象
            currentRow = self.__table.find_elements_by_tag_name("tr")[rowNo - 1]
            # 在找到的某行基础上,再找这行中的某一列,列号也从 0 开始
            currentCol = currentRow.find_elements_by_tag_name("td")[colNo - 1]
            # 返回找到的单元格对象
            return currentCol
        except Exception as err:
            raise err

    def getWebElementInCell(self, rowNo, colNo, by, value):
        # 获取表格中某行某列的单元格中某个页面元素对象,
        # by 表示定位页面元素的方法,比如 id,
        # value 表示定位表达式,比如 query
        try:
            currentRow = self.__table.find_elements_by_tag_name("tr")[rowNo - 1]
            currentCol = currentRow.find_elements_by_tag_name("td")[colNo - 1]
            # 获取具体某个单元格中的某个页面元素
            element = currentCol.find_element(by = by, value = value)
            # 返回找到的页面元素对象
            return element
        except Exception as err:
            raise err
```

OperTable.py 文件具体内容如下:

```
#encoding=utf-8
from selenium import webdriver
import unittest
import time
# 从当前文件所在目录下导入 Table.py 文件中的 Table 类
from Table import Table

class TestTableOpertion(unittest.TestCase):

    def setUp(self):
        # 启动 Firefox 浏览器
        self.driver = webdriver.Firefox(executable_path="c:\\geckodriver")

    def testTable(self):
        url = "D:\\book\\html.html"
        # 访问自定义的网页
        self.driver.get(url)
        # 获取被测试页面中的表格元素,并存储在 webTable 变量中
        webTable = self.driver.find_element_by_tag_name("table")
        # 使用 webTable 变量对 Table 类进行实例化
        table = Table(webTable)
        # 统计表格的行数
        print(table.getRowCount())
```

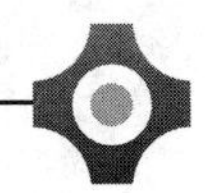

```
        # 统计表格的列数
        print(table.getColumnCount())
        # 获取表格中第二行第三列单元格对象
        cell = table.getCell(2, 3)
        # 断言获取的单元格文本内容是否是"第二行第三列"
        self.assertAlmostEqual("第二行第三列", cell.text)
        # 获取表格中第三行第二列单元格中的输入框对象
        cellInput = table.getWebElementInCell(3, 2, "tag name", "input")
        # 在找到的输入框中输入"第三行的第二列表格被找到"关键字内容
        cellInput.send_keys("第三行的第二列表格被找到")
        # 等待 3 秒,肉眼查看输入效果
        time.sleep(3)

    def tearDown(self):
        # 退出 Firefox 浏览器
        self.driver.quit()

if __name__ == '__main__':
    unittest.main()
```

执行 OperTable.py 文件,可以看到操作表格的公用类在测试过程中被成功调用。

OperTable.py 文件和 Table.py 文件在同一级目录中。

11.19　测试 HTML5 语言实现的视频播放器

目的:

能够获取 HTML5 语言实现的视频播放器视频文件的地址、时长,控制播放器进行播放或暂停播放等操作。

用于测试的网址:

http://www.w3school.com.cn/tiy/loadtext.asp?f=html5_video_simple

实例代码:

```
#encoding=utf-8
import unittest
from selenium import webdriver
import time

class TestDemo(unittest.TestCase):
    def setUp(self):
        # 获取浏览器驱动实例
        self.driver = webdriver.Firefox(executable_path = "c:\\geckodriver")

    def test_HTML5VideoPlayer(self):
        url = "https://www.17sucai.com/pins/demo-show?id=37427"
```

```
        # 访问 HTML5 语言实现的播放器网页
        self.driver.get(url)
        # 打印访问网页的页面源代码,供读者学习
        print(self.driver.page_source)
        self.driver.switch_to.frame(self.driver.find_element_by_id("iframe"))
        # 获取页面中的 video 标签元素对象
        videoPlayer = self.driver.find_element_by_tag_name("video")
        # 使用 JavaScript 语句,通过播放器内部的
        # currentSrc 属性获取视频文件的网络存储地址
        videoSrc = self.driver.execute_script\
            ("return arguments[0].currentSrc;", videoPlayer)
        # 打印网页中视频存放地址
        print(videoSrc)
        # 断言视频存放地址是否符合预期
        video_url = "https://blz-videos.nosdn.127.net/1/OverWatch/AnimatedShots/
Overwatch_AnimatedShot_Winston_Recall.mp4"
        self.assertEqual(videoSrc, video_url)
        # 使用 JavaScript 语句,通过播放器内部的
        # duration 属性获取视频文件的播放时长
        videoDuration = self.driver.execute_script\
            ("return arguments[0].duration;", videoPlayer)
        # 打印视频时长
        print(videoDuration)
        # 对获取到的视频时长取整,然后断言是否等于 3 秒
        self.assertEqual(int(videoDuration), 468)
        # 使用 JavaScript 语句,通过调用播放器内部的
        # play()方法来播放影片
        self.driver.find_element_by_xpath("//span[@class='sv-font sv-play']").click()
        time.sleep(2)
        # 播放 2 秒后,使用 JavaScript 语句,通过调用播放器
        # 内部的 pause 函数来暂停播放影片
        self.driver.execute_script("return arguments[0].pause();", videoPlayer)
        # 暂停 3 秒,以便人工确认视频是否已被暂停
        time.sleep(3)
        # 将暂停视频播放页面进行截屏,并保存为 D 盘的 videoPlay_pause.jpg 文件
        self.driver.save_screenshot("d:\\videoPlay_pause.jpg")

    def tearDown(self):
        # 退出浏览器
        self.driver.quit()

if __name__ == '__main__':
    unittest.main()
```

更多说明:

控制视频播放器的原理均需要使用 JavaScript 语句来调用视频播放器内部的属性和接口来完成我们想要做的操作。

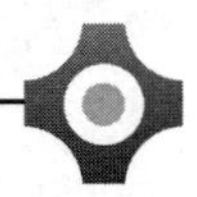

11.20　在 HTML5 的画布元素上进行绘画操作

目的：

能够在 HTML5 的画布上进行绘画操作。

用于测试的网址：

http://www.w3school.com.cn/tiy/loadtext.asp?f=html5_canvas_line

实例代码：

```
#encoding = utf-8
import unittest
from selenium import webdriver
import time

class TestDemo(unittest.TestCase):
    def setUp(self):
        # 获取浏览器驱动实例
        self.driver = webdriver.Firefox(executable_path = "c:\\geckodriver")

    def test_HTML5Canvas(self):
        url = "http://www.w3school.com.cn/tiy/loadtext.asp?f = html5_canvas_line"
        # 访问指定的网址
        self.driver.get(url)
        # 调用 JavaScript 语句,在页面画布上画一个红色的图案
        # getElementById('myCanvas'); 语句获取页面上的画布元素
        # var cxt = c.getContext('2d'); 设定画布为 2D
        # cxt.fillStyle = '#FF0000'; 设定填充色为 # FF0000 红色
        # cxt.fillRect(0, 0, 150, 150); 在画布上绘制矩形
        self.driver.execute_script("var c = document.getElementById('myCanvas');"
                + "var cxt = c.getContext('2d');"
                + "cxt.fillStyle = '#FF0000';"
                + "cxt.fillRect(0,0,150,150);")
        time.sleep(3)
        # 将绘制的红色矩形页面进行截屏,并保存为 D 盘的 HTML5Canvas.jpg
        self.driver.save_screenshot("d:\\HTML5Canvas.jpg")

    def tearDown(self):
        # 退出浏览器
        self.driver.quit()

if __name__ == '__main__':
    unittest.main()
```

11.21　操作 HTML5 存储对象

目的：

能够读取 HTML5 的 localStorage 和 sessionStorage 的内容，并删除存储的内容。

用于测试的网址：

localStorage：

http://www.w3school.com.cn/tiy/loadtext.asp?f=html5_webstorage_local

sessionStorage：

http://www.w3school.com.cn/tiy/loadtext.asp?f=html5_webstorage_session

实例代码：

```
#encoding = utf-8
from selenium import webdriver
import unittest, time

class Html5Storage(unittest.TestCase):

    def setUp(self):
        # 启动 Firefox 浏览器
        self.driver = webdriver.Firefox(executable_path = "c:\\geckodriver")

    def test_Html5localStorage(self):
        # 指定测试 localStorage 的网址
        localStorageUrl = \
            "http://www.w3school.com.cn/tiy/loadtext.asp?f = html5_webstorage_local"
        # 访问 localStorage 网址
        self.driver.get(localStorageUrl)
        time.sleep(2)
        # 通过 JavaScript 语句,获取存储在 localStorage 中的 lastname 的值
        lastName = self.driver.execute_script("return localStorage.lastname")
        print("lastName: ", lastName)
        # 断言获取的存储值是否为 Gates
        self.assertEquals("Gates", lastName)
        # 通过 JavaScript 语句"localStorage.clear();"
        # 清除所有存储在 localStorage 中的存储值
        self.driver.execute_script("localStorage.clear();")
        # 清除存储在 localStorage 中的存储值后再次查看 lastname 的值
        last_Name = self.driver.execute_script("return localStorage.lastname")
        # 断言获取的存储值是否为 None
        self.assertEquals(None, last_Name)

    def test_Html5SessionStorage(self):
        # 指定 sessionStorage 的网址
        sessionStorageUrl = \
          "http://www.w3school.com.cn/tiy/loadtext.asp?f = html5_webstorage_session"
        # 访问 sessionStorage 网址
        self.driver.get(sessionStorageUrl)
        time.sleep(2)
        # 单击页面上唯一的按钮,让单击次数计数增加一次
        self.driver.find_element_by_tag_name("button").click()
        # 通过 JavaScript 语句,获取存储在 sessionStorage 中的 clickCount 的值
        clickCount = self.driver.execute_script("return sessionStorage.clickcount")
        print("clickCount = ", clickCount)
```

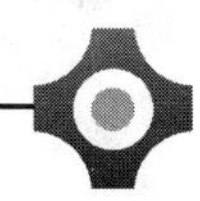

```
        # 断言获取的存储值的单击次数 clickCount 变量的值
        self.assertEquals(1, int(clickCount))
        # 通过 JavaScript 语句,清除存储在 sessionStorage 中的存储值
        self.driver.execute_script("sessionStorage.clear();")
        # 清除存储在 sessionStorage 中的存储值后再次查看 clickCount 的值
        click_count = self.driver.execute_script("return sessionStorage.clickcount")
        # 断言获取的存储值的单击次数 click_count 变量的值
        self.assertEquals(None, click_count)

    def tearDown(self):
        # 退出 Firefox 浏览器
        self.driver.quit()

if __name__ == '__main__':
    unittest.main()
```

11.22 使用 Chrome 浏览器自动将文件下载到指定路径

目的：

通过设定 Chrome 属性，实现脚本自动下载网页上的文件，并保存到指定路径下。

用于测试网址：

http://pypi.python.org/pypi/selenium

实例代码：

```
#encoding=utf-8
from selenium import webdriver
import unittest, time

class TestDemo(unittest.TestCase):

    def setUp(self):
        # 创建 Chrome 浏览器配置对象实例
        chromeOptions = webdriver.ChromeOptions()
        # 设定下载文件的保存目录为 C 盘的 iDownload 目录,
        # 如果该目录不存在,将会自动创建
        prefs = {"download.default_directory": "c:\\iDownload"}
        # 将自定义设置添加到 Chrome 配置对象实例中
        chromeOptions.add_experimental_option("prefs", prefs)
        # 启动带有自定义设置的 Chrome 浏览器
        self.driver = webdriver.Chrome(executable_path="c:\\chromedriver",\
                                       chrome_options=chromeOptions)

    def test_downloadFileByChrome(self):
        url = "http://pypi.python.org/pypi/selenium"
        # 访问将要下载文件的网址
        self.driver.get(url)
        # 找到要下载的文件链接页面元素,并单击进行下载
        self.driver.find_element_by_partial_link_text\
```

```
            ("selenium - 3.0.2.tar.gz").click()
        # 等待 100s,以便文件下载完成
        time.sleep(100)

    def tearDown(self):
        # 退出 Chrome 浏览器
        self.driver.quit()

if __name__ == '__main__':
    unittest.main()
```

11.23 使用 Firefox 浏览器自动下载文件到指定路径

目的：

在 Firefox 浏览器中单击下载一个文件时，通过配置 Firefox 浏览器设置项，来屏蔽弹出的询问你是直接打开该文件还是保存到指定文件的选择窗口，防止自动化实施过程中不能正确处理该弹出，导致自动化脚本执行错误或异常终止。

用于测试网址：

https://pypi.org/project/selenium/#files

实例代码：

```
#encoding = utf-8
from selenium import webdriver
import unittest, time

class TestDemo(unittest.TestCase):

    def setUp(self):
        # 创建 Firefox 浏览器配置对象实例
        profile = webdriver.FirefoxProfile()
        # 设置保证文件下载到指定文件设置选项
        profile.set_preference("browser.download.folderList", 2)
        # 指定文件下载的指定目录
        profile.set_preference("browser.download.dir", "D:\\driver")
        # 指定下载页面的 Content-type 值为文件类型(binary/octet-stream)
        profile.set_preference("browser.helperApps.neverAsk.saveToDisk",
                               "binary/octet-stream")
        # 启动带有自定义设置项的 Firefox 浏览器
        self.driver = webdriver.Firefox(executable_path = "D:\\driver\\geckodriver",
                                        firefox_profile = profile)

    def test_downloadFileByChrome(self):
        url = "https://pypi.org/project/selenium/#files"
        # 访问将要下载文件的网址
        self.driver.get(url)
        # 找到要下载的文件链接页面元素,并单击进行下载
        self.driver.find_element_by_partial_link_text\
```

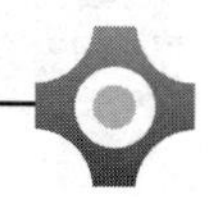

```
            (".tar.gz").click()
        # 等待 100s,以便文件下载完成
        time.sleep(100)

    def tearDown(self):
        # 退出浏览器
        self.driver.quit()

if __name__ == '__main__':
    unittest.main()
```

代码解释：

设置项 browser.download.folderList，表示设置文件下载后存放路径，0 表示下载到浏览器默认路径中，2 表示下载到指定路径中。

browser.helperApps.neverAsk.saveToDisk 设置项指定的是要下载文件的文件类型，如果不清楚当前要下载的文件的文件类型，可以打开浏览器，进入开发者工具——网络，然后手动单击要下载的文件，就会看到请求下载该文件时访问的消息体，在消息头中可以找到该文件的类型。

11.24 修改 Chrome 设置伪装成手机 M 站

目的：

通过更改 PC 端 Chrome 浏览器的属性值，将 PC 端 Chrome 浏览器设定为手机端尺寸的浏览器，以便模拟手机端的浏览器，并完成各种页面操作。

用于测试的网址：

http://www.baidu.com

实例代码：

```
#encoding = utf - 8
from selenium import webdriver
import unittest, time

class TestDemo(unittest.TestCase):

    def test_iPadChrome(self):
        options = webdriver.ChromeOptions()
        options.add_argument(
            '--user-agent=Mozilla/5.0 (iPad; CPU OS 5_0 like Mac OS X) \
            AppleWebKit/534.46 (KHTML, like Gecko) Version/5.1 \
            Mobile/9A334 Safari/7534.48.3')
        driver = webdriver.Chrome(executable_path="c:\\chromedriver",
                                  chrome_options=options)
        driver.get("http://www.baidu.com")
        # 暂停 3 秒,等待页面加载完成
        time.sleep(3)
        # 找到页面的搜索输入框,输入 iPad
        driver.find_element_by_id("kw").send_keys("iPad")
```

```
        # 等待 3 秒,人工查看效果
        time.sleep(1)
        # 通过在 Chrome 浏览器地址栏中输入 about:version,查看伪装效果
        driver.get("about:version")
        # 人工确认"用户代理"项配置信息是否跟设置一样
        time.sleep(10)
        driver.quit()

    def test_iPhoneChrome(self):
        options = webdriver.ChromeOptions()
        options.add_argument(
            '--user-agent=Mozilla/5.0 (iPhone; CPU iPhone OS 5_0 like Mac OS X) \
            AppleWebKit/534.46 (KHTML, like Gecko) Version/5.1 \
            Mobile/9A334 Safari/7534.48.3')
        driver = webdriver.Chrome(executable_path="c:\\chromedriver",
                                  chrome_options=options)
        driver.get("http://www.baidu.com")
        time.sleep(3)
        # 找到搜索输入框,输入 iPhone
        driver.find_element_by_id("kw").send_keys("iPhone")
        time.sleep(1)
        # 通过在 Chrome 浏览器地址栏中输入 about:version,查看伪装效果
        driver.get("about:version")
        # 人工确认"用户代理"项配置信息是否和设置一样
        time.sleep(10)
        driver.quit()

    def test_Android236Chrome(self):
        options = webdriver.ChromeOptions()
        options.add_argument(
            '--user-agent=Mozilla/5.0 (Linux; U; Android 2.3.6; en-us; \
            Nexus S Build/GRK39F) AppleWebKit/533.1 \
            (KHTML, like Gecko) Version/4.0 Mobile Safari/533.1')
        driver = webdriver.Chrome(executable_path="c:\\chromedriver",
                                  chrome_options=options)
        driver.get("http://www.baidu.com")
        time.sleep(3)
        # 找到搜索输入框,输入 Android 2.3.6
        driver.find_element_by_id("kw").send_keys("Android 2.3.6")
        time.sleep(1)
        # 通过在 Chrome 浏览器地址栏中输入 about:version,查看伪装效果
        driver.get("about:version")
        # 人工确认"用户代理"项配置信息是否和设置一样
        time.sleep(10)
        driver.quit()

    def test_Android402Chrome(self):
        options = webdriver.ChromeOptions()
        options.add_argument(
            '--user-agent=Mozilla/5.0 (Linux; U; Android 4.0.2; \
            en-us; Galaxy Nexus Build/ICL53F) AppleWebKit/534.30 \
```

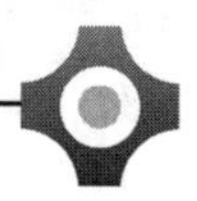

```
        (KHTML, like Gecko) Version/4.0 Mobile Safari/534.30')
    driver = webdriver.Chrome(executable_path = "c:\\chromedriver",
                            chrome_options = options)
    driver.get("http://www.baidu.com")
    time.sleep(3)
    # 找到搜索输入框,输入 Android 4.0.2
    driver.find_element_by_id("kw").send_keys("Android 4.0.2")
    time.sleep(1)
    # 通过在 Chrome 浏览器地址栏中输入 about:version,查看伪装效果
    driver.get("about:version")
    # 人工确认"用户代理"项配置信息是否和设置一样
    time.sleep(10)
    driver.quit()

if __name__ == '__main__':
    unittest.main()
```

代码解释：

通过--user-agent="xxx"来修改 HTTP 请求头部的 Agent 字符串，以便将 PC 端的 Chrome 浏览器伪装成手机浏览器。同时通过在 Chrome 地址栏中输入 about：version 来查看修改效果。

11.25 将 Firefox 浏览器伪装成手机 M 站

目的：

通过更改 PC 端 Chrome 浏览器的属性值，将 PC 端 Chrome 浏览器设定为手机端尺寸的浏览器，以便模拟手机端的浏览器，并完成各种页面操作。

用于测试的网址：

http://www.baidu.com

实例代码：

```
#encoding = utf - 8
from selenium import webdriver
import unittest, time

class TestDemo(unittest.TestCase):

    def test_iPadFirefox(self):
        profile = webdriver.FirefoxProfile()
        profile.set_preference("general.useragent.override",
            'Mozilla/5.0 (iPad; CPU OS 5_0 like Mac OS X) \
            AppleWebKit/534.46 (KHTML, like Gecko) Version/5.1 \
            Mobile/9A334 Safari/7534.48.3')
        driver = webdriver.Firefox(executable_path = "D:\\driver\\geckodriver",
                                firefox_profile = profile)
        driver.get("http://www.baidu.com")
        # 暂停 3 秒,等待页面加载完成
```

```
        time.sleep(3)
        # 找到页面的搜索输入框,输入 iPad
        driver.find_element_by_id("kw").send_keys("iPad")
        # 等待 3 秒,人工查看效果
        time.sleep(1)
        # 通过在 Firefox 浏览器地址栏中输入 about:support,查看伪装效果
        driver.get("about:support")
        # 人工确认"用户代理"项配置信息是否跟设置一样
        time.sleep(10)
        driver.quit()

    def test_Android404Firefox(self):
        profile = webdriver.FirefoxProfile()
        profile.set_preference("general.useragent.override",
            'Mozilla/5.0 (Linux; Android 4.0.4; \
            Galaxy Nexus Build/IMM76B)AppleWebKit/535.19 (KHTML, like Gecko) \
            Chrome/18.0.1025.133 MobileSafari/535.19')
        driver = webdriver.Firefox(executable_path = "D:\\driver\\geckodriver",
                                   firefox_profile = profile)
        driver.get("http://www.baidu.com")
        time.sleep(3)
        # 找到搜索输入框,输入 iPhone
        driver.find_element_by_id("index-kw").send_keys("Android4.0.4")
        time.sleep(1)
        # 通过在 Firefox 浏览器地址栏中输入 about:support,查看伪装效果
        driver.get("about:support")
        # 人工确认"用户代理"项配置信息是否和设置一样
        time.sleep(10)
        driver.quit()

    def test_IphoneFirefox(self):
        profile = webdriver.FirefoxProfile()
        profile.set_preference("general.useragent.override",
            'Mozilla/5.0 (iPhone; U; CPU iPhone OS 5_1_1 like Mac OS X; en) \
            AppleWebKit/534.46.0 (KHTML, like Gecko) \
            CriOS/19.0.1084.60 Mobile/9B206 Safari/7534.48.3')
        driver = webdriver.Firefox(executable_path = "D:\\driver\\geckodriver",
                                   firefox_profile = profile)
        driver.get("http://www.baidu.com")
        time.sleep(3)
        # 找到搜索输入框,输入 iPhone
        driver.find_element_by_id("index-kw").send_keys("iPhone")
        time.sleep(1)
        # 通过在 Firefox 浏览器地址栏中输入 about:suport,查看伪装效果
        driver.get("about:support")
        # 人工确认"用户代理"项配置信息是否和设置一样
        time.sleep(10)
        driver.quit()

    def test_Android402Firefox(self):
        profile = webdriver.FirefoxProfile()
```

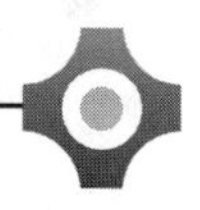

```
        profile.set_preference("general.useragent.override",
                               'Mozilla/5.0 (Linux; U; Android 4.0.2; \
            en-us; Galaxy Nexus Build/ICL53F) AppleWebKit/534.30 \
            (KHTML, like Gecko) Version/4.0 Mobile Safari/534.30')
        driver = webdriver.Firefox(executable_path = "D:\\driver\\geckodriver",
                                   firefox_profile = profile)
        driver.get("http://www.baidu.com")
        time.sleep(3)
        # 找到搜索输入框,输入 Android 4.0.2
        driver.find_element_by_id("index-kw").send_keys("Android 4.0.2")
        time.sleep(1)
        # 通过在 Firefox 浏览器地址栏中输入 about:support,查看伪装效果
        driver.get("about:support")
        # 人工确认"用户代理"项配置信息是否和设置一样
        time.sleep(10)
        driver.quit()

if __name__ == '__main__':
    unittest.main()
```

更多手机浏览器代理信息，请在相应手机查看相应浏览器的代理信息，比如在 Firefox 浏览器地址栏中访问 about：support 即可看到当前手机针对 Firefox 浏览器的代理信息。

11.26 屏蔽 Chrome 的--ignore-certificate-errors 提示及禁用扩展插件并实现窗口最大化

目的：

屏蔽 WebDriver 启动 Chrome 实例时总出现的--ignore-certificate-errors 提示信息，同时禁用 Chrome 浏览器的插件，并且让浏览器窗口最大化。

用于测试网址：

http://www.baidu.com

实例代码：

```
#encoding = utf-8
from selenium import webdriver
from selenium.webdriver.chrome.options import Options
import unittest, time

class TestDemo(unittest.TestCase):

    def setUp(self):
        # 创建 Chrome 浏览器的一个 Options 实例对象
        chrome_options = Options()
        # 向 Options 实例中添加禁用扩展插件的设置参数项
        chrome_options.add_argument("--disable-extensions")
        # 添加屏蔽--ignore-certificate-errors 提示信息的设置参数项
        chrome_options.add_experimental_option("excludeSwitches",
                    ["ignore-certificate-errors"])
```

```
        # 添加浏览器最大化的设置参数项,一启动就最大化
        chrome_options.add_argument('--start-maximized')
        # 启动带有自定义设置的 Chrome 浏览器
        self.driver = webdriver.Chrome(executable_path = "c:\\chromedriver",
                                       chrome_options = chrome_options)

    def test_extendedAttributesChrome(self):
        # 访问百度首页
        self.driver.get("http://www.baidu.com")
        # 暂停 3 秒,人工查看上面设置是否已生效
        time.sleep(3)
        # 找到页面的搜索输入框,输入"光荣之路自动化测试"
        self.driver.find_element_by_id("kw").send_keys("光荣之路自动化测试")
        time.sleep(2)

    def tearDown(self):
        # 退出 Chrome 浏览器
        self.driver.quit()

if __name__ == '__main__':
    unittest.main()
```

更多说明：

可以通过在 Chrome 浏览器地址栏中输入 chrome://extensions/后回车，查看已安装的扩展应用。--ignore-certificate-errors 提示只在 Chrome 部分版本中存在。

11.27 禁用 Chrome 浏览器的 PDF 和 Flash 插件

目的：

禁用 Chrome 浏览器的 PDF 和 Flash 插件，本实例包含了 11.23 小节的功能。

用于测试的网址：

http://www.iqiyi.com

实例代码

```
#encoding = utf-8
from selenium import webdriver
# 导入 Options 类
from selenium.webdriver.chrome.options import Options
import unittest, time

class TestDemo(unittest.TestCase):

    def setUp(self):
        # 创建 Chrome 浏览器的一个 Options 实例对象
        chrome_options = Options()
        # 设置 Chrome 浏览器禁用 PDF 和 Flash 插件
        profile = {"plugins.plugins_disabled":
            ['Chrome PDF Viewer', 'Adobe Flash Player']}
```

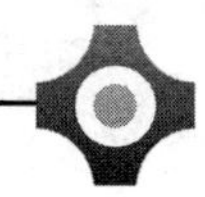

```
        chrome_options.add_experimental_option("prefs", profile)
        # 向 Options 实例中添加禁用扩展插件的设置参数项
        chrome_options.add_argument("--disable-extensions")
        # 添加屏蔽 --ignore-certificate-errors 提示信息的设置参数项
        chrome_options.add_experimental_option("excludeSwitches",
                                ["ignore-certificate-errors"])
        # 添加浏览器最大化的设置参数项,启动同时最大化窗口
        chrome_options.add_argument('--start-maximized')
        # 启动带有自定义设置的 Chrome 浏览器
        self.driver = webdriver.Chrome(executable_path = "c:\\chromedriver",
                                    chrome_options = chrome_options)

    def test_forbidPdfFlashChrome(self):
        # 访问爱奇艺首页
        self.driver.get("http://www.iqiyi.com")
        # 等待 50 秒,期间可以看到页面由于禁用了 Flash 插件,
        # 导致需要 Flash 支持的内容无法正常展示
        time.sleep(10)
        # 查看 PDF 和 Flash 插件禁用情况
        self.driver.get("chrome://plugins/")
        time.sleep(20)

    def tearDown(self):
        # 退出 Chrome 浏览器
        self.driver.quit()

if __name__ == '__main__':
    unittest.main()
```

更多说明：

本实例只提供了常用的禁用 PDF 和 Flash 插件项，如果读者有更多 Chrome 参数项需求，请自行上网搜索查看。

11.28 禁用 IE 的保护模式

目的：

关闭 IE 的保护模式，使自动化实施过程更方便。

实例代码：

```
#encoding = utf-8
from selenium import webdriver
from selenium.webdriver.common.desired_capabilities import DesiredCapabilities
import unittest, time

class TestDemo(unittest.TestCase):

    def setUp(self):
        caps = DesiredCapabilities.INTERNETEXPLORER
        # 将忽略 IE 保护模式的参数设置为 True
```

```
        caps['ignoreProtectedModeSettings'] = True
        # 启动带有自定义设置的 IE 浏览器
        self.driver = webdriver.Ie(executable_path = "c:\\IEDriverServer",
                                   capabilities = caps)

    def test_closeTheIEProtectedMode(self):
        # 访问百度首页
        self.driver.get("http://www.baidu.com")
        time.sleep(2)

    def tearDown(self):
        # 退出 IE 浏览器
        self.driver.quit()

if __name__ == '__main__':
    unittest.main()
```

11.29 禁用 Chrome 浏览器中的 Image 加载

目的：

禁用 Chrome 浏览器的图片加载，使访问的网页加载更迅速。

实例代码：

```
#encoding = utf-8
from selenium import webdriver
from selenium.webdriver.chrome.options import Options
import time,unittest

class TestDemo(unittest.TestCase):
    def setUp(self):
        # 创建 Chrome 浏览器的一个 Options 实例对象
        chrome_options = Options()
        # 设置 Chrome 禁用图片加载的设置
        prefs = {"profile.managed_default_content_settings.images": 2}

        # 添加屏蔽 Chrome 浏览器禁用图片的设置
        chrome_options.add_experimental_option("prefs", prefs)

        # 启动带有自定义设置的 Chrome 浏览器
        self.driver = webdriver.Chrome(executable_path = "c:\\chromedriver",
                     chrome_options = chrome_options)

    def test_forbidImageChrome(self):
        # 访问淘宝首页
        self.driver.get("https://www.taobao.com/")

        # 等待 50 秒，期间可以看到页面由于禁用了图片加载，
        # 导致图片均无法正常展示
        time.sleep(10)
        self.driver.quit()
```

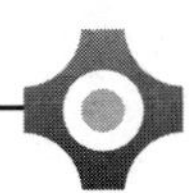

```
    def tearDown(self):
        # 退出 Chrome 浏览器
        self.driver.quit()

if __name__ == '__main__':
    unittest.main()
```

11.30　禁用 Firefox 浏览器中的 CSS、Flash 及 Image 加载

目的：

禁用 Firefox 浏览器的 CSS、Flash 及图片加载，使访问的网页加载更迅速。

实例代码：

```
#encoding = utf-8
from selenium import webdriver
import time, unittest

class TestDemo(unittest.TestCase):
    def setUp(self):
        # 创建 Firefox 浏览器的一个 Options 实例对象
        profile = webdriver.FirefoxProfile()

        # 禁用 CSS 加载
        profile.set_preference("permissions.default.stylesheet", 2)
        # 禁用 images 加载
        profile.set_preference("permissions.default.image", 2)
        # 禁用 Flash 插件
        profile.set_preference("dom.ipc.plugins.enabled.libflashplayer.so",
                               False)

        # 启动带有自定义设置的 Firefox 浏览器
        self.driver = webdriver.Firefox(executable_path = "c:\\geckodriver",
                          firefox_profile = profile)

    def test_forbidImageChrome(self):
        # 访问爱奇艺首页
        self.driver.get("http://www.iqiyi.com")

        # 等待 50 秒,期间可以看到页面由于禁用了 Flash 插件、图片及 CSS 加载,
        # 导致页面无法正常展示
        time.sleep(10)
        self.driver.quit()

    def tearDown(self):
        # 退出 Firefox 浏览器
        self.driver.quit()

if __name__ == '__main__':
    unittest.main()
```

第三篇 自动化测试框架搭建篇

第12章 数据驱动测试

数据驱动测试是自动化测试中的主流设计模式之一，属于中级自动化测试工程师必备知识，必须深入掌握数据驱动测试的工作原理和实现方法。

12.1 什么是数据驱动

相同的测试脚本使用不同的测试数据来执行，测试数据和测试行为完全分离，这样的测试脚本设计模式称为数据驱动。例如，测试网站的登录功能，自动化测试工程师想验证不同的用户名和密码在网站登录时对系统影响结果，就可以使用数据驱动模式来进行自动化测试。

实施数据驱动测试步骤如下：

(1) 编写测试脚本，脚本需要支持从程序对象、文件或数据库读入测试数据。

(2) 将测试脚本使用的测试数据存入程序对象、文件或数据库等外部介质中。

(3) 运行脚本过程中，循环调用存储在外部介质中的测试数据。

(4) 验证所有的测试结果是否符合预期结果。

12.2 数据驱动单元测试的环境准备

本书中 Python 数据驱动单元测试是将 unittest 和 ddt 模块结合起来实现的。下面介绍一下 ddt 模块的安装。

(1) 单击"开始"→"搜索程序和文件"框中输入 CMD 并回车，然后在打开的 CMD 窗口中输入 pip install ddt 并回车，进行 ddt 模块的安装。

(2) 等安装进度条走完，在 CMD 下输入 python 进入 python 交互模式，执行 import ddt，如果未报错，说明 ddt 模块已安装成功。

如果使用 pip 不能成功安装的读者，可以直接访问 https://pypi.python.org/pypi/ddt 下载 ddt 的源码包进行安装，如图 12-1 所示。

解压下载好的压缩包到某个目录，然后将 CMD 当前工作目录切换到 setup.py 文件所在目录，并执行 python setup.py install 进行 ddt 模块的安装，如图 12-2 所示。

等待安装完成，验证 ddt 模块是否安装成功，方法请参阅上面 pip 安装时的检测方法。

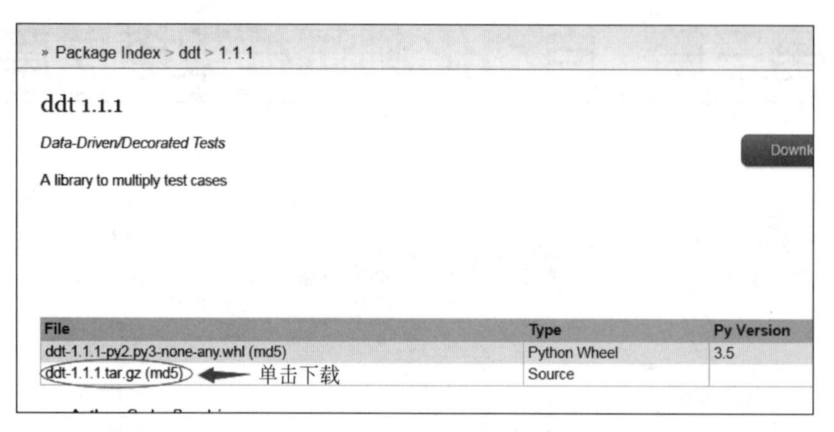

图 12-1

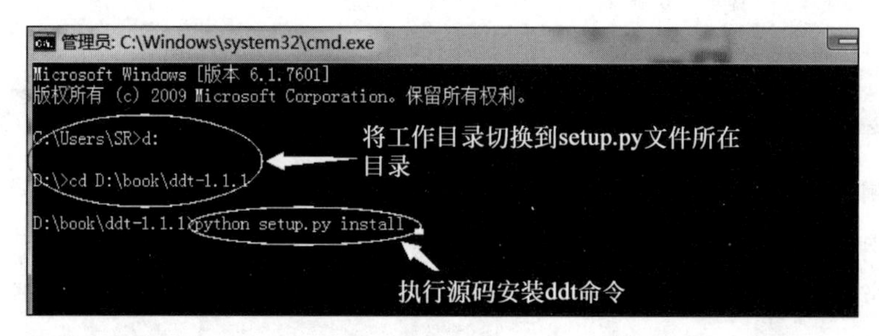

图 12-2

12.3 使用 unittest 和 ddt 进行数据驱动

测试逻辑：

(1) 打开百度首页。

(2) 在搜索框输入一个搜索关键词。

(3) 单击“搜索”按钮。

(4) 验证搜索结果页面是否包含预期关键字串，包含则认为测试执行通过，否则认为测试执行失败，并在测试过程中打印日志。

实例代码：

```
# encoding = utf - 8
from selenium import webdriver
import unittest, time
import logging, traceback
import ddt
from selenium.common.exceptions import NoSuchElementException

# 初始化日志对象
logging.basicConfig(
    # 日志级别
    level = logging.INFO,
```

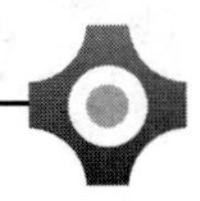

```
    # 日志格式
    # 时间、代码所在文件名、代码行号、日志级别名字、日志信息
    format = '%(asctime)s %(filename)s[line: %(lineno)d] %(levelname)s %(message)s',
    # 打印日志的时间
    datefmt = '%a, %d %b %Y %H: %M: %S',
    # 日志文件存放的目录(目录必须存在)及日志文件名
    filename = 'd:/DataDrivenTesting/report.log',
    # 打开日志文件的方式
    filemode = 'w'
)

@ddt.ddt
class TestDemo(unittest.TestCase):
    def setUp(self):
        self.driver = webdriver.Firefox(executable_path = "c:\\geckodriver")

    @ddt.data(["神奇动物在哪里", "叶茨"],
              ["疯狂动物城", "古德温"],
              ["大话西游之月光宝盒", "周星驰"])
    @ddt.unpack
    def test_dataDrivenByObj(self, testdata, expectdata):
        url = "http://www.baidu.com"
        # 访问百度首页
        self.driver.get(url)
        # 设置隐式等待时间为 10 秒
        self.driver.implicitly_wait(10)
        try:
            # 找到搜索输入框,并输入测试数据
            self.driver.find_element_by_id("kw").send_keys(testdata)
            # 找到搜索按钮,并单击
            self.driver.find_element_by_id("su").click()
            time.sleep(3)
            # 断言期望结果是否出现在页面源代码中
            self.assertTrue(expectdata in self.driver.page_source)
        except NoSuchElementException as err:
            logging.error("查找的页面元素不存在,异常堆栈信息:"
                          + str(traceback.format_exc()))
        except AssertionError as err:
            logging.info("搜索" %s",期望" %s",失败" %(testdata, expectdata))
        except Exception as err:
            logging.error("未知错误,错误信息:" + str(traceback.format_exc()))
        else:
            logging.info("搜索" %s",期望" %s"通过" %(testdata, expectdata))

    def tearDown(self):
        self.driver.quit()

if __name__ == '__main__':
    unittest.main()
```

执行结束后打印的日志文件内容：

```
Tue, 06 Dec 2016 16:59:56 dataDriver.py[line:54] INFO 搜索"神奇动物在哪里",期望"叶茨"通过
Tue, 06 Dec 2016 17:00:09 dataDriver.py[line:54] INFO 搜索"疯狂动物城",期望"古德温"通过
Tue, 06 Dec 2016 17:00:24 dataDriver.py[line:54] INFO 搜索"大话西游之月光宝盒",期望"周星
驰"通过
```

代码解释：

在 unittest 中结合 ddt 实现数据驱动，首先是在头部导入 ddt 模块(import ddt)，其次在测试类 TestDemo 前声明使用 ddt(@ddt. ddt)，然后在测试方法前使用@ddt. data()添加该测试方法需要的测试数据，@ddt. data 接收一个可迭代的类型，以此来判断需要执行的次数。多组测试数据间以逗号隔开(比如@ddt. ddt(3,1,5,6))。如果每组测试数据存在多个测试数据，需要将每组数据存于列表中，比如本节实例中的@ddt. data(["神奇动物在哪里", "叶茨"],["疯狂动物城", "古德温"],["大话西游之月光宝盒", "周星驰"])，表示存在三组数据，每组数据中的数据与测试方法中定义的形参个数及顺序一一对应。最后使用@unpack 进行修饰，也就是在测试方法被调用过程中，对测试数据进行解包，将每组测试数据中的第一个数据传给 testdata 形参，将每组测试数据中的第二个测试数据传给 expectdata 形参。

由此在脚本执行过程中，会根据提供的三组测试数据三次打开浏览器，分别输入三个不同的搜索词进行查询，并在三次搜索结果中断言是否出现期望的结果数据。

12.4 使用数据文件进行数据驱动

测试逻辑：

(1) 打开百度首页。

(2) 从扩展名为 json 的文件中读出测试相关数据，并将要查询的数据输入搜索框中。

(3) 单击“搜索”按钮。

(4) 断言搜索结果页面是否包含文本文件中提供的预期关键字串，包含则认为测试执行通过，否则认为测试执行失败，并生成自定义的 HTML 测试报告。

测试数据文件准备：

Pycharm 中新建一个名叫 DataDrivenProject 的 Python 工程，在该工程下新建 test_data_list. json、ReportTemplate. py 以及 DataDrivenTest. py 三文件。

文件 test_data_list. json 用于存放测试所需要的测试数据，具体内容如下：

```
[
    "邓肯||蒂姆",
    "乔丹||迈克尔",
    "库里||斯蒂芬",
    "杜兰特||凯文",
    "詹姆斯||勒布朗"
]
```

查询关键词和期望出现的关键词用“||”分隔，多组数据，以逗号隔开存于列表中。如果直接用文本编辑器创建该文件，需要将文件保存为 utf-8 编码的格式，否则中文会出现乱码。

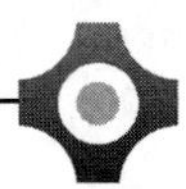

实例代码：

ReportTemplate.py 文件，用于生成自定义 HTML 测试报告，具体内容如下：

```
# encoding = utf - 8
def htmlTemplate(trData):
    htmlStr = '''<! DOCTYPE HTML >
    < html >
    < head >
    < title >单元测试报告</title >
    < style >
    body {
        width: 80 % ;                          /* 整个 body 区域占浏览器的宽度百分比 */
        margin: 40px auto;           /* 整个 body 区域相对浏览器窗口摆放位置(左右,上下) */
        font - weight: bold;                   /* 整个 body 区域的字体加粗 */
        font - family: 'trebuchet MS', 'Lucida sans', SimSun;     /* 表格中文字的字体类型 */
        font - size: 18px;                     /* 表格中文字字体大小 */
        color: #000;                           /* 整个 body 区域字体的颜色 */
    }
    table {
        * border - collapse: collapse;         /* 合并表格边框 */
        border - spacing: 0;                   /* 表格的边框宽度 */
        width: 100 % ;                         /* 整个表格相对父元素的宽度 */
    }
    .tableStyle {
        /* border: solid #ggg 1px; */
        border - style: outset;                /* 整个表格外边框样式 */
        border - width: 2px;                   /* 整个表格外边框宽度 */
        /* border: 2px; */
        border - color: blue;                  /* 整个表格外边框颜色 */
    }
    .tableStyle tr:hover {
        background: rgb(173,216,230);      /* 鼠标滑过一行时,动态显示的颜色 146,208,80 */
    }

    .tableStyle td,.tableStyle th {
        border - left: solid 1px rgb(146,208,80); /* 表格的竖线颜色 */
        border - top: 1px solid rgb(146,208,80); /* 表格的横线颜色 */
        padding: 15px;                         /* 表格内边框尺寸 */
        text - align: center;                  /* 表格内容显示位置 */
    }
    .tableStyle th {
        padding: 15px;                         /* 表格标题栏,字体的尺寸 */
        background - color: rgb(146,208,80);   /* 表格标题栏背景颜色 */
        /* 表格标题栏设置渐变颜色 */
         background - image: - webkit - gradient (linear, left top, left bottom, from (#
92D050), to(#A2D668));
        /* rgb(146,208,80) */
    }
    </style >
    </head >
```

```
    <body>
        <center><h1>测试报告</h1></center><br />
        <table class = "tableStyle">
            <thead>
            <tr>
            <th> Search Words </th>
            <th> Assert Words </th>
            <th> Start Time </th>
            <th> Waste Time(s)</th>
            <th> Status </th>
            </tr>
            </thead>'''
    endStr = u'''
        </table>
    </body>
    </html>'''
    # 拼接完整的测试报告 HTML 页面代码
    html = htmlStr + trData + endStr
    print(html)
    # 生成.html 文件
    with open(r"D:\book\testTemplate.html", "w") as fp:
        fp.write(html)
```

DataDrivenTest.py 文件,用于编写数据驱动测试脚本,具体内容如下:

```
# encoding = utf - 8
from selenium import webdriver
import unittest, time
import logging, traceback
import ddt
from ReportTemplate import htmlTemplate
from selenium.common.exceptions import NoSuchElementException

# 初始化日志对象
logging.basicConfig(
    # 日志级别
    level = logging.INFO,
    # 日志格式
    # 时间、代码所在文件名、代码行号、日志级别名字、日志信息
    format = '%(asctime)s %(filename)s[line: %(lineno)d] %(levelname)s %(message)s',
    # 打印日志的时间
    datefmt = '%a, %Y-%m-%d %H:%M:%S',
    # 日志文件存放的目录(目录必须存在)及日志文件名
    filename = 'd:/DataDrivenTesting/report.log',
    # 打开日志文件的方式
    filemode = 'w'
)

@ddt.ddt
class TestDemo(unittest.TestCase):
```

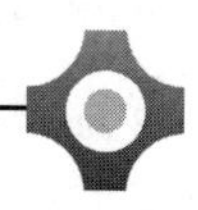

```
@classmethod
def setUpClass(cls):
    # 整个测试过程只被调用一次
    TestDemo.trStr = ""

def setUp(self):
    self.driver = webdriver.Firefox(executable_path = "c:\\geckodriver")
    status = None                    # 用于存放测试结果状态,失败为 fail,成功为 pass
    flag = 0 # 数据驱动测试结果的标志,失败置 0,成功置 1

@ddt.file_data("test_data_list.json")
def test_dataDrivenByFile(self, value):
    # 决定测试报告中状态单元格中内容的颜色
    flagDict = {0: 'red', 1: '#00AC4E'}

    url = "http://www.baidu.com"
    # 访问百度首页
    self.driver.get(url)
    print(value)
    # 将从.json 文件中读取出的数据用"||"分隔成测试数据
    # 和期望数据
    testdata, expectdata = tuple(value.strip().split("||"))
    # 设置隐式等待时间为 10 秒
    self.driver.implicitly_wait(10)
    # 获取当前的时间戳,用于后面计算查询耗时用
    start = time.time()
      # 获取当前时间的字符串,表示测试开始时间
      startTime = time.strftime("%Y-%m-%d %H:%M:%S", time.localtime())

    try:
        # 找到搜索输入框,并输入测试数据
        self.driver.find_element_by_id("kw").send_keys(testdata)
        # 找到搜索按钮,并单击
        self.driver.find_element_by_id("su").click()
        time.sleep(3)
        # 断言期望结果是否出现在页面源代码中
        self.assertTrue(expectdata in self.driver.page_source)
    except NoSuchElementException as err:
        logging.error("查找的页面元素不存在,异常堆栈信息:"
                      + str(traceback.format_exc()))
        status = 'fail'
        flag = 0
    except AssertionError as err:
        logging.info("搜索"%s",期望"%s",失败" % (testdata, expectdata))
        status = 'fail'
        flag = 0
    except Exception as err:
        logging.error("未知错误,错误信息:" + str(traceback.format_exc()))
        status = 'fail'
        flag = 0
    else:
```

```
                logging.info("搜索" % s",期望" % s"通过" % (testdata, expectdata))
                status = 'pass'
                flag = 1
            # 计算耗时,从将测试数据输入到输入框中到断言期望结果之间所耗时
            # wasteTime = time.time() - start - 3 减去强制等待的 3 秒
            # 每一组数据测试结束后,都将其测试结果信息插入表格行
            # 的 HTML 代码中,并将这些行 HTML 代码拼接到变量 trStr 变量中,
            # 等所有测试数据都被测试结束后,传入 htmlTemplate()函数中
            # 生成完整测试报告的 HTML 代码
            TestDemo.trStr += '''
            <tr>
                <td>%s</td>
                <td>%s</td>
                <td>%s</td>
                <td>%.2f</td>
                <td style="color:%s">%s</td>
            </tr><br />''' % (testdata, expectdata,startTime,
                                wasteTime, flagDict[flag], status)

    def tearDown(self):
        self.driver.quit()

    @classmethod
    def tearDownClass(cls):
        # 写自定义的 HTML 测试报告
        # 整个测试过程只被调用一次
        htmlTemplate(TestDemo.trStr)

if __name__ == '__main__':
    unittest.main()
```

执行结束后打印的日志文件内容:

```
Wed, 2016-12-07 10:07:51 dataDriver.py[line:53] INFO 搜索"邓肯",期望"蒂姆"通过
Wed, 2016-12-07 10:08:06 dataDriver.py[line:53] INFO 搜索"乔丹",期望"迈克尔"通过
Wed, 2016-12-07 10:08:20 dataDriver.py[line:53] INFO 搜索"库里",期望"斯蒂芬"通过
Wed, 2016-12-07 10:08:34 dataDriver.py[line:53] INFO 搜索"杜兰特",期望"凯文"通过
Wed, 2016-12-07 10:08:48 dataDriver.py[line:53] INFO 搜索"詹姆斯",期望"勒布朗"通过
```

生成的 HTML 测试报告内容如图 12-3 所示。

测试报告

Search Words	Assert Words	Start Time	Waste Time(s)	Status
邓肯	蒂姆	2016-12-07 11:34:45	0.47	pass
乔丹	迈克尔	2016-12-07 11:34:59	0.87	pass
库里	斯蒂芬	2016-12-07 11:35:15	0.40	pass
杜兰特	凯文	2016-12-07 11:35:29	0.42	pass
詹姆斯	勒布朗	2016-12-07 11:35:44	1.02	pass

图 12-3

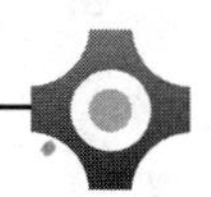

代码解释：

在测试方法 test_dataDrivenByFile 上使用@ddt.file_data()装饰该测试方法，目的是为测试方法提供存放在文件 test_data_list.json 里的测试数据，文件里的数据通过列表包裹，而列表中每一个元素代表一组测试数据，比如“邓肯||蒂姆”属于第一组测试数据，也就是程序第一次调用 test_dataDrivenByFile 测试方法时传入的数据。测试脚本依次从 test_data_list.json 文件中的列表中取出需要测试的数据，并记录每次测试开始时间、耗时以及测试结果，最后将测试结果信息组装成 HTML 代码进行展示，如图 12-3 所示。

更多说明：

存放测试数据的文件也可以是.txt 等类型的文件，在文件中除了将测试数据包裹在列表中，还可以使用字典对象进行包装，详细使用方法请参看 http://ddt.readthedocs.io/en/latest/example.html。

备选方法：

```
# encoding = utf - 8
from selenium import webdriver
import unittest, time
import logging, traceback
import ddt
import HTMLTestRunner
from selenium.common.exceptions import NoSuchElementException

# 初始化日志对象
logging.basicConfig(
    # 日志级别
    level = logging.INFO,
    # 日志格式
    # 时间、代码所在文件名、代码行号、日志级别名字、日志信息
    format = '%(asctime)s %(filename)s[line: %(lineno)d] %(levelname)s %(message)s',
    # 打印日志的时间
    datefmt = '%a, %Y-%m-%d %H: %M: %S',
    # 日志文件存放的目录(目录必须存在)及日志文件名
    filename = 'd:/DataDrivenTesting/report.log',
    # 打开日志文件的方式
    filemode = 'w'
)

@ddt.ddt
class TestDemo(unittest.TestCase):
    def setUp(self):
        self.driver = webdriver.Firefox(executable_path = "c:\\geckodriver")

    @ddt.file_data("test_data_list.json")
    def test_dataPickerByRightKey(self, value):
        url = "http://www.baidu.com"
        # 访问百度首页
        self.driver.get(url)
        print(value)
```

```
        # 将从.json文件中读取出的数据用"||"进行分隔成测试数据
        # 和期望数据
        testdata, expectdata = tuple(value.strip().split("||"))
        # 设置隐式等待时间为10秒
        self.driver.implicitly_wait(10)
        try:
            # 找到搜索输入框,并输入测试数据
            self.driver.find_element_by_id("kw").send_keys(testdata)
            # 找到搜索按钮,并单击
            self.driver.find_element_by_id("su").click()
            time.sleep(3)
            # 断言期望结果是否出现在页面源代码中
            self.assertTrue(expectdata in self.driver.page_source)
        except NoSuchElementException as err:
            logging.error("查找的页面元素不存在,异常堆栈信息:"
                          + str(traceback.format_exc()))
        except AssertionError as err:
            logging.info("搜索"%s",期望"%s",失败" %(testdata, expectdata))
        except Exception as err:
            logging.error("未知错误,错误信息:" + str(traceback.format_exc()))
        else:
            logging.info("搜索"%s",期望"%s"通过" %(testdata, expectdata))

    def tearDown(self):
        self.driver.quit()

if __name__ == '__main__':
    # unittest.main()
    suite1 = unittest.TestLoader().loadTestsFromTestCase(TestDemo)
    suite = unittest.TestSuite(suite1)
    filename = "d:\\test.html"
    fp = open(filename, 'wb')
    runner = HTMLTestRunner.HTMLTestRunner(stream=fp, title='Report_title', description=
'Report_description')
    runner.run(suite)                        # 运行测试集合
```

执行结束后打印的日志文件内容：

```
Wed, 2016-12-07 10:07:51 dataDriver.py[line:53] INFO 搜索"邓肯",期望"蒂姆"通过
Wed, 2016-12-07 10:08:06 dataDriver.py[line:53] INFO 搜索"乔丹",期望"迈克尔"通过
Wed, 2016-12-07 10:08:20 dataDriver.py[line:53] INFO 搜索"库里",期望"斯蒂芬"通过
Wed, 2016-12-07 10:08:34 dataDriver.py[line:53] INFO 搜索"杜兰特",期望"凯文"通过
Wed, 2016-12-07 10:08:48 dataDriver.py[line:53] INFO 搜索"詹姆斯",期望"勒布朗"通过
```

代码解释：

在测试方法test_dataDrivenByFile上使用@ddt.file_data()装饰该测试方法，目的是为测试方法提供存放在文件test_data_list.json里的测试数据，文件里的数据通过列表包裹，而列表中每一个元素代表一组测试数据，比如“邓肯||蒂姆”属于第一组测试数据，也就是程序第一次调用test_dataDrivenByFile测试方法时传入的数据。测试脚本依次从test_data_list.json文件的列表取出需要测试的数据，测试结束后使用HTMLTestRunner模块提供

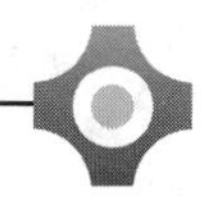

的 HTML 报告模板生成测试报告。

12.5 使用 Excel 进行数据驱动测试

测试逻辑：

（1）打开百度首页，从 Excel 文件中读取测试数据作为搜索关键词。

（2）在搜索输入框中输入读取出的搜索关键词。

（3）单击“搜索”按钮。

（4）断言搜索结果页面中是否出现 Excel 文件中提供的预期内容，包含则认为测试执行成功，否则认为失败。

环境准备：

（1）按下 Win＋R 组合键，再调出的运行窗口输入框中输入 CMD 并按下回车键，以便调出 Windows 的 CMD 窗口。

（2）在弹出的 CMD 窗口中输入 pip install openpyxl＝＝2.5.10，安装 Python 解析 Excel 2007 及以上版本的模块。

（3）如果上述方法安装失败，读者可以直接访问 https://pypi.python.org/pypi/openpyxl 下载 openpyxl 源码安装包，后续安装方法及检验安装成功与否，请参看本章的 12.2 小节。

本书中有关 Excel 操作的实例，笔者都是在 openpyxl(2.5.10)版本上开发的，为防止高版本 openpyxl 的 API 不兼容低版本的 API，导致本书中有关 Excel 解析的实例代码不能成功执行，这里需要使用 pip install openpyxl＝＝2.5.10 命令指定版本安装。

测试数据准备：

在本地磁盘 D:\DataDrivenTesting 目录中新建“测试数据.xlsx”，工作表名为“搜索数据表”的 Excel 文件，“搜索数据表”工作表内容如表 12-1 所示。

表　12-1

序　　号	搜　索　词	期望结果
1	邓肯	蒂姆
2	乔丹	迈克尔
3	库里	斯蒂芬

实例代码：

在 Pycharm 中新建一名叫 ExcelDataDrivenProject 的 Python 工程，在工程下新建两个文件，文件名分别为 ExcelUtil.py 和 DataDriven.py。

ExcelUtil.py 文件用于编写读取 Excel 的脚本，具体内容如下：

```
#encoding = utf - 8
from openpyxl import load_workbook
```

```
class ParseExcel(object):

    def __init__(self, excelPath, sheetName):
        # 将要读取的 Excel 工作表加载到内存
        self.wb = load_workbook(excelPath)
        # 通过工作表名称获取该工作表对象
        self.sheet = self.wb.get_sheet_by_name(sheetName)
        # 获取工作表中存在数据的区域的最大行号
        self.maxRowNum = self.sheet.max_row

    def getDatasFromSheet(self):
        # 用于存放从工作表中读取出来的数据
        dataList = []
        # 获取表中所有行数据,生成器类型数据
        rows_data = list(self.sheet.rows)
        # 因为工作表中的第一行是标题行,所以需要去掉
        for line in rows_data[1:]:
            # 遍历工作表中数据区域的每一行,
            # 并将每行中各个单元格的数据取出存于列表 tmpList 中,
            # 然后再将存放一行数据的列表添加到最终数据列表 dataList 中
            tmpList = []
            tmpList.append(line[1].value)
            tmpList.append(line[2].value)
            dataList.append(tmpList)
        # 将获取工作表中的所有数据的迭代对象返回
        return dataList

if __name__ == '__main__':
    excelPath = 'D:\\DataDrivenTesting\\测试数据.xlsx'
    sheetName = "搜索数据表"
    pe = ParseExcel(excelPath, sheetName)
    for i in pe.getDatasFromSheet():
        print(i[0], i[1])
```

DataDriven.py 文件用于编写数据驱动测试脚本代码,具体内容如下:

```
# encoding = utf-8
from selenium import webdriver
import unittest, time
import logging, traceback
import ddt
from ExcelUtil import ParseExcel
from selenium.common.exceptions import NoSuchElementException

# 初始化日志对象
logging.basicConfig(
    # 日志级别
    level = logging.INFO,
    # 日志格式
    # 时间、代码所在文件名、代码行号、日志级别名字、日志信息
    format = '%(asctime)s %(filename)s[line: %(lineno)d] %(levelname)s %(message)s',
```

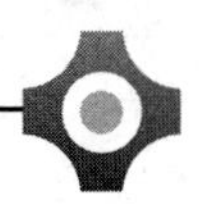

```
    # 打印日志的时间
    datefmt = '%a, %Y-%m-%d %H:%M:%S',
    # 日志文件存放的目录(目录必须存在)及日志文件名
    filename = 'd:/DataDrivenTesting/dataDriveRreport.log',
    # 打开日志文件的方式
    filemode = 'w'
)

excelPath = 'D:\\DataDrivenTesting\\测试数据.xlsx'
sheetName = "搜索数据表"
# 创建 ParseExcel 类的实例对象
excel = ParseExcel(excelPath, sheetName)

@ddt.ddt
class TestDemo(unittest.TestCase):

    def setUp(self):
        self.driver = webdriver.Firefox(executable_path="c:\\geckodriver")

    @ddt.data(*excel.getDatasFromSheet())
    def test_dataDrivenByFile(self, data):
        testData, expectData = tuple(data)
        url = "http://www.baidu.com"
        # 访问百度首页
        self.driver.get(url)
        # 将浏览器窗口最大化
        self.driver.maximize_window()
        print(testData, expectData)
        # 设置隐式等待时间为 10 秒
        self.driver.implicitly_wait(10)

        try:
            # 找到搜索输入框,并输入测试数据
            self.driver.find_element_by_id("kw").send_keys(testData)
            # 找到搜索按钮,并单击
            self.driver.find_element_by_id("su").click()
            time.sleep(3)
            # 断言期望结果是否出现在页面源代码中
            self.assertTrue(expectData in self.driver.page_source)
        except NoSuchElementException as err:
            logging.error("查找的页面元素不存在,异常堆栈信息:"
                          + str(traceback.format_exc()))
        except AssertionError as err:
            logging.info("搜索"%s",期望"%s",失败" %(testData, expectData))
        except Exception as err:
            logging.error("未知错误,错误信息:" + str(traceback.format_exc()))
        else:
            logging.info("搜索"%s",期望"%s"通过" %(testData, expectData))

    def tearDown(self):
        self.driver.quit()
```

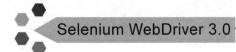

```
if __name__ == '__main__':
    unittest.main()
```

执行结束后日志文件内容如下：

```
Thu, 2016-12-08 11:02:59 dataDriver.py[line:67] INFO 搜索"邓肯",期望"蒂姆"通过
Thu, 2016-12-08 11:03:12 dataDriver.py[line:67] INFO 搜索"乔丹",期望"迈克尔"通过
Thu, 2016-12-08 11:03:27 dataDriver.py[line:67] INFO 搜索"库里",期望"斯蒂芬"通过
```

代码解释：

@ddt.data 从 excel.getDatasFromSheet()方法中接收一个可迭代的数组对象，以此来判断需要执行的次数。如果@ddt.data()括号中传的是一个方法，方法前需要加星号(*)修饰。

12.6 使用 XML 进行数据驱动测试

测试逻辑：

(1) 打开百度首页，从 XML 文件中读取测试数据作为搜索关键词。

(2) 在搜索输入框中输入读取出的搜索关键词。

(3) 单击“搜索”按钮。

(4) 断言搜索结果页面中是否出现 XML 文件中提供的预期内容，包含则认为测试执行成功，否则认为失败。

实例代码：

在 Pycharm 中新建 XMLDataDrivenProject 的 Python 工程，并在工程下新建三个文件，文件名分别为 XmlUtil.py、TestData.xml 以及 DataDrivenByXML.py。

TestData.xml 文件用于存放测试数据，具体内容如下：

```
<?xml version="1.0" encoding="utf-8"?>
<bookList type="technology">
    <book>
        <name>Selenium WebDriver 实战宝典</name>
        <author>吴晓华</author>
    </book>
    <book>
        <name>HTTP 权威指南</name>
        <author>古尔利</author>
    </book>
    <book>
        <name>探索式软件测试</name>
        <author>惠特克</author>
    </book>
</bookList>
```

XmlUtil.py 文件用于解析 XML 文件，获取测试数据，具体内容如下：

```
#encoding=utf-8
```

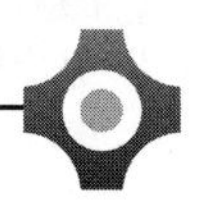

```
from xml.etree import ElementTree

class ParseXML(object):
    def __init__(self, xmlPath):
        self.xmlPath = xmlPath

    def getRoot(self):
        # 打开将要解析的 XML 文件
        tree = ElementTree.parse(self.xmlPath)
        # 获取 XML 文件的根节点对象,也就是树的根
        # 然后返回给调用者
        return tree.getroot()

    def findNodeByName(self, parentNode, nodeName):
        # 通过节点的名字,获取节点对象
        nodes = parentNode.findall(nodeName)
        return nodes

    def getNodeOfChildText(self, node):
        # 获取节点 node 下所有子节点的节点名作为 key,
        # 文本节点作为 value 组成的字典对象
        childrenTextDict = {i.tag: i.text for i in list(node.iter())[1:]}
        # 上面代码等价于下面代码
        '''
        childrenTextDict = {}
        for i in list(node.iter())[1:]:
            childrenTextDict[i.tag] = i.text
        '''
        return childrenTextDict

    def getDataFromXml(self):
        # 获取 XML 文档树的根节点对象
        root = self.getRoot()
        # 获取根节点下所有名叫 book 的节点对象
        books = self.findNodeByName(root, "book")
        dataList = []
        # 遍历获取到的所有 book 节点对象,
        # 取得需要的测试数据
        for book in books:
            childrenText = self.getNodeOfChildText(book)
            dataList.append(childrenText)
        return dataList

if __name__ == '__main__':
    xml = ParseXML(r"D:\PythonProject\Calc\TestData.xml")
    datas = xml.getDataFromXml()
    for i in datas:
        print(i["name"], i["author"])
```

DataDrivenByXML.py 文件用于编写数据驱动测试脚本,具体内容如下:

```
# encoding = utf - 8
from selenium import webdriver
import unittest, time, os
import logging, traceback
import ddt
from XmlUtil import ParseXML
from selenium.common.exceptions import NoSuchElementException

# 初始化日志对象
logging.basicConfig(
    # 日志级别
    level = logging.INFO,
    # 日志格式
    # 时间、代码所在文件名、代码行号、日志级别名字、日志信息
    format = '%(asctime)s %(filename)s[line: %(lineno)d] %(levelname)s %(message)s',
    # 打印日志的时间
    datefmt = '%a, %Y-%m-%d %H:%M:%S',
    # 日志文件存放的目录(目录必须存在)及日志文件名
    filename = 'd:/DataDrivenTesting/dataDriveRreport.log',
    # 打开日志文件的方式
    filemode = 'w'
)
# 获取当前文件所在父目录的绝对路径
currentPath = os.path.dirname(os.path.abspath(__file__))
# 获取数据文件的绝对路径
dataFilePath = os.path.join(currentPath, "TestData.xml")

# 创建 ParseXML 类实例对象
xml = ParseXML(dataFilePath)

@ddt.ddt
class TestDemo(unittest.TestCase):

    def setUp(self):
        self.driver = webdriver.Firefox(executable_path = "c:\\geckodriver")

    @ddt.data( * xml.getDataFromXml())
    def test_dataDrivenByXML(self, data):
        testData, expectData = data["name"], data["author"]
        url = "http://www.baidu.com"
        # 访问百度首页
        self.driver.get(url)
        # 将浏览器窗口最大化
        self.driver.maximize_window()
        print(testData, expectData)
        # 设置隐式等待时间为 10 秒
        self.driver.implicitly_wait(10)

        try:
            # 找到搜索输入框,并输入测试数据
            self.driver.find_element_by_id("kw").send_keys(testData)
```

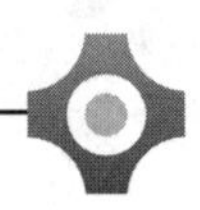

```
            # 找到搜索按钮,并单击
            self.driver.find_element_by_id("su").click()
            time.sleep(3)
            # 断言期望结果是否出现在页面源代码中
            self.assertTrue(expectData in self.driver.page_source)
        except NoSuchElementException as err:
            logging.error("查找的页面元素不存在,异常堆栈信息:"
                          + str(traceback.format_exc()))
        except AssertionError as err:
            logging.info("搜索"%s",期望"%s",失败" %(testData, expectData))
        except Exception as err:
            logging.error("未知错误,错误信息:" + str(traceback.format_exc()))
        else:
            logging.info("搜索"%s",期望"%s"通过" %(testData, expectData))

    def tearDown(self):
        self.driver.quit()

if __name__ == '__main__':
    unittest.main()
```

执行结束后日志文件内容如下:

```
Thu, 2016-12-08 14:52:17 DataDrivenByXML.py[line:64] INFO 搜索"Selenium WebDriver 实战宝典",期望"吴晓华"通过
Thu, 2016-12-08 14:52:32 DataDrivenByXML.py[line:64] INFO 搜索"HTTP 权威指南",期望"古尔利"通过
Thu, 2016-12-08 14:52:47 DataDrivenByXML.py[line:64] INFO 搜索"探索式软件测试",期望"惠特克"通过
```

12.7　使用 MySQL 数据库进行数据驱动测试

测试逻辑:

(1) 打开百度首页,从 MySQL 数据库中获取到测试过程中需要的测试数据。

(2) 在搜索输入框中输入查询关键词测试数据。

(3) 单击搜索按钮。

(4) 断言搜索结果页面中是否出现数据库中提供的预期内容,包含则认为测试执行成功,否则认为失败。

环境准备:

(1) 从 http://dev.mysql.com/downloads/mysql/5.5.html#downloads 下载 MySQL5.5 安装文件(扩展名为.msi 的文件),请读者根据自己操作系统位数选择相应的安装文件进行下载。

(2) 在安装 MySQL5.5 过程中,请读者设置好登录数据库的用户名 root 的密码并记住,如图 12-4 所示,并将数据库默认字符集修改成 utf8,如图 12-5 所示。

(3) 按 Ctrl+R 组合键调出运行窗口,在打开输入框中输入 SERVICES.msc 回车,打开计算机服务管理界面,在该界面找到 MySQL 服务并启动,如果在安装过程中选择了启动 MySQL 服务选项,这一步可以略过。

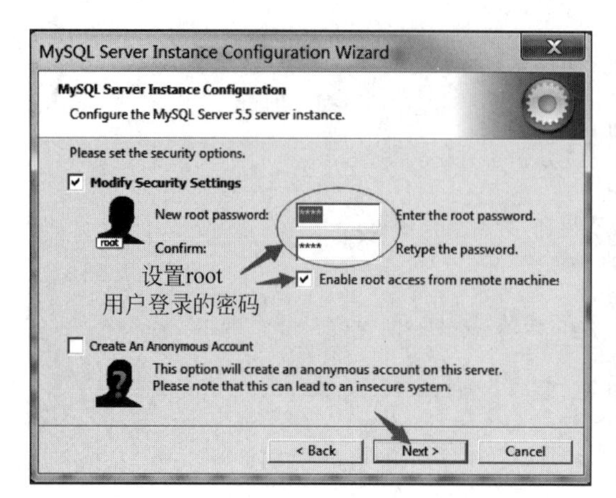

图 12-4

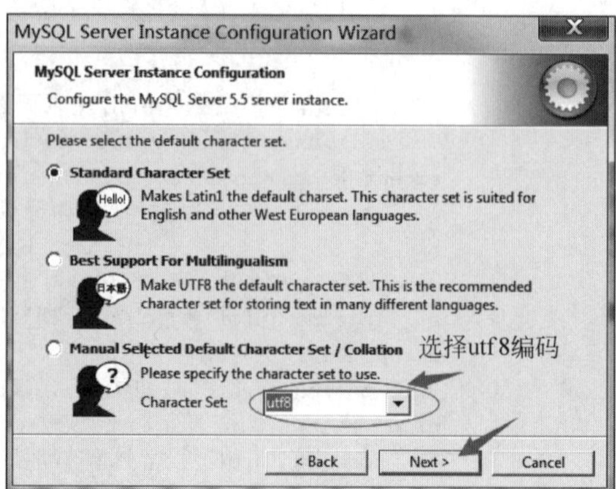

图 12-5

（4）单击“开始”→“所有程序”→MySQL→MySQL Server 5.5→MySQL 5.5 Command Line Client，启动 MySQL 数据库，在弹出的命令窗口输入 MySQL 登录密码进行登录，后出现如图 12-6 所示结果，表示 MySQL 数据库安装成功。

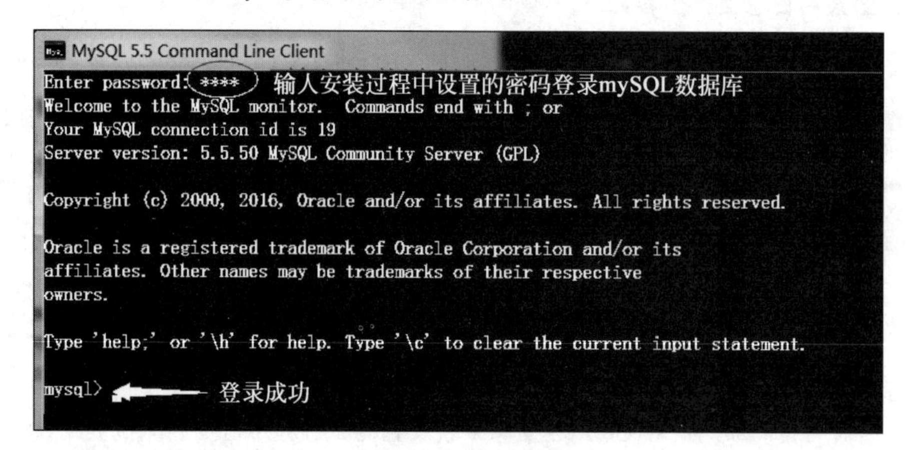

图 12-6

（5）安装 Python 连接 MySQL 数据库的连接器安装，直接通过 pip 命令安装即可，pip install PyMySQL，未报错说明安装成功。

实例代码：

在 Pycharm 工具中新建一名为 DataBaseDataDrivenProject 的 Python 工程，并在该工程下新建四个文件，文件名分别为 Sql.py、DatabaseInit.py、MysqlUtil.py 和 DataDrivenByMySQL.py。

Sql.py 文件用于编写创建数据库及数据表的 SQL 语句，具体内容如下：

```
#encoding = utf-8
# 创建 gloryroad 数据库 SQL 语句
create_database = 'CREATE DATABASE IF NOT EXISTS gloryroad DEFAULT CHARSET utf8 COLLATE utf8_
general_ci;'
drop_table = "drop table if exists testdata;"
# 创建 testdata 表
create_table = """
```

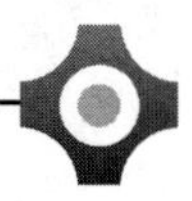

```
    create table testdata(
        id int not null auto_increment comment '主键',
        bookname varchar(40) unique not null comment '书名',
        author varchar(30) not null comment '作者',
        primary key(id)
    )engine = innodb character set utf8 comment '测试数据表';
"""
```

DatabaseInit.py 文件用于编写初始化数据库的脚本，具体内容如下：

```
# encoding = utf - 8
import pymysql
from Sql import *

class DataBaseInit(object):
    # 本类用于完成初始化数据操作
    # 创建数据库,创建数据表,向表中插入测试数据
    def __init__(self, host, port, dbName, username, password, charset):
        self.host = host
        self.port = port
        self.db = dbName
        self.user = username
        self.password = password
        self.charset = charset

    def create(self):
        try:
            # 连接 MySQL 数据库
            conn = pymysql.connect(
                host = self.host,
                port = self.port,
                user = self.user,
                password = self.passwd,
                charset = self.charset
            )
            # 获取数据库游标
            cur = conn.cursor()
            cur.execute(drop_table)
            # 创建数据库
            cur.execute(create_database)
            # 选择创建好的 gloryroad 数据库
            conn.select_db("gloryroad")
            # 创建测试表
            cur.execute(create_table)
        except Exception as err:
            raise err
        else:
            # 关闭游标
            cur.close()
            # 提交操作
            conn.commit()
```

```
            # 关闭连接
            conn.close()
            print("创建数据库及表成功")

    def insertDatas(self):
        try:
            # 连接 MySQL 数据库中具体某个库
            conn = pymysql.connect(
                host = self.host,
                port = self.port,
                database = self.db,
                user = self.user,
                password = self.password,
                charset = self.charset
            )
            cur = conn.cursor()
            # 向测试表中插入测试数据
            sql = "insert into testdata(bookname, author) values(%s, %s);"
            res = cur.executemany(sql, [('Selenium WebDriver 实战宝典', '吴晓华'),
                                        ('HTTP 权威指南', '古尔利'),
                                        ('探索式软件测试', '惠特克'),
                                        ('暗时间', '刘未鹏')])
        except Exception as err:
            raise err
        else:
            conn.commit()
            print("初始数据插入成功")
            # 确认插入数据成功
            cur.execute("select * from testdata;")
            for i in cur.fetchall():
                print(i[1], i[2])
            cur.close()
            conn.close()


if __name__ == '__main__':
    db = DataBaseInit(
        host = "localhost",
        port = 3306,
        dbName = "gloryroad",
        username = "root",
        password = "root",
        charset = "utf8"
    )
    db.create()
    db.insertDatas()
    print("数据库初始化结束")
```

MysqlUtil.py 文件用于从数据库中获取测试数据，具体代码如下：

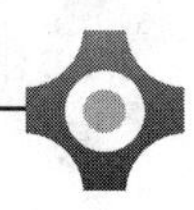

```
# encoding = utf-8
import pymsql
from DatabaseInit import DataBaseInit

class MyMySQL(object):
    def __init__(self, host, port, dbName, username, password, charset):
        # 进行数据库初始化
        dbInit = DataBaseInit(host, port, dbName, username, password, charset)
        dbInit.create()
        dbInit.insertDatas()
        self.conn = pymsql.connect(
            host = host,
            port = port,
            database = dbName,
            user = username,
            password = password,
            charset = charset
        )
        self.cur = self.conn.cursor()

    def getDataFromDataBases(self):
        # 从 testdata 表中获取需要的测试数据
        # bookname 作为搜索关键词,author 作为预期关键词
        self.cur.execute("select bookname, author from testdata;")
        # 从查询区域取回所有查询结果
        datasTuple = self.cur.fetchall()
        return datasTuple

    def closeDatabase(self):
        # 数据库后期清理工作
        self.cur.close()
        self.conn.commit()
        self.conn.close()

if __name__ == '__main__':
    db = MyMySQL(
        host = "localhost",
        port = 3306,
        dbName = "gloryroad",
        username = "root",
        password = "root",
        charset = "utf8"
    )
    print(db.getDataFromDataBases())
    db.closeDatabase()
```

DataDrivenByMySQL.py 文件用于编写执行数据驱动测试脚本，具体内容如下：

```
# encoding = utf-8
from selenium import webdriver
import unittest, time
```

```
import logging, traceback
import ddt
from MysqlUtil import MyMySQL
from selenium.common.exceptions import NoSuchElementException

# 初始化日志对象
logging.basicConfig(
    # 日志级别
    level = logging.INFO,
    # 日志格式
    # 时间、代码所在文件名、代码行号、日志级别名字、日志信息
    format = '%(asctime)s %(filename)s[line:%(lineno)d] %(levelname)s %(message)s',
    # 打印日志的时间
    datefmt = '%a, %Y-%m-%d %H:%M:%S',
    # 日志文件存放的目录(目录必须存在)及日志文件名
    filename = 'd:/DataDrivenTesting/dataDriveRreport.log',
    # 打开日志文件的方式
    filemode = 'w'
)

def getTestDatas():
    db = MyMySQL(
        host = "localhost",
        port = 3306,
        dbName = "gloryroad",
        username = "root",
        password = "root",
        charset = "utf8"
    )
    # 从数据库测试表中获取测试数据
    testData = db.getDataFromDataBases()
    # 关闭数据库连接
    db.closeDatabase()
    return testData

@ddt.ddt
class TestDemo(unittest.TestCase):

    def setUp(self):
        self.driver = webdriver.Firefox(executable_path = "c:\\geckodriver")

    @ddt.data( * getTestDatas())
    def test_dataDrivenByDatabase(self, data):
        # 对获得的数据进行解包
        testData, expectData = data
        url = "http://www.baidu.com"
        # 访问百度首页
        self.driver.get(url)
        # 将浏览器窗口最大化
        self.driver.maximize_window()
        print(testData, expectData)
```

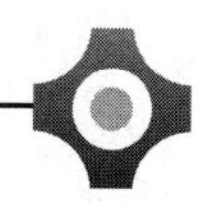

```
        # 设置隐式等待时间为 10 秒
        self.driver.implicitly_wait(10)
        try:
            # 找到搜索输入框,并输入测试数据
            self.driver.find_element_by_id("kw").send_keys(testData)
            # 找到搜索按钮,并单击
            self.driver.find_element_by_id("su").click()
            time.sleep(3)
            # 断言期望结果是否出现在页面源代码中
            self.assertTrue(expectData in self.driver.page_source)
        except NoSuchElementException as err:
            logging.error("查找的页面元素不存在,异常堆栈信息:"\
                          + str(traceback.format_exc()))
        except AssertionError as err:
            logging.info("搜索" %s",期望" %s",失败" %(testData, expectData))
        except Exception as err:
            logging.error("未知错误,错误信息:" + str(traceback.format_exc()))
        else:
            logging.info("搜索" %s",期望" %s"通过" %(testData, expectData))

    def tearDown(self):
        self.driver.quit()

if __name__ == '__main__':
    unittest.main()
```

执行结束后打印的日志文件内容如下:

```
Fri, 2016-12-09 14:38:15 DataDrivenByMySQL.py[line:70] INFO 搜索"Selenium WebDriver 实战宝典",期望"吴晓华"通过
Fri, 2016-12-09 14:38:28 DataDrivenByMySQL.py[line:70] INFO 搜索"HTTP 权威指南",期望"古尔利"通过
Fri, 2016-12-09 14:38:42 DataDrivenByMySQL.py[line:70] INFO 搜索"探索式软件测试",期望"惠特克"通过
Fri, 2016-12-09 14:38:55 DataDrivenByMySQL.py[line:70] INFO 搜索"暗时间",期望"刘未鹏"通过
```

说明:

由于官方的 MySQLdb 模块目前只支持到 Python3.4 版本,再高版本想连接数据库需要安装 PyMySQL 模块,并且也不支持一次性执行多条独立的 SQL 语句。

第13章 行为驱动测试

行为驱动测试方法已经在敏捷开发模式中普遍使用，通过使用标准化的语言将客户需求人员、开发人员和测试人员关联在一起，让产品开发相关人员在沟通上保持一致。请参与敏捷开发项目的读者仔细阅读本章内容，充分理解行为驱动测试原则、机制和实践方法。

13.1 行为驱动开发和 lettuce 简介

行为驱动开发是一种敏捷软件开发技术，它的英文全称是 Behavior Driven Development，英文缩写为 BDD。BDD 最初由 Dan North 在 2003 年命名，它包括验收测试和客户测试驱动等极限编程实践，作为对测试驱动开发的回应。它鼓励软件项目中的开发者、QA、非技术人员或商业参与者之间进行协作。在过去数年里，BDD 开发模式得到了很大的发展，BDD 的流行已然无法逆转。

lettuce 是实现 BDD 开发模式的一种测试框架，实现了使用自然语言来执行相关联测试的代码的需求。lettuce 是基于 Cucumber 的一款非常易于使用的 BDD 工具，可以执行纯文本的功能描述。lettuce 使用 Gherkin 语言来描述测试功能、测试场景、测试步骤和测试结果，Gherkin 语言支持超过 40 种自然语言，包括英文和中文。Gherkin 语言使用的主要英文关键词有 Scenario、Given、When、And、Then 和 But 等，这些关键词也可以转换为中文关键词，例如“场景”“如果”“当”和“那么”。根据用户故事，需求人员或测试人员使用 Gherkin 语言编写好测试场景的每个执行步骤，lettuce 就会一步一步地解析关键词右侧的自然语言并执行相应的代码。

关键词的含义如下：

（1）Feature：特性，将多个测试用例的集合到一起，对应于 unittest 中的 test suite（测试用例集）。

（2）Scenario：情景，用于描述一个用例，对应于 unittest 中的 test case（测试用例）。

（3）Given：如果，用例开始执行前的一个前置条件，类似于 unittest 中 setup 方法中的一些步骤。

（4）When：当，用例开始执行时的一些关键操作步骤，类似于 unittest 中的以 test 开头的方法，比如执行一个单击元素的操作。

（5）Then：那么，验证结果，就是平时用例中的验证步骤，比如 assert 方法。

（6）And：和，一个步骤中如果存在多个 Given 操作，后面的 Given 可以用 And 替代。

（7）But：一个步骤中如果存在多个 Then 操作，第二个开始后面的 Then 可以用 But 替代。

使用 Gherkin 语言编写测试场景的执行步骤，并将执行步骤保存在扩展名为 feature 的文件中，每个.feature 文件都要开始于 Feature（功能）关键词，Feature 之后的描述可以自定

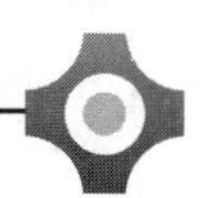

义，直到出现 Scenario（场景）关键词。一个.feature 文件中可以有多个 Scenario，每个 Scenario 包含步骤（step）列表，不同步骤使用 Given、When、Then、But、And 这些关键词进行区分。

BDD 开发模式的好处在于，可以将用户故事（敏捷开发中的 User Story）或者需求和测试用例建立起一一对应的映射关系，保证开发和测试的目标与范围严格地和需求保持一致，可以更好地让需求方、开发者以及测试人员用 唯一的需求进行相关开发工作，防止对需求理解的不一致，并且 BDD 框架的测试结果很容易被参与者所理解。

lettuce 的工作流程如图 13-1 所示。

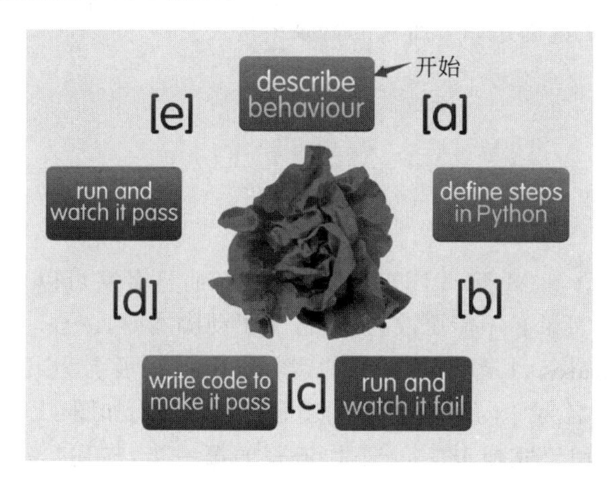

图　13-1

13.2　行为驱动测试的环境准备

（1）在“开始”→“搜索程序和文件”框中输入 CMD 并回车，然后在打开的 CMD 窗口中输入 pip install lettuce 并回车，进行 lettuce 模块的安装。

（2）等安装进度条走完，在 CMD 下输入 python 进入 Python 交互模式，执行 import lettuce，如果未报错，说明 lettuce 模块已安装成功。

如果使用 pip 不能成功安装的读者，可以直接访问 https://pypi.python.org/pypi/lettuce/0.2.23 下载 lettuce 的源码包进行安装，如图 13-2 所示，源码安装方法请参看第 6 章 selenium3 的安装方法。

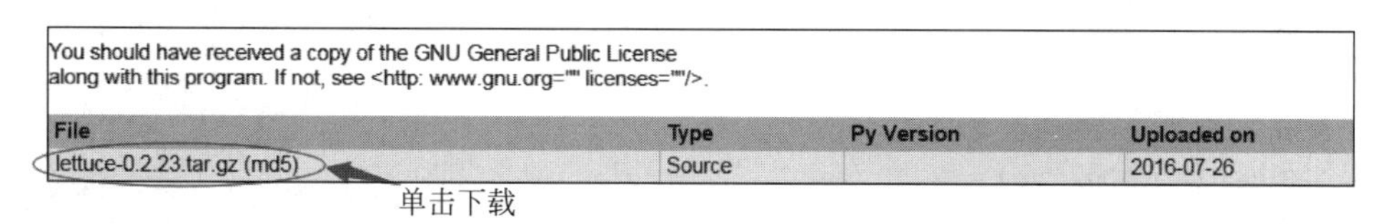

图　13-2

由于 lettuce 官网未出针对 Python3 的安装包，所以本章仍保留的是 Python2 版本的功能。

13.3 第一个英文语言行为驱动测试

测试逻辑：

（1）从 lettuce 全局变量命名空间 world 中取得一个整数。

（2）计算该整数的阶乘。

（3）断言计算结果的正确性。

BDD 实施步骤：

（1）在 Pycharm 工具创建如下所示的目录结构及文件。

```
|lettuce
    |MyFirstBDD
        |features
            - zero.feature
            - steps.py
```

前两层目录 lettuce 和 MyFirstBDD 不是必须的，而且名字可以自定义；但 features 目录是必须存在的，并且目录名不能更改，执行行为驱动脚本时，lettuce 首先寻找的就是具有这个名字的目录；features 目录下存放的执行场景文件(扩展名为.feature 的文件)和描述行为的脚本文件(扩展名为.py 的文件)。创建好的目录结构如图 13-3 所示。

（2）zero.feature 用于完成 lettuce 工作流程的第一步，描述测试场景的行为，具体内容如下：

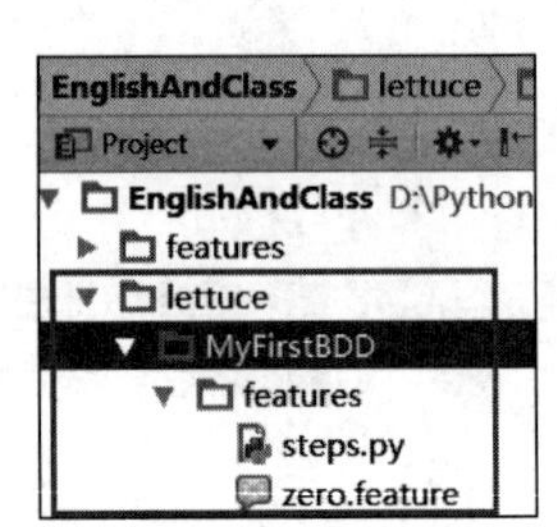

图 13-3

```
Feature: Compute factorial
  In order to play with Lettuce
  As beginners
  We'll implement factorial

  Scenario: Factorial of 0
    Given I have the number 0
    When I compute its factorial
    Then I see the number 1

  Scenario: Factorial of 1
    Given I have the number 1
    When I compute its factorial
    Then I see the number 1

  Scenario: Factorial of 2
    Given I have the number 2
    When I compute its factorial
    Then I see the number 2

  Scenario: Factorial of 3
    Given I have the number 3
    When I compute its factorial
    Then I see the number 6
```

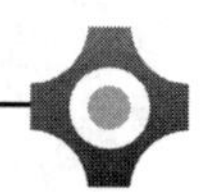

注意　如果.feature 文件想直接通过记事本等文本编辑器进行创建，必须将该文件保存为 utf-8 编码。

(3) steps.py 里面使用 Python 语言编写行为步骤，并且提供检测执行结果代码，具体内容如下：

```
#encoding=utf-8
from lettuce import *

# 用于计算整数的阶乘函数
def factorial(number):
    number = int(number)
    if (number == 0) or (number == 1):
        return 1
    else:
        return reduce(lambda x, y: x * y, range(1, number + 1))

@step('I have the number (\d+)')
def have_the_number(step, number):
    # 将通过正则表达式匹配的数字存于全局变量 world 中
    world.number = int(number)

@step('I compute its factorial')
def compute_its_factorial(step):
    # 从全局变量 world 中取出匹配的数字,
    # 计算其阶乘,并将结果再存回 world 中
    world.number = factorial(world.number)

@step('I see the number (\d+)')
def check_number(step, expected):
    # 通过正则匹配到预期数字
    expected = int(expected)
    # 断言计算阶乘结果是否等于预期
    assert world.number == expected, "Got %d" %world.number
```

(4) 在 PyCharm 工具的 Terminal(终端)中，将当前工作目录切换到 features 目录所在目录(操作方法同 Windows 的 CMD)，然后执行命令 lettuce 启动行为驱动测试，如图 13-4 所示。

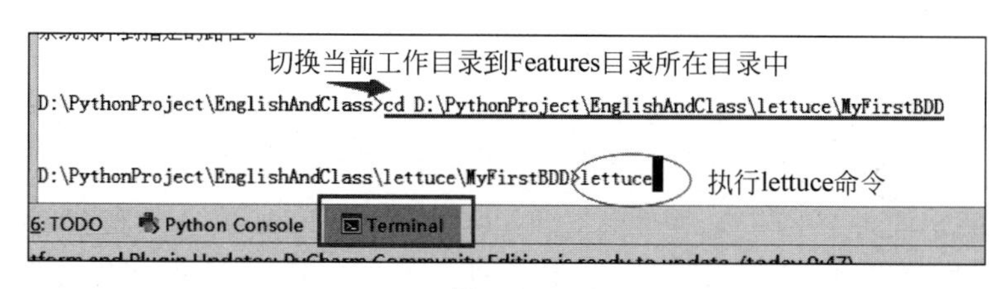

图　13-4

执行结果：

执行结束后，会在 Terminal 中打印结果信息，如下：

```
Feature: Compute factorial              # \features\zero.feature:1
  In order to play with Lettuce         # \features\zero.feature:2
  As beginners                          # \features\zero.feature:3
  We'll implement factorial             # \features\zero.feature:4

  Scenario: Factorial of 0              # \features\zero.feature:6
    Given I have the number 0           # \features\steps.py:13
    When I compute its factorial        # \features\steps.py:17
    Then I see the number 1             # \features\steps.py:21

  Scenario: Factorial of 1              # \features\zero.feature:11
    Given I have the number 1           # \features\steps.py:13
    When I compute its factorial        # \features\steps.py:17
    Then I see the number 1             # \features\steps.py:21

  Scenario: Factorial of 2              # \features\zero.feature:16
    Given I have the number 2           # \features\steps.py:13
    When I compute its factorial        # \features\steps.py:17
    Then I see the number 2             # \features\steps.py:21

  Scenario: Factorial of 3              # \features\zero.feature:21
    Given I have the number 3           # \features\steps.py:13
    When I compute its factorial        # \features\steps.py:17
    Then I see the number 6             # \features\steps.py:21

1 feature (1 passed)
4 scenarios (4 passed)
12 steps (12 passed)
```

代码解释：

执行结果中 1 feature (1 passed)表示有一个 features 被执行通过了。4 scenarios (4 passed)表示有 4 个 scenarios(场景，对应 zero.feature 文件中的四个 scenarios，每个场景均会被执行)被执行通过了。12 steps (12 passed)表示 4 个 scenarios 总共有 12 步，均被执行通过。

一个 scenarios 中有三步，分别为 Given、When 和 Then 标注的步骤，执行它们的方法分别对应 steps.py 文件中的@step('I have the number (\d+)')、@step('I compute its factorial')和@step('I see the number (\d+)')修饰的方法，其中(\d+)是一个正则表达式，\d 表示匹配一个数字，+号表示匹配的数字至少有一个或多个，@step('I have the number (\d+)')步骤中(\d+)数字来自 zero.feature 文件中的 Given 后面步骤描述中的数字，比如 Given I have the number 0 中的 0，@step('I see the number (\d+)')步骤中的(\d+)来自 zero.feature 文件中的 Then 后面步骤描述中的数字，比如 I see the number 1 中的 1。这些数字通过正则表达式匹配出来以后存储在 lettuce 全局变量命名空间 world 中，对应实例代码 world.number = int(number)，后续测试过程中需要使用时直接从 world 中取即可。

更多说明：

关于行为驱动测试的执行方式除了本例所提到的方式以外，还可以直接通过 CMD 进

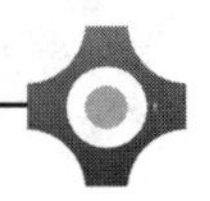

行执行。首先按下 Win + R 组合键，在弹出的运行框中输入 cmd 回车，弹出 CMD 窗口，然后将当前的工作目录切换到要执行的测试用例集 features 目录所在的目录中，最后执行 lettuce 命令即可，如图 13-5 所示。

```
C:\Users\SR>d:

D:\>cd D:\PythonProject\EnglishAndClass\lettuce\MyFirstBDD

D:\PythonProject\EnglishAndClass\lettuce\MyFirstBDD>lettuce

Feature: Compute factorial        # \features\zero.feature:1
  In order to play with Lettuce   # \features\zero.feature:2
  As beginners                    # \features\zero.feature:3
  We'll implement factorial       # \features\zero.feature:4

  Scenario: Factorial of 0        # \features\zero.feature:6
    Given I have the number 0     # \features\steps.py:13
    When I compute its factorial  # \features\steps.py:18
    Then I see the number 1       # \features\steps.py:24

  Scenario: Factorial of 1        # \features\zero.feature:11
    Given I have the number 1     # \features\steps.py:13
    When I compute its factorial  # \features\steps.py:18
    Then I see the number 1       # \features\steps.py:24

  Scenario: Factorial of 2        # \features\zero.feature:16
    Given I have the number 2     # \features\steps.py:13
    When I compute its factorial  # \features\steps.py:18
    Then I see the number 2       # \features\steps.py:24

  Scenario: Factorial of 3        # \features\zero.feature:21
    Given I have the number 3     # \features\steps.py:13
    When I compute its factorial  # \features\steps.py:18
    Then I see the number 6       # \features\steps.py:24

1 feature (1 passed)
4 scenarios (4 passed)
12 steps (12 passed)
```

执行lettuce命令

输出结果

图　13-5

13.4　通过类模式实现英文行为驱动

测试逻辑：

(1) 将测试步骤方法封装到类中，并从全局变量中获取需要的计算阶乘的整数。

(2) 计算该整数的阶乘。

(3) 断言计算结果的正确性。

BDD 的实施步骤：

(1) 在 PyCharm 工具创建如下所示的目录结构及文件。

```
|lettuce
    |ClassBDD
        |features
            - zero.feature
            - steps.py
```

(2) zero.feature 文件具体内容如下：

```
Feature: Compute factorial
  In order to play with Lettuce
```

```
  As beginners
  We'll implement factorial

  Scenario: Factorial of 0
    Given I have the number 0
    When I compute its factorial
    Then I see the number 1

  Scenario: Factorial of 1
    Given I have the number 1
    When I compute its factorial
    Then I see the number 1

  Scenario: Factorial of 2
    Given I have the number 2
    When I compute its factorial
    Then I see the number 2

  Scenario: Factorial of 3
    Given I have the number 3
    When I compute its factorial
    Then I see the number 6
```

（3）steps.py 文件具体内容如下：

```
#encoding = utf-8
from lettuce import world, steps

def factorial(number):
  number = int(number)
  if (number == 0) or (number == 1):
    return 1
  else:
      return reduce(lambda x, y: x * y, range(1, number + 1))

@steps
class FactorialSteps(object):
    """Methods in exclude or starting with _ will not be considered as step"""

    exclude = ['set_number', 'get_number']

    def __init__(self, environs):
        # 初始全局变量
        self.environs = environs

    def set_number(self, value):
        # 设置全局变量中的 number 变量的值
        self.environs.number = int(value)
```

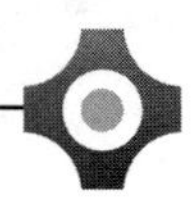

```
    def get_number(self):
        # 从全局变量中取出 number 的值
        return self.environs.number

    def _assert_number_is(self, expected, msg = "Got %d"):
        number = self.get_number()
        # 断言
        assert number == expected, msg % number

    def have_the_number(self, step, number):
        '''I have the number (\d+)'''
        # 上面的三引号引起的代码必须写,并且必须是三引号引起
        # 表示从场景步骤中获取需要的数据
        # 并将获得数据存到全局变量 number 中
        self.set_number(number)

    def i_compute_its_factorial(self, step):
        number = self.get_number()
        # 调用 factorial 方法进行阶乘结算,
        # 并将结算结果存于全局变量中的 number 中
        self.set_number(factorial(number))

    def check_number(self, step, expected):
        '''I see the number (\d+)'''
        # 上面的三引号引起的代码必须写,并且必须是三引号引起
        # 表示从场景步骤中获取需要的数据以便断言测试结果
        self._assert_number_is(int(expected))

FactorialSteps(world)
```

（4）在 PyCharm 工具的 Terminal(终端)中，或者 CMD 中，将当前的工作目录切换到 features 目录所在目录中，然后执行 lettuce > testReport.log。

执行结果：

执行结束后，将会在 features 目录作者目录下生产一个 testReport.log 文件，里面内容即为直接通过 lettuce 命令执行时打印到屏幕上的测试过程信息，将其重定向到文件中，方便后续查看，文件内如图 13-6 所示，但这样得到的输出结果中，每一步都被打印了两次。

代码解释：

本实例将 13.3 小节中的步骤方法封装到类 FactorialSteps 中，并且使用@steps 注解进行修饰类 FactorialSteps，表示此类中集合了多个步骤，同时内部修饰@steps 和__init__方法组成闭包，表示__init__方法将和@steps 共存亡。

类模式的行为驱动中，关键字 exclude 列表中列出的方法名或以下画线“_”开始的方法将不会被认为是测试步骤，行为驱动测试执行过程中不会被 lettuce 主动执行，比如本实例中的 exclude = ['set_number', 'get_number']中的两个方法以及_assert_number_is 方法均不会被当成行为驱动测试步骤而被 lettuce 执行。

```
Feature: Compute factorial          # \features\zero.feature:1
  In order to play with Lettuce     # \features\zero.feature:2
  As beginners                      # \features\zero.feature:3
  We'll implement factorial         # \features\zero.feature:4

  Scenario: Factorial of 0          # \features\zero.feature:6
    Given I have the number 0       # \features\steps.py:35
    Given I have the number 0       # \features\steps.py:35
    When I compute its factorial    # \features\steps.py:42
    When I compute its factorial    # \features\steps.py:42
    Then I see the number 1         # \features\steps.py:48
    Then I see the number 1         # \features\steps.py:48

  Scenario: Factorial of 1          # \features\zero.feature:11
    Given I have the number 1       # \features\steps.py:35
    Given I have the number 1       # \features\steps.py:35
    When I compute its factorial    # \features\steps.py:42
    When I compute its factorial    # \features\steps.py:42
    Then I see the number 1         # \features\steps.py:48
    Then I see the number 1         # \features\steps.py:48

  Scenario: Factorial of 2          # \features\zero.feature:16
    Given I have the number 2       # \features\steps.py:35
    Given I have the number 2       # \features\steps.py:35
    When I compute its factorial    # \features\steps.py:42
    When I compute its factorial    # \features\steps.py:42
    Then I see the number 2         # \features\steps.py:48
    Then I see the number 2         # \features\steps.py:48

  Scenario: Factorial of 3          # \features\zero.feature:21
    Given I have the number 3       # \features\steps.py:35
    Given I have the number 3       # \features\steps.py:35
    When I compute its factorial    # \features\steps.py:42
    When I compute its factorial    # \features\steps.py:42
    Then I see the number 6         # \features\steps.py:48
    Then I see the number 6         # \features\steps.py:48

1 feature (1 passed)
4 scenarios (4 passed)
12 steps (12 passed)
```

执行结果概述

图 13-6

13.5 lettuce 框架的步骤数据表格

在自动化测试实施过程中，经常会遇到在测试用例执行过程中的某一步或某几步需要往文件或数据库中添加大量不同数据的场景，同时还会检查一下这些数据添加后的新状态，lettuce 框架也为此种情况提供了很好的支持，只需要使用 lettuce 框架的步骤数据表格即可很容易实现向数据或文件等中写入大量的不同数据。

实例：

（1）在 PyCharm 工具中创建如下所示的目录结构及文件。

```
|lettuce
    |StepDataTables
        |features
            - student.feature
            - steps.py
```

（2）student.feature 文件具体内容如下所示。

```
Feature: bill students alphabetically
```

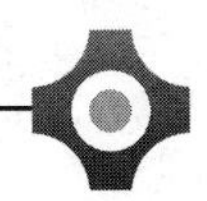

```
In order to bill students properly
As a financial specialist
I want to bill those which name starts with some letter

Scenario: Bill students which name starts with "G"
Given I have the following students in my database:
    | name    | monthly_due | billed |
    | Anton   | $ 500       | no     |
    | Jack    | $ 400       | no     |
    | Gabriel | $ 300       | no     |
    | Gloria  | $ 442.65    | no     |
    | Ken     | $ 907.86    | no     |
    | Leonard | $ 742.84    | no     |
When I bill names starting with "G"
Then I see those billed students:
    | name    | monthly_due | billed |
    | Gabriel | $ 300       | no     |
    | Gloria  | $ 442.65    | no     |
And those that weren't:
    | name    | monthly_due | billed |
    | Anton   | $ 500       | no     |
    | Jack    | $ 400       | no     |
    | Ken     | $ 907.86    | no     |
    | Leonard | $ 742.84    | no     |
```

Given、Then 及 And 步骤下都存在步骤数据表格，数据间以“|”进行分隔，数据表的第一行表示数据表的列名，不作为数据存在。

（3）steps.py 文件，用于编写获取 student.feature 文件中的数据，并提供后续操作，具体代码如下。

```
# encoding = utf - 8
from lettuce import *

@step('I have the following students in my database:')
def students_in_database(step):
    if step.hashes:
        # 如果存在步骤表格数据,则继续后续步骤
        print type(step.hashes)
        assert step.hashes == [
            {
                'name': 'Anton',
                'monthly_due': '$ 500',
                'billed': 'no'
            },
            {
                'name': 'Jack',
                'monthly_due': '$ 400',
                'billed': 'no'
            },
            {
```

```
                'name': 'Gabriel',
                'monthly_due': '$ 300',
                'billed': 'no'
            },
            {
                'name': 'Gloria',
                'monthly_due': '$ 442.65',
                'billed': 'no'
            },
            {
                'name': 'Ken',
                'monthly_due': '$ 907.86',
                'billed': 'no'
            },
            {
                'name': 'Leonard',
                'monthly_due': '$ 742.84',
                'billed': 'no'
            },
        ]

@step('I bill names starting with "(.*)"')
def match_starting(step, startAlpha):
    # 将通过正则表达式匹配步骤中最后一个字母,
    # 并存于全局变量 startAlpha 中
    world.startAlpha = startAlpha
    print step.hashes

@step('I see those billed students:')
def get_starting_with_G_student(step):
    # 遍历步骤数据表中的数据
    for i in step.hashes:
        # 断言学生的名字是否以 world.startAlpha 变量存取的字母开头
        assert i["name"].startswith(world.startAlpha)

@step("those that weren't:")
def result(step):
    for j in step.hashes:
        # 断言学生名字不以 world.startAlpha 变量存取的字母开头
        assert world.startAlpha not in j["name"][0]
```

执行结果：

```
Feature: bill students alphabetically                            # \features\data.feature:1
  In order to bill students properly                             # \features\data.feature:2
  As a financial specialist                                      # \features\data.feature:3
  I want to bill those which name starts with some letter        # \features\data.feature:4

  Scenario: Bill students which name starts with "G"             # \features\data.feature:6
    Given I have the following students in my database:          # \features\steps.py:5
    Given I have the following students in my database:          # \features\steps.py:5
```

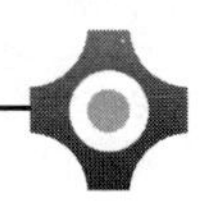

```
    | name     | monthly_due | billed |
    | Anton    | $ 500       | no     |
    | Jack     | $ 400       | no     |
    | Gabriel  | $ 300       | no     |
    | Gloria   | $ 442.65    | no     |
    | Ken      | $ 907.86    | no     |
    | Leonard  | $ 742.84    | no     |
  When I bill names starting with "G"                 # \features\steps.py:41
  When I bill names starting with "G"                 # \features\steps.py:41
  Then I see those billed students:                   # \features\steps.py:48
    | name     | monthly_due | billed |
    | Gabriel  | $ 300       | no     |
    | Gloria   | $ 442.65    | no     |
  And those that weren't:                             # \features\steps.py:55
    | name     | monthly_due | billed |
    | Anton    | $ 500       | no     |
    | Jack     | $ 400       | no     |
    | Ken      | $ 907.86    | no     |
    | Leonard  | $ 742.84    | no     |

1 feature (1 passed)
1 scenario (1 passed)
4 steps (4 passed)
```

代码解释：

执行测试脚本中的步骤函数时，lettuce 框架自动将 Given、Then 及 And 步骤中的数据表格中的数据转换成一个可迭代的数据对象 list，list 中的每一个元素都是数据表中每一行数据组成的以列名为 key，列中的数据为 value 的字典对象，然后直接通过访问 step.hashes 属性就可以取到存储的这些数据，以便提供给后续提取数据的使用。

行为驱动场景中的每一步骤都可以有自己的步骤数据表格。

13.6　使用 WebDriver 进行英文的行为数据驱动测试

测试逻辑：

(1) 访问 http://www.sogou.com。

(2) 依次搜索几个球星的英文名字中的一部分。

(3) 在搜索结果页面断言搜索的球星的全名。

实例：

(1) 在 Pycharm 工具创建如下所示的目录结构及文件。

```
|lettuce
    |BddDataDrivenByEnglish
        |features
            - sogou.feature
            - sogou.py
            - terrain.py
```

（2）sogou.feature 文件用于存放数据驱动所需要的数据，具体内容如下：

```
Feature: Search in Sogou website
  In order to Search in Sogou  website
  As a visitor
  We'll search the NBA best player

  Scenario: Search NBA player
    Given I have the english name "<search_name>"
    When  I search it in Sogou website
    Then  I see the entire name "<search_result>"

  Examples:
    | search_name   | search_result  |
    | Jordan        | Michael        |
    | Curry         | Stephen        |
    | Kobe          | Bryant         |
```

Examples 下面是一个场景数据表，数据间以“|”进行分隔，数据表的第一行表示列名，其跟场景中的变量名对应，比如 Given I have the english name "<search_name>"语句中的 search_name 对应数据表中的 search_name 列。

（3）sogou.py 文件编写实施结合行为驱动的数据驱动测试，具体内容如下：

```
#encoding = utf-8
from lettuce import *
from selenium import webdriver
import time

@step('I have the english name "(.*)"')
def have_the_searchWord(step, searchWord):
    world.searchWord = str(searchWord)
    print world.searchWord

@step('I search it in Sogou website')
def search_in_sogou_website(step):
    world.driver = webdriver.Firefox(executable_path = "c:\\geckodriver")
    world.driver.get("http://www.sogou.com")
    world.driver.find_element_by_id("query").send_keys(world.searchWord)
    world.driver.find_element_by_id("stb").click()
    time.sleep(3)

@step('I see the entire name "(.*)"')
def check_result_in_sogou(step, searchResult):
    assert searchResult in world.driver.page_source, "got word: %s" % searchResult
    world.driver.quit()
```

（4）terrain.py 文件用于在测试过程中和测试结束后打印日志，具体内容如下：

```
#encoding = utf-8
from lettuce import *
import logging
```

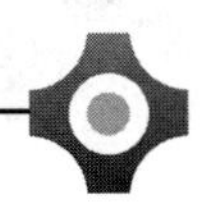

```
# 初始化日志对象
logging.basicConfig(
    # 日志级别
    level = logging.INFO,
    # 日志格式
    # 时间、代码所在文件名、代码行号、日志级别名字、日志信息
    format = '%(asctime)s %(filename)s[line: %(lineno)d] %(levelname)s %(message)s',
    # 打印日志的时间
    datefmt = '%a, %Y-%m-%d %H: %M: %S',
    # 日志文件存放的目录(目录必须存在)及日志文件名
    filename = 'D:/lettuce/BddDataDrivenByEnglish/BddDataDriveRreport.log',
    # 打开日志文件的方式
    filemode = 'w'
)
# 在所有场景执行前执行
@before.all
def say_hello():
    logging.info("Lettuce will start to run tests right now...")
    print "Lettuce will start to run tests right now..."

# 在每个 secnario 开始执行前执行
@before.each_scenario
def setup_some_scenario(scenario):
    # 每个 Scenario 开始前,打印场景的名字
    print 'Begin to execute scenario name:' + scenario.name
    # 将开始执行的场景信息打印到日志
    logging.info('Begin to execute scenario name:' + scenario.name)

# 每个 step 开始前执行
@before.each_step
def setup_some_step(step):
    run = "running step %r, defined at %s" % (
        step.sentence,                  # 执行的步骤
        step.defined_at.file            # 步骤定义在哪个文件
    )
    # 将每个场景的每一步信息打印到日志
    logging.info(run)

# 每个 step 执行后执行
@after.each_step
def teardown_some_step(step):
    logging.info("End of the '%s'" % step.sentence)

# 在每个 secnario 执行结束执行
@after.each_scenario
def teardown_some_scenario(scenario):
    print 'finished, scenario name:' + scenario.name
    logging.info('finished, scenario name:' + scenario.name)

# 在所有场景执行结束后执行
```

```
@after.all #默认获取执行结果的对象作为 total 参数
def say_goodbye(total):
    result = "Congratulations, %d of %d scenarios passed!" % (
        total.scenarios_ran,          #一共多少场景运行了
        total.scenarios_passed        #一共多少场景运行成功了
    )
    print result
    logging.info(result)
    # 将测试结果写入日志文件
    logging.info("Goodbye!")
    print "------ Goodbye! ------"
```

执行结果日志文件内容如图 13-7 所示。

```
Fri, 2017-02-10 13:38:52 terrain.py[line:22] INFO Lettuce will start to run tests right now...
Fri, 2017-02-10 13:38:52 terrain.py[line:31] INFO Begin to execute scenario name:Search NBA player
Fri, 2017-02-10 13:38:52 terrain.py[line:41] INFO running step u'Given I have the english name "Jordan"', defined at \features\sogou.py
Fri, 2017-02-10 13:38:52 terrain.py[line:46] INFO End of the 'Given I have the english name "Jordan"'
Fri, 2017-02-10 13:38:52 terrain.py[line:41] INFO running step u'When  I search it in Sogou website', defined at \features\sogou.py
Fri, 2017-02-10 13:39:07 terrain.py[line:46] INFO End of the 'When  I search it in Sogou website'
Fri, 2017-02-10 13:39:07 terrain.py[line:41] INFO running step u'Then  I see the entire name "Michael"', defined at \features\sogou.py
Fri, 2017-02-10 13:39:07 terrain.py[line:46] INFO End of the 'Then  I see the entire name "Michael"'
Fri, 2017-02-10 13:39:07 terrain.py[line:41] INFO running step u'Given I have the english name "Curry"', defined at \features\sogou.py
Fri, 2017-02-10 13:39:07 terrain.py[line:46] INFO End of the 'Given I have the english name "Curry"'
Fri, 2017-02-10 13:39:07 terrain.py[line:41] INFO running step u'When  I search it in Sogou website', defined at \features\sogou.py
Fri, 2017-02-10 13:39:22 terrain.py[line:46] INFO End of the 'When  I search it in Sogou website'
Fri, 2017-02-10 13:39:22 terrain.py[line:41] INFO running step u'Then  I see the entire name "Stephen"', defined at \features\sogou.py
Fri, 2017-02-10 13:39:23 terrain.py[line:46] INFO End of the 'Then  I see the entire name "Stephen"'
Fri, 2017-02-10 13:39:23 terrain.py[line:41] INFO running step u'Given I have the english name "Kobe"', defined at \features\sogou.py
Fri, 2017-02-10 13:39:23 terrain.py[line:46] INFO End of the 'Given I have the english name "Kobe"'
Fri, 2017-02-10 13:39:23 terrain.py[line:41] INFO running step u'When  I search it in Sogou website', defined at \features\sogou.py
Fri, 2017-02-10 13:39:38 terrain.py[line:46] INFO End of the 'When  I search it in Sogou website'
Fri, 2017-02-10 13:39:38 terrain.py[line:41] INFO running step u'Then  I see the entire name "Bryant"', defined at \features\sogou.py
Fri, 2017-02-10 13:39:38 terrain.py[line:46] INFO End of the 'Then  I see the entire name "Bryant"'
Fri, 2017-02-10 13:39:38 terrain.py[line:52] INFO finished, scenario name:Search NBA player
Fri, 2017-02-10 13:39:38 terrain.py[line:62] INFO Congratulations, 3 of 3 scenarios passed!
Fri, 2017-02-10 13:39:38 terrain.py[line:64] INFO Goodbye!
```

图 13-7

13.7 使用 WebDriver 进行中文语言的行为数据驱动测试

测试逻辑：

(1) 访问 http://www.baidu.com。

(2) 依次搜索几本中文名字的书。

(3) 在搜索结果页面断言是否出现书的预期作者。

实例：

(1) 在 PyCharm 工具创建如下所示的目录结构及文件。

```
|lettuce
    |BddDataDrivenByChinese
        |features
            - baidu.feature
            - baidu.py
            - terrain.py
            - log.py
```

(2) baidu.feature 文件内容如下：

```
#encoding = utf - 8
# language: zh - CN
```

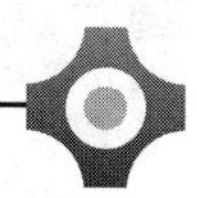

```
特性: 在百度网址搜索 IT 相关书籍
    能够搜索到书的作者,比如吴晓华

    场景: 在百度网站搜索 IT 相关书籍
        如果将搜索词设定为书的名字"<书名>"
        当打开百度网站
        和在搜索输入框中输入搜索的关键词,并单击搜索按钮后
        那么在搜索结果中可以看到书的作者"<作者>"

例如:

| 书名                          | 作者   |
| Selenium WebDriver 实战宝典   | 吴晓华 |
| HTTP 权威指南                 | 古尔利 |
| Python 核心编程               | 丘恩   |
```

(3) baidu.py 文件内容如下:

```
#encoding = utf-8
# language: zh-CN
from lettuce import *
from selenium import webdriver
import time

@step(u'将搜索词设定为书的名字"(.*)"')
def have_the_searchWord(step, searchWord):
    world.searchWord = searchWord
    print world.searchWord

@step(u'打开百度网站')
def visit_baidu_website(step):
    world.driver = webdriver.Firefox(executable_path = "c:\\geckodriver")
    world.driver.get("http://www.baidu.com")

@step(u'在搜索输入框中输入搜索的关键词,并单击搜索按钮后')
def search_in_sogou_website(step):
    world.driver.find_element_by_id("kw").send_keys(world.searchWord)
    world.driver.find_element_by_id("su").click()
    time.sleep(3)

@step(u'在搜索结果中可以看到书的作者"(.*)"')
def check_result_in_sogou(step, searchResult):
    assert searchResult in world.driver.page_source, "not got words: %s" % searchResult
    world.driver.quit()
```

(4) terrain.py 文件内容如下:

```
#encoding = utf-8
from lettuce import *
from log import *

# 在所有场景执行前执行
```

```
@before.all
def say_hello():
    logging.info(u"开始执行行为数据驱动测试...")

# 在每个 secnario 开始执行前执行
@before.each_scenario
def setup_some_scenario(scenario):
    # 将开始执行的场景信息打印到日志
    logging.info(u'开始执行场景" %s"' % scenario.name)

# 每个 step 开始前执行
@before.each_step
def setup_some_step(step):
    world.stepName = step.sentence
    run = u"执行步骤" %s", 定义在" %s"文件" % (
        step.sentence,                    # 执行的步骤
        step.defined_at.file              # 步骤定义在哪个文件
    )
    # 将每个场景的每一步信息打印到日志
    logging.info(run)

# 每个 step 执行后执行
@after.each_step
def teardown_some_step(step):
    logging.info(u"步骤" %s"执行结束" % world.stepName)

# 在每个 secnario 执行结束执行
@after.each_scenario
def teardown_some_scenario(scenario):
    logging.info(u'场景" %s"执行结束' % scenario.name)

# 在所有场景执行结束后执行
@after.all #默认获取执行结果的对象作为 total 参数
def say_goodbye(total):
    result = u"恭喜, %d 个场景运行, %d 个场景运行成功" % (
        total.scenarios_ran,              #一共多少场景运行了
        total.scenarios_passed            #一共多少场景运行成功了
    )
    logging.info(result)
    # 将测试结果写入日志文件
    logging.info(u"本次行为数据驱动执行结束")
```

（5）log.py 文件用于编写初始化日志对象的程序，具体内容如下：

```
#encoding = utf-8
import logging

# 初始化日志对象
logging.basicConfig(
    # 日志级别
    level = logging.INFO,
```

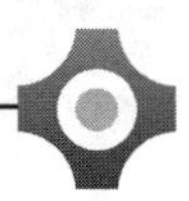

```
    # 日志格式
    # 时间、代码所在文件名、代码行号、日志级别名字、日志信息
    format = '%(asctime)s %(filename)s[line:%(lineno)d] %(levelname)s %(message)s',
    # 打印日志的时间
    datefmt = '%a, %Y-%m-%d %H:%M:%S',
    # 日志文件存放的目录(目录必须存在)及日志文件名
    filename = 'D:/lettuce/BddDataDrivenByChinese/BddDataDriveRreport.log',
    # 打开日志文件的方式
    filemode = 'w'
)
```

执行结果：

执行结束后日志文件 BddDataDriveRreport. log 内容如图 13-8 所示。

```
dDataDriveRreport.log
Fri, 2016-12-16 12:31:57 terrain.py[line:9] INFO 开始执行行为数据驱动测试...
Fri, 2016-12-16 12:31:57 terrain.py[line:15] INFO 开始执行场景"在百度网站搜索IT相关书籍"
Fri, 2016-12-16 12:31:57 terrain.py[line:26] INFO 执行步骤"如果将搜索词设定为书的名字"Selenium WebDriver实战宝典"", 定义在"\features\baidu.py"文件
Fri, 2016-12-16 12:31:57 terrain.py[line:31] INFO 步骤"如果将搜索词设定为书的名字"Selenium WebDriver实战宝典""执行结束
Fri, 2016-12-16 12:31:57 terrain.py[line:26] INFO 执行步骤"当打开百度网站", 定义在"\features\baidu.py"文件
Fri, 2016-12-16 12:32:07 terrain.py[line:31] INFO 步骤"当打开百度网站"执行结束
Fri, 2016-12-16 12:32:07 terrain.py[line:26] INFO 执行步骤"和在搜索输入框中输入搜索的关键词, 并点击搜索按钮后", 定义在"\features\baidu.py"文件
Fri, 2016-12-16 12:32:10 terrain.py[line:31] INFO 步骤"和在搜索输入框中输入搜索的关键词, 并点击搜索按钮后"执行结束
Fri, 2016-12-16 12:32:11 terrain.py[line:26] INFO 执行步骤"那么在搜索结果中可以看到书的作者"吴晓华"", 定义在"\features\baidu.py"文件
Fri, 2016-12-16 12:32:11 terrain.py[line:31] INFO 步骤"那么在搜索结果中可以看到书的作者"吴晓华""执行结束
Fri, 2016-12-16 12:32:11 terrain.py[line:26] INFO 执行步骤"如果将搜索词设定为书的名字"HTTP权威指南"", 定义在"\features\baidu.py"文件
Fri, 2016-12-16 12:32:11 terrain.py[line:31] INFO 步骤"如果将搜索词设定为书的名字"HTTP权威指南""执行结束
Fri, 2016-12-16 12:32:11 terrain.py[line:26] INFO 执行步骤"当打开百度网站", 定义在"\features\baidu.py"文件
Fri, 2016-12-16 12:32:22 terrain.py[line:31] INFO 步骤"当打开百度网站"执行结束
Fri, 2016-12-16 12:32:22 terrain.py[line:26] INFO 执行步骤"和在搜索输入框中输入搜索的关键词, 并点击搜索按钮后", 定义在"\features\baidu.py"文件
Fri, 2016-12-16 12:32:25 terrain.py[line:31] INFO 步骤"和在搜索输入框中输入搜索的关键词, 并点击搜索按钮后"执行结束
Fri, 2016-12-16 12:32:25 terrain.py[line:26] INFO 执行步骤"那么在搜索结果中可以看到书的作者"古尔利"", 定义在"\features\baidu.py"文件
Fri, 2016-12-16 12:32:26 terrain.py[line:31] INFO 步骤"那么在搜索结果中可以看到书的作者"古尔利""执行结束
Fri, 2016-12-16 12:32:26 terrain.py[line:26] INFO 执行步骤"如果将搜索词设定为书的名字"Python核心编程"", 定义在"\features\baidu.py"文件
Fri, 2016-12-16 12:32:26 terrain.py[line:31] INFO 步骤"如果将搜索词设定为书的名字"Python核心编程""执行结束
Fri, 2016-12-16 12:32:26 terrain.py[line:26] INFO 执行步骤"当打开百度网站", 定义在"\features\baidu.py"文件
Fri, 2016-12-16 12:32:37 terrain.py[line:31] INFO 步骤"当打开百度网站"执行结束
Fri, 2016-12-16 12:32:37 terrain.py[line:26] INFO 执行步骤"和在搜索输入框中输入搜索的关键词, 并点击搜索按钮后", 定义在"\features\baidu.py"文件
Fri, 2016-12-16 12:32:40 terrain.py[line:31] INFO 步骤"和在搜索输入框中输入搜索的关键词, 并点击搜索按钮后"执行结束
Fri, 2016-12-16 12:32:40 terrain.py[line:26] INFO 执行步骤"那么在搜索结果中可以看到书的作者"丘恩"", 定义在"\features\baidu.py"文件
Fri, 2016-12-16 12:32:41 terrain.py[line:31] INFO 步骤"那么在搜索结果中可以看到书的作者"丘恩""执行结束
Fri, 2016-12-16 12:32:41 terrain.py[line:36] INFO 场景"在百度网站搜索IT相关书籍"执行结束
Fri, 2016-12-16 12:32:41 terrain.py[line:45] INFO 恭喜, 3个场景运行, 3个场景运行成功   ← 执行结果概述
Fri, 2016-12-16 12:32:41 terrain.py[line:47] INFO 本次行为数据驱动执行结束
```

图　13-8

控制台输出如图 13-9 所示。

```
特性: 在百度网址搜索IT相关书籍                          # \features\baidu.feature:4
  能够搜索到书的作者, 比如吴晓华                        # \features\baidu.feature:5

  场景模板: 在百度网站搜索IT相关书籍                    # \features\baidu.feature:7
    如果将搜索词设定为书的名字"<书名>"                    # \features\baidu.py:8
Selenium WebDriver实战宝典
    当打开百度网站                                      # \features\baidu.py:13
    和在搜索输入框中输入搜索的关键词, 并点击搜索按钮后  # \features\baidu.py:18
    那么在搜索结果中可以看到书的作者"<作者>"            # \features\baidu.py:24

  例如:
    | 书名                         | 作者   |
    | Selenium WebDriver实战宝典   | 吴晓华 |
HTTP权威指南
    | HTTP权威指南                 | 古尔利 |
Python核心编程
    | Python核心编程               | 丘恩   |

1 feature (1 passed)
3 scenarios (3 passed)
12 steps (12 passed)
```

图　13-9

代码解释：

lettuce 支持中文编写的测试场景，是通过在场景文件以及测试场景的脚本文件的顶部

添加# language：zh-CN，声明使用的语言为中文，同时.feature 文件的编码必须保存为utf-8。

在 baidu.feature 文件中，关键词与后面的描述信息间的冒号（:）必须是英文字符的冒号。

13.8 批量执行行为驱动用例集

lettuce 支持一次执行多个用例，也就是放到 features 目录下的多个.feature 文件。

实例：

在 Pycharm 工具创建如下所示的目录结构及文件。

```
|lettuce
    |MultipleFeatures
        |features
            - Login_Chinese.feature
            - Login_Chinese.py
            - Login_English.feature
            - Login_English.py
        - terrain.py
```

Login_Chinese.feature 文件具体内容如下：

```
# encoding = utf - 8
# language: zh - CN

特性: 登录 126 邮箱和退出 126 邮箱登录

    场景: 成功登录 126 邮箱
        假如启动一个浏览器
        当用户访问 http://mail.126.com 网址
        当用户输入用户名"xxx"和密码"xxx"
        那么页面会出现"未读邮件"关键字

    场景: 成功退出 126 邮箱
        当用户从页面单击退出链接
        那么页面显示"您已成功退出网易邮箱"关键内容
```

Login_Chinese.py 文件内容如下：

```
# encoding = utf - 8
# language: zh - CN
from lettuce import *
from selenium import webdriver
from selenium.webdriver.common.keys import Keys
import time

@step(u'启动一个浏览器')
```

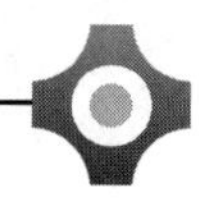

```
def open_browser(step):
    try:
        # 创建 Chrome 浏览器的 driver 实例,并存于全局对象 world 中,
        # 供后续场景或步骤函数使用
        world.driver = webdriver.Chrome(executable_path = "c:\\chromedriver")
    except Exception, e:
        raise e

@step(u'用户访问(. * )网址')
def visit_url(step, url):
    print url
    world.driver.get(url)

@step(u'用户输入用户名"(. * )"和密码"(. * )"')
def user_enters_UserName_and_Password(step, username, password):
    print username, password
    # 浏览器窗口最大化
    world.driver.maximize_window()
    time.sleep(3)
    # 切换进 frame 控件
    world.driver.switch_to.frame("x - URS - iframe")
    # 获取用户名输入框
    userName = world.driver.find_element_by_xpath('//input[@name = "email"]')
    # 输入用户名
    userName.send_keys(username)
    # 获取密码输入框
    pwd = world.driver.find_element_by_xpath("//input[@name = 'password']")
    # 输入密码
    pwd.send_keys(password)
    # 发送一个回车键
    pwd.send_keys(Keys.RETURN)
    # 等待 15 秒,以便登录后成功进入登录后的页面
    time.sleep(15)

@step(u'页面会出现"(. * )"关键字')
def message_displayed_Login_Successfully(step, keywords):
    # print world.driver.page_source.encode('utf - 8')
    # 断言登录成功后,页面是否出现预期的关键字
    assert keywords in world.driver.page_source
    # 断言成功后,打印登录成功信息
    print "Login Success"

@step(u'用户从页面单击退出链接')
def LogOut_from_the_Application(step):
    print " ==== ",world.driver
    # time.sleep(5)
    # 单击退出按钮,退出登录
    world.driver.find_element_by_link_text(u"退出").click()
    time.sleep(4)

@step(u'页面显示"(. * )"关键内容')
```

```
def displayed_LogOut_Successfully(step, keywords):
    # 断言退出登录后,页面是否出现退出成功关键内容
    assert keywords in world.driver.page_source
    print u"Logout Success"
    # 退出浏览器
    world.driver.quit()
```

Login_English. feature 文件内容如下：

```
#encoding = utf - 8

Feature: login and logout

   Scenario: Successful Login with Valid Credentials
      Given Launch a browser
      When User visit to http://mail.126.com Page
      And User enters UserName"xxx" and Password"xxx"
      Then Message displayed Login Successfully

    Scenario: Successful LogOut
        When User LogOut from the Application
        Then Message displayed LogOut Successfully
```

Login_English. py 文件内容如下：

```
#encoding = utf - 8
from lettuce import *
from selenium import webdriver
from selenium.webdriver.common.keys import Keys
import time

@step('Launch a browser')
def open_browser(step):
    try:
        world.driver = webdriver.Chrome(executable_path = "c:\\chromedriver")
    except Exception, e:
        raise e

@step('User visit to (.*) Page')
def visit_url(step, url):
    world.driver.get(url)

@step('User enters UserName"(.*)" and Password"(.*)"')
def user_enters_UserName_and_Password(step, username, password):
    world.driver.maximize_window()
    time.sleep(3)
    world.driver.switch_to.frame("x-URS-iframe")
    userName = world.driver.find_element_by_xpath('//input[@name = "email"]')
    userName.send_keys(username)
    pwd = world.driver.find_element_by_xpath("//input[@name = 'password']")
    pwd.send_keys(password)
    pwd.send_keys(Keys.RETURN)
```

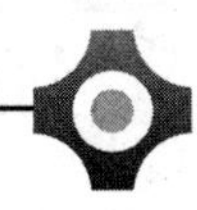

```
    time.sleep(15)

@step('Message displayed Login Successfully')
def message_displayed_Login_Successfully(step):
    # print world.driver.page_source.encode('utf - 8')
    assert u"未读邮件" in world.driver.page_source
    print "Login Success"

@step('User LogOut from the Application')
def LogOut_from_the_Application(step):
    print " ==== ",world.driver
    # time.sleep(15)
    world.driver.find_element_by_link_text(u"退出").click()
    time.sleep(4)

@step('Message displayed LogOut Successfully')
def displayed_LogOut_Successfully(step):
    assert u"您已成功退出网易邮箱" in world.driver.page_source
    print u"Logout Success"
    world.driver.quit()
```

terrain.py 文件内容用于统计各个场景执行结果信息，具体内容如下：

```
# encoding = utf - 8
from lettuce import *

# 在所有场景执行前执行
@before.all
def say_hello():
    print u"开始执行行为数据驱动测试..."

# 在每个 secnario 开始执行前执行
@before.each_scenario
def setup_some_scenario(scenario):
    print u'开始执行场景" % s"' % scenario.name

# 在每个 secnario 执行结束后执行
@after.each_scenario
def teardown_some_scenario(scenario):
    print u'场景" % s"执行结束' % scenario.name

# 在所有场景执行结束后执行
@after.all # 默认获取执行结果的对象作为 total 参数
def say_goodbye(total):
    result = u"恭喜, % d 个场景被运行, % d 个场景运行成功" % (
        total.scenarios_ran,            # 一共多少场景运行了
        total.scenarios_passed          # 一共多少场景运行成功了
    )
    print result
```

将 Login_Chinese.feature 和 Login_English.feature 文件中的 xxx 字符替换成有效的 126 邮箱登录账号及密码，然后在 Pycharm 工具的 Terminal（终端）中，或者 CMD 中，将当前的工作目录切换到 features 目录所在目录中，然后执行 lettuce 命令，启动行为测驱动测试。

结果说明：

lettuce 可以一次性执行多个.feature 文件，用户只需要将写好的多个.feature 文件放入 features 目录中即可，至于 terrain.py 文件，可以放在 features 目录中，也可以放在跟 features 目录同级目录中。步骤函数中的 print 语句执行的结果，只能在执行过程中在控制台中能看到，但不会被保留在输出结果信息中。

更多说明：

此程序在 Selenium3.x 和 Firefox 浏览器上运行会抛出 WebDriverException：Message：can't access dead object 异常，出现此异常的原因目前还不是特别清楚，可能是 Selenium3.x 版本的 bug，也可能是高版本 Firefox 浏览器的保护机制导致的，如果读者确实想在 Firefox 浏览器上执行此程序，可以将 Selenium 降到 2.x 版本（比如 2.53.6，Firefox 可以用 47.0.2 或者 43.0.4 版本）。

13.9 解决中文描述的场景输出到控制台乱码

在默认编码为 GBK 的 Windows 系统中执行场景使用中文描述的行为驱动测试时，打印到控制台的场景等信息，中文会出现乱码，这是由于 lettuce 框架将输出到控制台的场景描述信息转成 UTF8 编码的字符导致的。下面针对 lettuce（0.2.23）版本给出具体解决方法。

（1）进入 Python 安装目录中 lettuce 安装路径中的 plugins 目录中，比如笔者的本地路径为 C:\Python27\Lib\site-packages\lettuce\plugins。

（2）找到该目录下的 colored_shell_output.py 文件，如图 13-10 所示。

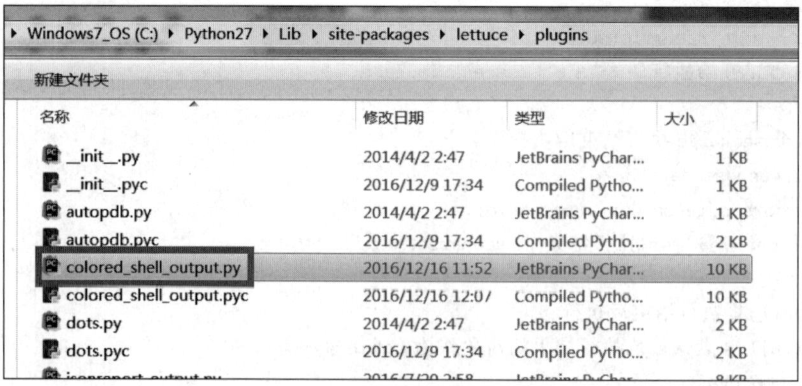

▸ Windows7_OS (C:) ▸ Python27 ▸ Lib ▸ site-packages ▸ lettuce ▸ plugins

新建文件夹

名称	修改日期	类型	大小
__init__.py	2014/4/2 2:47	JetBrains PyChar...	1 KB
__init__.pyc	2016/12/9 17:34	Compiled Pytho...	1 KB
autopdb.py	2014/4/2 2:47	JetBrains PyChar...	1 KB
autopdb.pyc	2016/12/9 17:34	Compiled Pytho...	2 KB
colored_shell_output.py	2016/12/16 11:52	JetBrains PyChar...	10 KB
colored_shell_output.pyc	2016/12/16 12:07	Compiled Pytho...	10 KB
dots.py	2014/4/2 2:47	JetBrains PyChar...	2 KB
dots.pyc	2016/12/9 17:34	Compiled Pytho...	2 KB

图 13-10

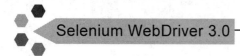

（3）打开该文件，找到该文件的第 32 行代码 what = what.encode('utf-8')，将其改成 what = what#.encode('utf-8')，或者将包裹 what = what.encode('utf-8')代码的整个 if 语句块一起去掉，如图 13-11 所示。

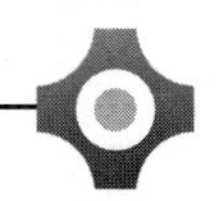

```
colored_shell_output.py
25  from lettuce.terrain import after
26  from lettuce.terrain import before
27  from lettuce.terrain import world
28
29
30  def wrt(what):
31      if isinstance(what, unicode):
32          what = what#.encode('utf-8')
33      sys.stdout.write(what)
34
35
```

删除该行或者按上述方式修改

图　13-11

（4）然后保存对该文件的修改，再次执行场景使用中文描述的行为驱动测试时，就可以看到控制台中打印的中文正常显示，如 13.7 小节结果截图所示。

第14章　Selenium Grid的使用

Selenium Grid 组件专门用于远程分布式测试或并发测试，通过并发执行测试用例的方式可以提高测试用例的执行速度和效率，解决界面自动化测试执行速度过慢的问题。此章为高级自动化测试人员的必修内容。

14.1 Selenium Grid 简介

前面章节介绍的自动化测试代码均是在本地计算机上运行的，在自动化测试用例不多的情况下，一台计算机可以在较短的时间内运行完所有的测试用例，并给出测试报告。但是随着计算机行业的快速发展，越来越多的大型项目横空出世，需要测试人员在较短的时间内完成成百上千个测试用例的运行，此时仅依靠一台计算机来完成这么多用例的执行肯定无法满足实际的测试要求。另外，越来越多的项目对浏览器的兼容性要求也越来越高，需要在多种操作系统和各种流行的浏览器以及不同版本间进行兼容性测试，单台计算机很难满足这样的需求。由此我们可以借助分布式的方式来并行执行大量的测试用例，以满足缩短测试时间和兼容性测试的要求。

Selenium Grid 的产生就是为了解决分布式运行自动化测试用例的需求。使用此组件可以在一台计算机上给多台计算机(不同操作系统和不同版本浏览器环境)分发多个测试用例从而并发执行，大大提高了测试用例执行效率，基本满足大型项目自动化测试的时限要求和兼容性要求。

Selenium Grid 目前有两个版本，一个是 1.0 版本，一个是 2.0 版本。Selenium Grid 和 Selenium RC 进行了合并，现在下载一个 selenium-server-standalone-3.xx.x.jar 文件就可以使用了。Selenium Grid 2 集成了 Apache Ant，可以同时支持 Selenium RC 的脚本和 WebDriver 脚本，最多可以远程控制 5 个浏览器。

SeleniumGrid 使用 Hub 和 Node 模式，一台计算机作为 Hub(管理中心)管理其他多个 Node(节点)计算机，Hub 负责将测试用例分发给多台 Node 计算机执行，并收集多台 Node 计算机执行结果的报告，汇总后提交一份总的测试报告，如图 14-1 所示。

Hub：

- 在分布式测试模式中，只能有一台作为 Hub 的计算机。
- Hub 负责管理测试脚本，并负责发送脚本给其他 Node 节点。
- 所有的 Node 节点计算机必须先在作为 Hub 的计算机中进行注册，注册成功后再和 Hub 计算机通信，Node 节点计算机会

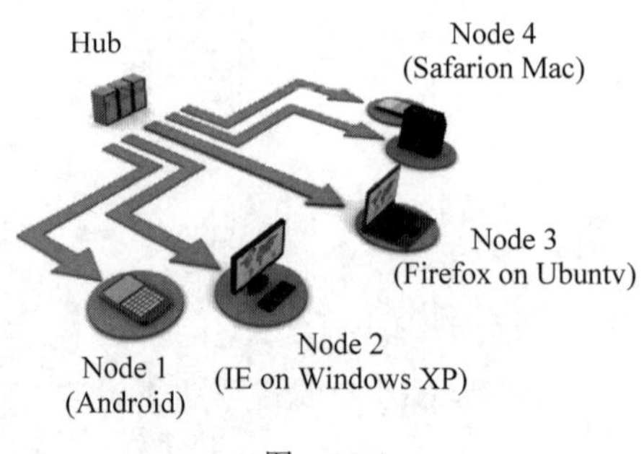

图　14-1

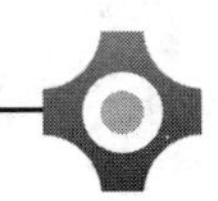

告之 Hub 自己的相关信息，例如，Node 节点的操作系统和浏览器相关版本。

- Hub 计算机可以给自己分配执行测试用例的任务。
- Hub 计算机分发的测试用例任务会在各个 Node 计算机节点执行。

Node：

- 在分布式测试模式中，可以有一个或者多个 Node 节点。
- Node 节点会打开本地的浏览器完成测试任务并返回测试结果给 Hub。
- Node 节点的操作系统和浏览器版本无须和 Hub 保持一致。
- 在 Node 节点上可以同时打开多个浏览器并行执行测试任务。

14.2 分布式自动化测试环境准备

安装 Java JDK，配置 Java 环境

1. 下载 JDK 1.8 安装文件

具体操作步骤如下：

（1）访问 https://profile.oracle.com/myprofile/account/create-account.jspx，先注册 Oracle 用户，注册过程请参阅页面提示信息，需要使用个人有效邮箱进行验证才能激活正式的注册用户。

（2）注册成功后，在页面 http://www.oracle.com/index.html 进行用户登录。

（3）登录成功后，访问 http://www.oracle.com/technetwork/java/javase/archive-139210.html，单击 Java SE8，进入 JDK 1.8 版本下载页面，如图 14-2 所示。

（4）在下载页面单击下载地址列表最上方链接，如图 14-3 所示。

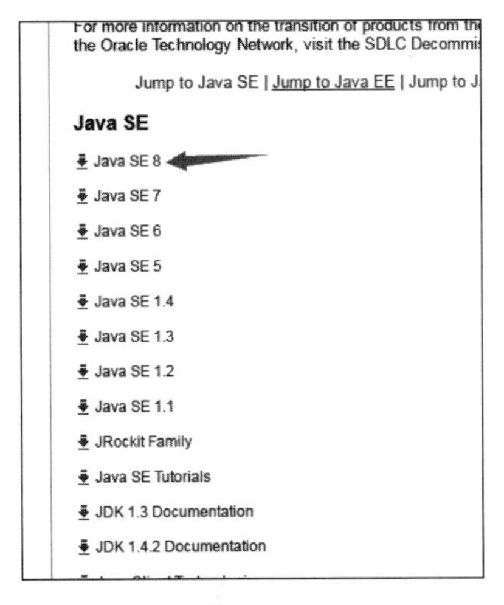

图　14-2

The JDK includes tools useful for developing and testing programs written in the Java programming language and running on the Java™ platform.

WARNING: These older versions of the JRE and JDK are provided to help developers debug issues in older systems. They are not updated with the latest security patches and are not recommended for use in production.

For production use Oracle recommends downloading the latest JDK and JRE versions and allowing auto-update.

Only developers and Enterprise administrators should download these releases.

Downloading these releases requires an oracle.com account. If you don't have an oracle.com account you can use the links on the top of this page to learn more about it and register for one for free.

For current Java releases, please consult the Oracle Software Download page.

- Java SE Development Kit 8u102　单击此链接
- Java SE Runtime Environment 8u102
- Server JRE (Java SE Runtime Environment) 8u102
- Java SE Development Kit 8u101
- Java SE Runtime Environment 8u101
- Server JRE (Java SE Runtime Environment) 8u101
- Java SE Development Kit 8u92

图　14-3

（5）跳转到下载版本选择页面，勾选 Accept License Agreement 后，选择和当前操作系统位数对应的 JDK 1.8 进行下载，如图 14-4 所示。

2. 安装 JDK 与配置环境变量

具体操作步骤如下：

(1) 双击下载好的 JDK 1.8 安装包进行安装操作，安装过程中只需要一直单击“下一步”即可。

(2) 安装完成 JDK 后，右击 Windows 7 操作系统桌面上的“计算机”图标，在弹出的快捷菜单中选择“属性”命令，如图 14-5 所示。

- Server JRE (Java SE Runtime Environment) 8

Java SE Development Kit 8u102

You must accept the Oracle Binary Code License Agreement for Java SE to download this software.

Thank you for accepting the Oracle Binary Code License Agreement for Java SE; you may now download this software.

Product / File Description	File Size	Download
Linux ARM 32 Hard Float ABI	77.78 MB	jdk-8u102-linux-arm32-vfp-hflt.tar.gz
Linux ARM 64 Hard Float ABI	74.73 MB	jdk-8u102-linux-arm64-vfp-hflt.tar.gz
Linux x86	160.35 MB	jdk-8u102-linux-i586.rpm
Linux x86	175.03 MB	jdk-8u102-linux-i586.tar.gz
Linux x64	158.35 MB	jdk-8u102-linux-x64.rpm
Linux x64	173.03 MB	jdk-8u102-linux-x64.tar.gz
Mac OS X	227.35 MB	jdk-8u102-macosx-x64.dmg
Solaris SPARC 64-bit	139.59 MB	jdk-8u102-solaris-sparcv9.tar.Z
Solaris SPARC 64-bit	98.98 MB	jdk-8u102-solaris-sparcv9.tar.gz
Solaris x64	140.02 MB	jdk-8u102-solaris-x64.tar.Z
Solaris x64	96.24 MB	jdk-8u102-solaris-x64.tar.gz
Windows x86	189.2 MB	jdk-8u102-windows-i586.exe
Windows x64	194.68 MB	jdk-8u102-windows-x64.exe

32位

64位

Back to top

Java SE Runtime Environment 8u102

图 14-4

图 14-5

(3) 在弹出的 Windows 窗口右边单击“高级系统设置”，如图 14-6 所示。

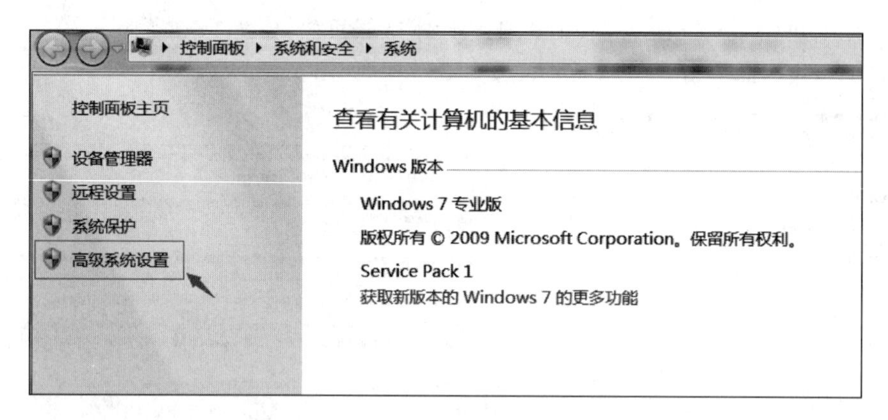

图 14-6

(4) 在弹出的“系统属性”对话框，单击“高级”标签栏，然后单击“环境变量”按钮，如图 14-7 所示。

(5) 在弹出的“环境变量”对话框中，单击系统变量的下方“新建”按钮，并在弹出的对话框的变量名输入框中输入 JAVA_HOME，变量值输入框中输入 JDK 1.8 的安装目录(比如笔者的安装目录为 C:\Program Files\Java\jdk1.8.0_111)，如图 14-8 所示，然后单击“确定”按钮保存，回到“环境变量”对话框。

(6) 在系统环境变量中找到 Path 变量行，双击该行，在弹出的 Path 环境变量的编辑界面的变量值输入框的最前面添加“%JAVA_HOME%\bin;%JAVA_HOME%\jre\bin;”，

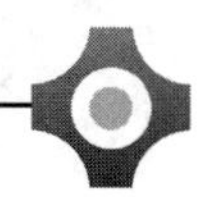

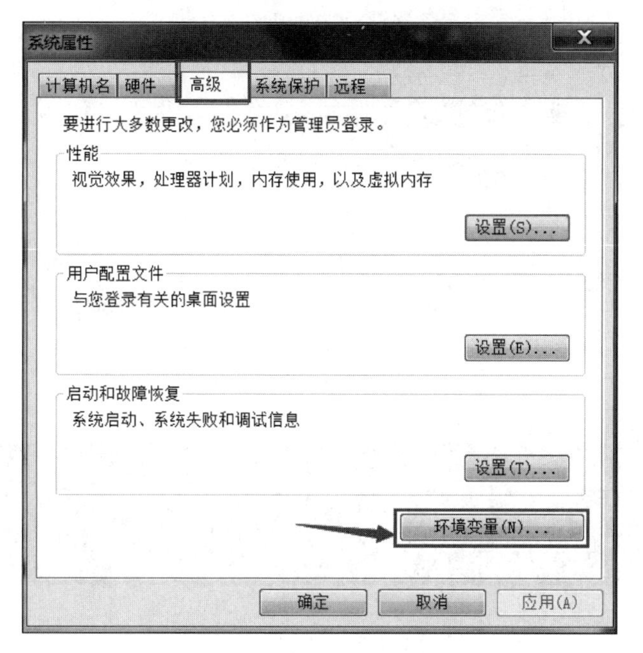

图　14-7

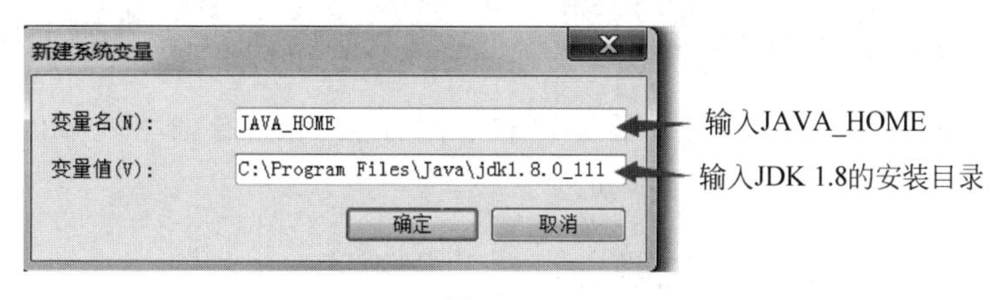

图　14-8

如图 14-9 所示，然后单击“确定”保存修改并且返回“环境变量”对话框。

(7) 再次单击系统变量下面的“新建”按钮，在弹出的“新建系统变量”对话框中，在“变量名”输入框中输入 CLASSPATH，在“变量值”输入框中输入.；%JAVA_HOME%\lib；%JAVA_HOME%lib\tools.jar(注意最前面有一点)，如图 14-10 所示，然后单击“确定”完成 Java 环境变量的配置步骤。

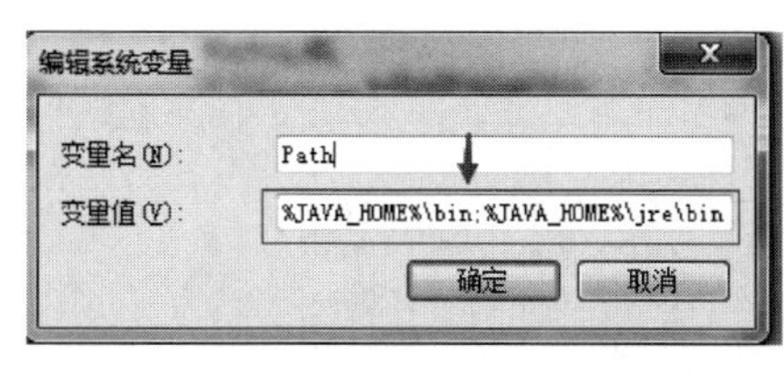

图　14-9

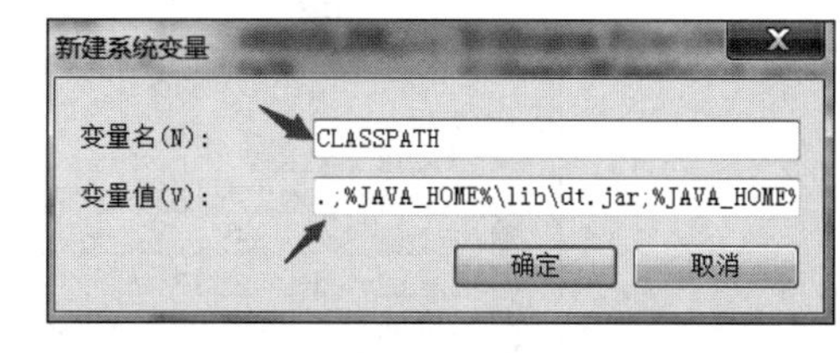

图　14-10

(8) 按下 WIN + R 组合键，在弹出的“运行”输入框中输入 cmd，如图 14-11 所示，然后按下回车键，调出 CMD 对话框，并在 CMD 界面输入 java -version，按回车键后显示出 Java 的版本信息，如图 14-12 所示，表示 Java 环境配置成功。

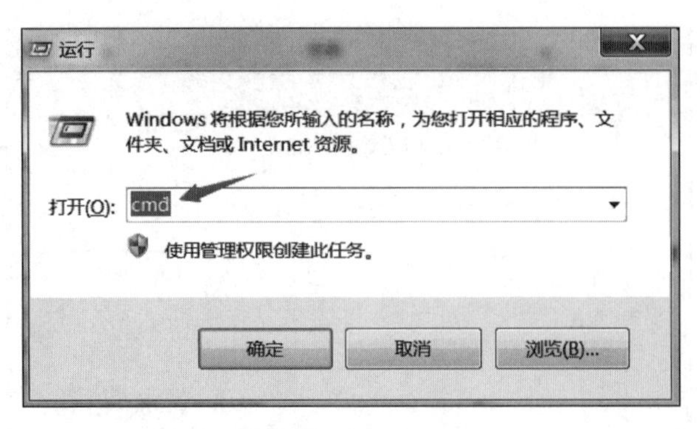

图 14-11

```
C:\Users\SR>java -version
java version "1.8.0_73"        JAVA的版本信息
Java(TM) SE Runtime Environment (build 1.8.0_73-b02)
Java HotSpot(TM) 64-Bit Server VM (build 25.73-b02, mixed mode)
```

图 14-12

14.3 Selenium Grid 的使用方法

本节将详细讲解 Selenium Grid 的配置和使用方法，请读者根据本节内容在本地计算机网络内尝试搭建分布式测试执行环境。

14.3.1 远程调用 Firefox 浏览器进行自动化测试

具体操作步骤如下：

(1) 找到两台 Windows 系统的计算机 A 和 B，A 计算机作为 Hub，B 计算机作为 Node

(2) 两台计算机均访问 http://selenium-release.storage.googleapis.com/index.html，并找到与你使用的 Selenium 同版本的下载链接下载 selenium-server-standalone-x.xx.x.jar，如图 14-13 所示，并保存在两台计算机 C 盘根目录中。

Index of /3.14/

Name	Last modified	Size
Parent Directory		
IEDriverServer_Win32_3.14.0.zip	2018-08-02 16:46:20	1.02MB
IEDriverServer_x64_3.14.0.zip	2018-08-02 16:46:20	1.13MB
selenium-dotnet-3.14.0.zip	2018-08-02 16:46:21	5.20MB
selenium-dotnet-strongnamed-3.14.0.zip	2018-08-02 16:46:22	5.21MB
selenium-html-runner-3.14.0.jar	2018-08-02 20:15:26	14.21MB
selenium-java-3.14.0.zip	2018-08-02 20:15:15	8.40MB
selenium-server-3.14.0.zip	2018-08-02 20:15:09	20.92MB
selenium-server-standalone-3.14.0.jar	2018-08-02 20:14:54	22.17MB

图 14-13

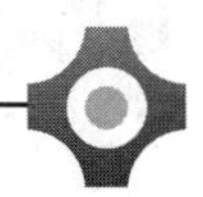

本书中下载的文件名为 selenium-server-standalone-3.14.0.jar。

(3) 在机器 A 上打开 CMD 窗口，将当前工作目录切换到 C 盘根目录(也就是 selenium-server-standalone-3.14.0.jar 文件所在目录)，然后执行如下语句：

```
java -jar selenium-server-standalone-3.14.0.jar -role hub
```

-role hub：启动一个 hub 服务，作为分布式管理中心，等待 WebDriver 客户端进行注册和请求，默认接收注册的地址为 http://localhost:4444/grid/register，默认启动端口为 4444。

此行语句表示使用 JAVA 命令把 JAR 文件作为程序执行，并将 role 参数值传递给 JAR 文件的函数，以此启动管理中心，如图 14-14 所示。

```
C:\WINDOWS\system32\cmd.exe - java -jar selenium-server-standalone-3.14.0.jar -role hub

D:\driver>java -jar selenium-server-standalone-3.14.0.jar -role hub
15:14:56.099 INFO [GridLauncherV3.launch] - Selenium build info: version: '3.14.0', revision: 'aacccce0'
15:14:56.099 INFO [GridLauncherV3$2.launch] - Launching Selenium Grid hub on port 4444
2018-12-16 15:14:56.599:INFO::main: Logging initialized @963ms to org.seleniumhq.jetty9.util.log.StdErrLog
15:14:57.270 INFO [Hub.start] - Selenium Grid hub is up and running
15:14:57.270 INFO [Hub.start] - Nodes should register to http://192.168.0.107:4444/grid/register/
15:14:57.270 INFO [Hub.start] - Clients should connect to http://192.168.0.107:4444/wd/hub
```

图 14-14

(4) 在机器 A(假如 IP 地址为 192.168.1.107)中的 Firefox 浏览器地址栏中访问 http://localhost:4444/grid/console，如果访问的网页中显示出 view config 的链接，表示 Hub 已经成功启动。默认情况下 Selenium Grid 使用 4444 作为服务端口号。在机器 B 上也可以访问此网址，只需要将 localhost 换成 A 机器的 IP 地址 192.168.1.107 即可，访问地址为 http://192.168.1.107:4444/grid/console，成功界面如图 14-15 所示。

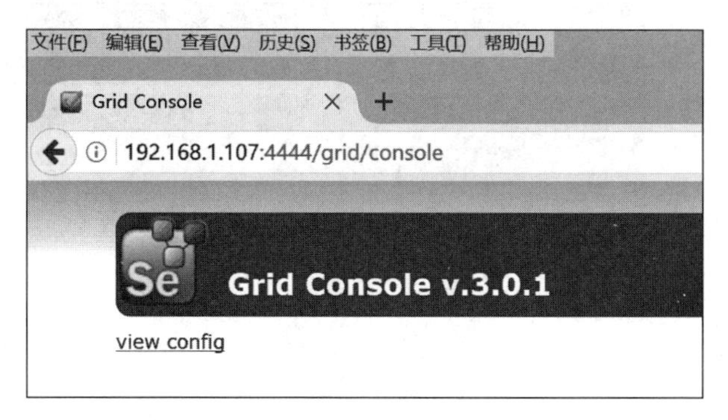

图 14-15

(5) 在机器 B(IP 地址为 192.168.1.113)中打开 CMD 窗口，并将当前工作目录切换到 C 盘根目录(也就是 selenium-server-standalone-3.14.0.jar 文件所在目录)，输入如下命令：

```
java -jar selenium-server-standalone-3.14.0.jar -role webdriver -hub http://192.168.
1.107:4444/grid/register -Dwebdriver.firefox.driver="C:\geckodriver.exe" -port 6655 -
maxSession 5 -browser browserName="firefox",maxInstances=5
```

参数说明：

- role：参数值 webdriver 表示 Node 节点的名字。

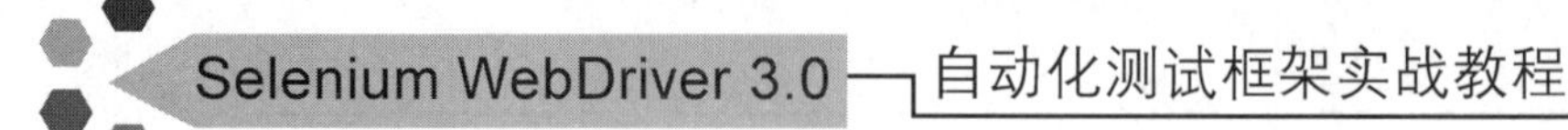

- hub：参数值表示管理中心的 URL 地址，Node 会连接这个地址进行节点注册。
- port：参数值表示 Node 节点服务的端口号为 6655，建议使用大于 5000 的端口号。

命令执行后会看到如图 14-16 所示的结果。

```
C:\WINDOWS\system32\cmd.exe - java -jar selenium-server-standalone-3.14.0.jar -role webdriver -hub http://192.168.1.107:4444/grid/regi...

D:\driver>java -jar selenium-server-standalone-3.14.0.jar -role webdriver -hub http://192.168.1.107:4444/grid/register-D
webdriver.firefox.driver="C:\geckodriver.exe" -port 6655 -maxSession 5 -browser browserName="firefox",maxInstances=5
15:18:07.897 INFO [GridLauncherV3.launch] - Selenium build info: version: '3.14.0', revision: 'aacccce0'
15:18:07.913 INFO [GridLauncherV3$3.launch] - Launching a Selenium Grid node on port 6655
2018-12-16 15:18:08.654:INFO::main: Logging initialized @1237ms to org.seleniumhq.jetty9.util.log.StdErrLog
15:18:08.842 INFO [SeleniumServer.boot] - Selenium Server is up and running on port 6655
15:18:08.842 INFO [GridLauncherV3$3.launch] - Selenium Grid node is up and ready to register to the hub
15:18:09.076 INFO [SelfRegisteringRemote$1.run] - Starting auto registration thread. Will try to register every 5000 ms.

15:18:09.076 INFO [SelfRegisteringRemote.registerToHub] - Registering the node to the hub: http://192.168.1.107:4444/gri
d/register
15:18:31.339 INFO [SelfRegisteringRemote$1.run] - Couldn't register this node: Error sending the registration request: F
ailed to connect to /192.168.1.107:4444
15:18:57.383 INFO [SelfRegisteringRemote$1.run] - Couldn't register this node: The hub is down or not responding: Failed
 to connect to /192.168.1.107:4444
15:19:23.445 INFO [SelfRegisteringRemote$1.run] - Couldn't register this node: The hub is down or not responding: Failed
 to connect to /192.168.1.107:4444
15:19:49.490 INFO [SelfRegisteringRemote$1.run] - Couldn't register this node: The hub is down or not responding: Failed
 to connect to /192.168.1.107:4444
15:20:15.495 INFO [SelfRegisteringRemote$1.run] - Couldn't register this node: The hub is down or not responding: Failed
 to connect to /192.168.1.107:4444
15:20:41.500 INFO [SelfRegisteringRemote$1.run] - Couldn't register this node: The hub is down or not responding: Failed
 to connect to /192.168.1.107:4444
15:21:07.510 INFO [SelfRegisteringRemote$1.run] - Couldn't register this node: The hub is down or not responding: Failed
 to connect to /192.168.1.107:4444
```

图 14-16

(6) 再次访问网址 http://192.168.1.107:4444/grid/console，验证 Node 节点是否已在 Hub 上注册成功，注册成功会看到如图 14-17 所示的信息。

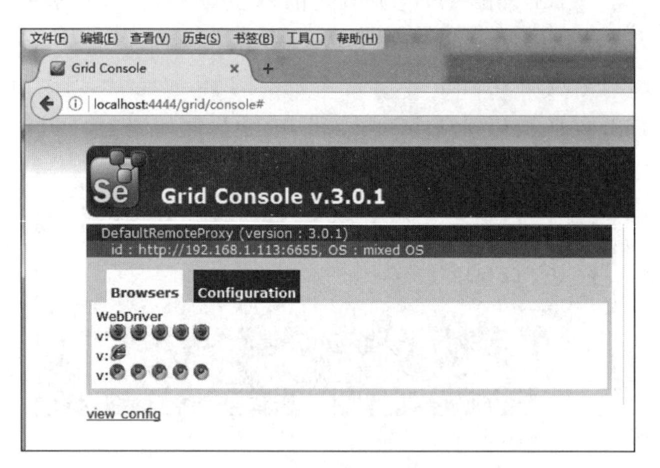

图 14-17

从此页面可以获取到节点计算机允许不同种类的浏览器打开多少个实例，验证节点计算机执行命令行的正确性。

(7) 编写分布式执行的测试脚本。

测试逻辑：

使用 Firefox 浏览器访问 sogou 首页，进行关键词"webdriver 实战宝典"的搜索，并验证搜索结果页面源码中是否出现"吴晓华"关键内容。

测试脚本：

```
# encoding = utf - 8
```

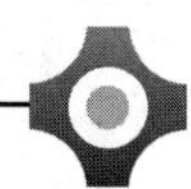

```
from selenium import webdriver
from selenium.webdriver.common.keys import Keys
import time

driver = webdriver.Remote(
  # 设定 Node 节点的 URL 地址,后续将通过访问这个地址连接到 Node 计算机
  command_executor = 'http://192.168.1.113:6655/wd/hub',
  desired_capabilities = {
        # 指定远程计算机执行使用的浏览器为 Firefox
        "browserName": "firefox",
        "video": "True",
        # 远程计算机的平台
        "platform": "WINDOWS"
  })
print("Video: " + "http://www.baidu.com" + driver.session_id)

try:
    driver.implicitly_wait(30)
    driver.maximize_window()
    driver.get("http://www.sogou.com")
    assert "搜狗" in driver.title
    elem = driver.find_element_by_id("query")
    elem.send_keys("webdriver 实战宝典")
    elem.send_keys(Keys.RETURN)
    time.sleep(3)
    assert "吴晓华" in driver.page_source
    print('done!')
finally:
    driver.quit()
```

测试结果:

在机器 B 上可以看到,计算机会自动启动 Firefox 浏览器执行测试脚本,执行完毕后浏览器会自动退出。在机器 A 上可以看到自动化测试的执行结果,如图 14-18 所示。

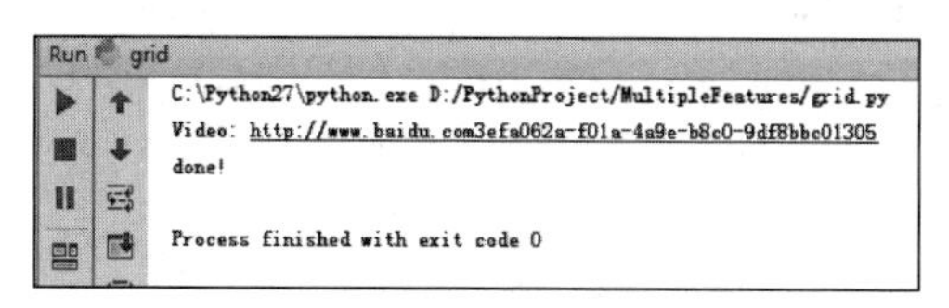

图　14-18

作为 Hub 启动时部分参数的含义如表 14-1 所示。

表　14-1

参数名称	参数含义
-rolehub	启动一个 Hub 服务,等待 Node 进行注册和请求
-hubConfig [filename]	设置一个符合 Selenium Grid2 规则的 JSON 格式的 Hub 配置文件
-port	指定 Hub 监听的端口
-host	ip 或 host,指定 Hub 机的 ip 或者 host 值

续表

参数名称	参数含义
-newSessionWaitTimeout	指定一个新的测试 session 等待执行的间隔时间,即一个代理节点上前后两个测试间的时间间隔,单位毫秒。默认为－1,即没有超时
-browserTimeout	浏览器无响应的超时时间
-servlets xxx	在 Hub 上注册一个新的 serlet,访问地址为/grid/admin/xxx

作为 Node 节点启动时部分参数的含义如表 14-2 所示。

表 14-2

参数名称	参数含义
-port	节点计算机提供远程连接服务的端口号,也是 Hub 监听的端口
-role [node|wd|rc]	为 Node 值时表示注册的 RC 可以支持所有版本的 Selenium 为 wd 值时表示注册的 RC 不支持 Selenium 1,也可以写成 webdriver。 为 rc 值时表示注册的 RC 仅支持 Selenium 1
-hub url_to_hub	url_to_hubz 值为 Hub 启动的注册地址,默认为 http://hub_ip:4444/grid/register,该选项包含了-hubHost 和-hubPort 两个选项
-timeout	Hub 在无法收到 Node 节点的任何请求后,在等待 timeout 设定的时间后会自动释放和 Node 节点的连接 注意:此参数不是 WebDriver 定位页面元素最大的等待时间
-maxSession	在一台节点计算机中,允许同时最多打开多少个浏览器窗口
-browser	设定节点计算机允许使用的浏览器信息,例如: `browserName = firefox, version = 3. 6, firefox _ binary = /home/myhomedir/firefox36/firefox, maxInstances = 3, platform = LINUX` version:设定浏览器版本版本号,当多个版本号共存的时候,要明确使用哪个版本进行测试 platform:设定节点操作系统属性为 Linux。可使用的值有 Windows、Linux 和 Mac firefox_binary:设定 Firefox 浏览器启动路径 maxInstances:最多允许同时启动 3 个 Firefox 浏览器窗口
-registerCycle	设定节点计算机间隔多少毫秒去注册一下 Hub 管理中心,以便重启 Hub 时不需要再重新启动所有的代理节点
-browserTimeout	浏览器无响应的超时时间
-nodeTimeout	客户端超时时间
-cleanupCycle	代理节点检查超时的周期
-nodeConfig jsonFile	一个符合 Selenium Grid2 规则的 json 格式的 Node 配置文件
-proxy 代理类	默认指向 org. openqa. grid. selenium. proxy. DefaultRemoteProxy

14.3.2 远程调用 IE 浏览器进行自动化测试

远程调用 IE 浏览器进行自动化测试的步骤和远程调用 Firefox 浏览器的步骤基本相同,只是节点注册 Hub 的命令行参数和测试程序有一些变化。

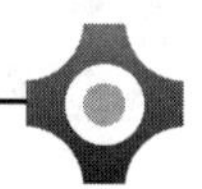

操作步骤：

（1）在机器 A 上启动 Hub，切换 CMD 当前工作目录到 selenium-server-standalone-3.14.0.jar 文件所在目录，然后执行如下命令：

```
java -jar selenium-server-standalone-3.14.0.jar -role hub
```

（2）在机器 B 上进行 Node 节点注册，切换 CMD 当前工作目录到 selenium-server-standalone-3.14.0.jar 文件所在目录，然后执行如下命令：

```
java -jar selenium-server-standalone-3.14.0.jar -role webdriver -hub http://192.168.
1.107:4444/grid/register -Dwebdriver.ie.driver="C:\IEDriverServer.exe" -port 6655 -
maxSession 5 -browser browserName="internet explorer",maxInstances=5
```

与 Firefox 浏览器相比，启动 Node 节点 hub 服务的命令只需要将浏览器名 firefox 修改成 internet explorer，浏览器驱动路径及文件名 C:\geckodriver.exe 修改为 C:\IEDriverServer.exe 即可。

测试程序：

```
#encoding=utf-8
from selenium import webdriver
from selenium.webdriver.common.keys import Keys
import time

driver = webdriver.Remote(
  # 设定 Node 节点的 URL 地址，后续将通过访问这个地址连接到 Node 计算机
  command_executor = 'http://192.168.1.113:6655/wd/hub',
  desired_capabilities = {
        # 指定远程计算机执行使用的浏览器为 IE
        # IE8 以下版本写 ie,IE8 写 iehta,IE11 写 internet explorer
        "browserName": "internet explorer",
        "video": "True",
        # 远程计算机的平台
        "platform": "WINDOWS" # 或者写 XP
  })
print("Video: " + "http://www.baidu.com" + driver.session_id)

try:
    driver.implicitly_wait(30)
    driver.maximize_window()
    driver.get("http://www.sogou.com")
    assert "搜狗" in driver.title
    elem = driver.find_element_by_id("query")
    elem.send_keys("WebDriver 实战宝典")
    elem.send_keys(Keys.RETURN)
    time.sleep(3)
    assert "吴晓华" in driver.page_source
    print('done!')
finally:
    driver.quit()
```

14.3.3 远程调用 Chrome 浏览器进行自动化测试

远程调用 Chrome 浏览器进行自动化测试的步骤和远程调用 Firefox 浏览器的步骤基本相同，只是节点注册 Hub 的命令行参数和测试程序有一些变化。

操作步骤：

(1) 在机器 A 上启动 Hub，切换 CMD 当前工作目录到 selenium-server-standalone-3.14.0.jar 文件所在目录，然后执行如下命令：

```
java -jar selenium-server-standalone-3.14.0.jar -role hub
```

(2) 在机器 B 上进行 Node 节点注册，切换 CMD 当前工作目录到 selenium-server-standalone-3.14.0.jar 文件所在目录，然后执行如下命令：

```
java -jar selenium-server-standalone-3.14.0.jar -role webdriver -hub http://192.168.
1.107:4444/grid/register -Dwebdriver.chrome.driver="C:\chromedriver.exe" -port 8855 -
maxSession 5 -browser browserName="chrome",maxInstances=5
```

测试脚本：

```
#encoding=utf-8
from selenium import webdriver
from selenium.webdriver.common.keys import Keys
import time

driver = webdriver.Remote(
  # 设定 Node 节点的 URL 地址,后续将通过访问这个地址连接到 Node 计算机
  command_executor = 'http://192.168.1.113:8855/wd/hub',
  desired_capabilities = {
        # 指定远程计算机执行使用的浏览器为 IE
        "browserName": "chrome",
        "video": "True",
        # 远程计算机的平台
        "platform": "WINDOWS" # 或者写 XP
  })
print("Video: " + "http://www.baidu.com" + driver.session_id)

try:
    driver.implicitly_wait(30)
    driver.maximize_window()
    driver.get("http://www.sogou.com")
    assert "搜狗" in driver.title
    elem = driver.find_element_by_id("query")
    elem.send_keys("WebDriver 实战宝典")
    elem.send_keys(Keys.RETURN)
    time.sleep(3)
    assert "吴晓华" in driver.page_source
    print('done!')
finally:
    driver.quit()
```

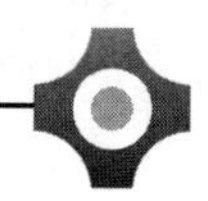

14.3.4 同时支持多个浏览器进行自动化测试

Selenium Grid 支持同时将 Node 节点计算机上多个浏览器注册到 Hub 上，以便满足测试过程中对不同浏览器的需求。

操作步骤：

(1) 在机器 A 上启动 Hub，切换 CMD 当前工作目录到 selenium-server-standalone-3.14.0.jar 文件所在目录，然后执行如下命令：

```
java -jar selenium-server-standalone-3.14.0.jar -role hub
```

(2) 在机器 B 上进行 Node 节点注册，切换 CMD 当前工作目录到 selenium-server-standalone-3.14.0.jar 文件所在目录，然后执行如下命令：

```
java -jar selenium-server-standalone-3.14.0.jar -role webdriver -hub http://192.168.1.107:4444/grid/register -Dwebdriver.chrome.driver="C:\chromedriver.exe" -Dwebdriver.ie.driver="C:\IEDriverServer.exe" -Dwebdriver.firefox.driver="C:\geckodriver.exe" -port 6666 -maxSession 5 -browser browserName="internet explorer",maxInstances=5 -browser browserName="chrome",maxInstances=5 -browser browserName="firefox",maxInstance=5
```

测试脚本：

```python
#encoding=utf-8
from selenium import webdriver
from selenium.webdriver.common.keys import Keys
import time,random

# 节点主机的访问地址
host="http://192.168.1.113:6666/wd/hub"
browsers = ["firefox", "chrome", "internet explorer"]
driver = webdriver.Remote(
  # 设定 Node 节点的 URL 地址，后续将通过访问这个地址连接到 Node 计算机
  command_executor = host,
  desired_capabilities = {
        # 在 browsers 列表中随机选择一个浏览器
        "browserName": random.choice(browsers),
        "platform": "WINDOWS"
  })

try:
    driver.implicitly_wait(30)
    driver.maximize_window()
    driver.get("http://www.sogou.com")
    assert "搜狗" in driver.title
    elem = driver.find_element_by_id("query")
    elem.send_keys("WebDriver 实战宝典")
    elem.send_keys(Keys.RETURN)
    time.sleep(3)
    assert "吴晓华" in driver.page_source
    print('done!')
finally:
    driver.quit()
```

14.4 结合 uittest 完成分布式自动化测试

操作步骤：

(1) 在机器 A 和 B 的 C 盘根目录下均放入 selenium-server-standalone-xx. xx. xx. jar 文件。

(2) 在机器 A 中打开 CMD，并将 CMD 当前工作目录切换到 C 盘根目录，然后执行如下命令启动 Hub 服务：

```
java -jar selenium-server-standalone-3.14.0.jar -role hub
```

(3) 在机器 B 中打开 CMD，并将 CMD 当前工作目录切换到 C 盘根目录，然后执行如下命令启动机器 B 节点服务，并进行注册：

```
java -jar selenium-server-standalone-3.14.0.jar -role webdriver -hub http://192.168.
1.107:4444/grid/register -Dwebdriver.firefox.driver="C:\geckodriver.exe" -port 6655 -
maxSession 5 -browser browserName="firefox",firefox_binary="C:\Program Files\Mozilla
Firefox\firefox.exe",platform="WINDOWS",maxInstances=5
```

上述命令中-browser 参数后面的值均是本次远程调用的 Firefox 浏览器的属性，其中 firefox_binary 用于设置浏览器的安装路径。

测试脚本：

```
#encoding=utf-8
from selenium import webdriver
from selenium.webdriver.common.keys import Keys
import unittest, time
from HTMLTestRunner import HTMLTestRunner

class SeleniumGridTest(unittest.TestCase):
    def setUp(self):
        self.driver = webdriver.Remote(
            # 设定 Node 节点的 URL 地址，后续将通过访问这个地址连接到 Node 计算机
            command_executor='http://192.168.1.113:6655/wd/hub',
            desired_capabilities={
                # 在 browsers 列表中随机选择一个浏览器
                "browserName": "firefox",
                "video": "True",
                "platform": "WINDOWS"
            }
        )
        self.driver.implicitly_wait(30)
        print("Video: " + "http://www.baidu.com" + self.driver.session_id)

    def testSogou(self):
        self.driver.maximize_window()
        self.driver.get("http://www.sogou.com")
        assert "搜狗" in self.driver.title
        elem = self.driver.find_element_by_id("query")
```

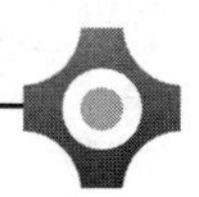

```
        elem.send_keys("WebDriver 实战宝典")
        elem.send_keys(Keys.RETURN)
        time.sleep(3)
        assert "吴晓华" in self.driver.page_source
        print 'done!'

    def tearDown(self):
        self.driver.quit()

def suite():
    suite1 = unittest.TestLoader().loadTestsFromTestCase(SeleniumGridTest)
    return unittest.TestSuite(suite1)

def run(suite, report = "d:\\seleniumGridTest.html"):
    # 以二进制方式打开文件,准备写
    with open(report, "wb") as fp:
        # 使用 HTMLTestRunner 配置参数,输出报告路径、报告标题、描述,均可以配置
        runner = HTMLTestRunner(
            stream = fp,
            verbosity = 2,
            title = '分布式测试结果',
            description = '测试报告描述')
        runner.run(suite)                    # 运行测试集合

if __name__ == '__main__':
    run(suite())                             # 运行测试集合
```

执行结束后生成的测试报告如图 14-19 所示。

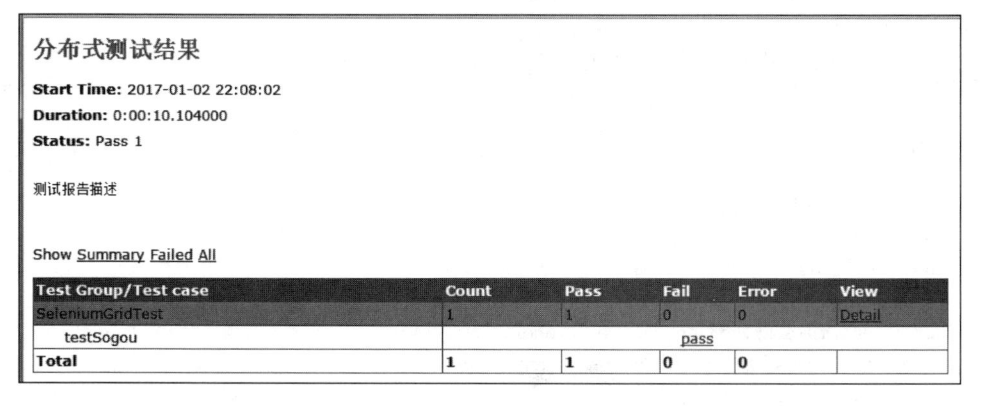

分布式测试结果

Start Time: 2017-01-02 22:08:02

Duration: 0:00:10.104000

Status: Pass 1

测试报告描述

Show Summary Failed All

Test Group/Test case	Count	Pass	Fail	Error	View
SeleniumGridTest	1	1	0	0	Detail
testSogou	pass				
Total	1	1	0	0	

图　14-19

14.5　实现并发的分布式自动化测试

Selenium Grid 不仅支持在同一台机器上同时启动好几个浏览器并发跑测试用例,也支持同时在多台不同机器上启动不同的浏览器并发跑测试用例,本小节主要针对第二种方法进行讲解。

操作步骤：

（1）在机器 A、B 和 C 的 C 盘根目录放入 selenium-server-standalone-xx. xx. xx. jar 文件。

（2）在机器 A 中打开 CMD，并将 CMD 当前工作目录切换到 C 盘根目录，然后执行如下命令启动 Hub 服务：

```
java -jar selenium-server-standalone-3.14.0.jar -role hub
```

再重新打开一个 CMD，将 CMD 当前工作目录切换到 C 盘根目录，然后执行如下命令进行节点注册：

```
java -jar selenium-server-standalone-3.14.0.jar -role webdriver -hub http://192.168.
31.26:4444/grid/register -Dwebdriver.ie.driver = "C:\IEDriverServer.exe" -port 6655 -
maxSession 5 -browser browserName = "internet explorer",maxInstances = 5
```

（3）在机器 B 中打开 CMD，并将 CMD 当前工作目录切换到 C 盘根目录，然后执行如下命令启动机器 B 节点服务，并进行注册：

```
java -jar selenium-server-standalone-3.14.0.jar -role webdriver -hub http://192.168.
31.26:4444/grid/register -Dwebdriver.firefox.driver = "C:\geckodriver.exe" -port 6666 -
maxSession 5 -browser browserName = "firefox",platform = "WINDOWS",maxInstances = 5
```

（4）在机器 C 中打开 CMD，并将 CMD 当前工作目录切换到 C 盘根目录，然后执行如下命令启动机器 B 节点服务，并进行注册：

```
java -jar selenium-server-standalone-3.14.0.jar -role webdriver -hub http://192.168.
32.26:4444/grid/register -Dwebdriver.chrome.driver = "C:\chromedriver.exe" -port 8889 -
maxSession 5 -browser browserName = "chrome",platform = "WINDOWS",maxInstances = 5
```

注册的 A、B 和 C 机器节点的端口号必须不一致。

测试脚本：

```python
#encoding = utf-8
from multiprocessing import Pool
import os, time
from selenium import webdriver
from selenium.webdriver.common.keys import Keys
from multiprocessing import Manager, current_process

def node_task(name, lock, arg, successTestCases, failTestCases):
    # 获取当前进程名
    procName = current_process().name
    print(procName)
    time.sleep(1.2)
    print(arg['node'])
    print(arg["browerName"])
    print('Run task %s (%s)...\n' % (name, os.getpid()))
    start = time.time()
    driver = webdriver.Remote(
```

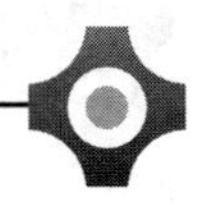

```
            command_executor = "%s" % arg['node'],
            desired_capabilities = {
                "browserName": "%s" % arg["browerName"],
                "video": "True",
                "platform": "WINDOWS"})
    try:
        driver.implicitly_wait(30)
        driver.maximize_window()
        driver.get("http://www.sogou.com")
        assert "搜狗" in driver.title
        elem = driver.find_element_by_id("query")
        elem.send_keys("webdriver 实战宝典")
        time.sleep(2)
        elem.send_keys(Keys.RETURN)
        assert "吴晓华" not in driver.page_source
        # 请求获取共享资源的锁
        lock.acquire()
        # 向进程间共享列表 successTestCases 中添加执行成功的用例名称
        successTestCases.append("TestCase" + str(name))
        # 释放共享资源的锁,以便其他进程能获取到此锁
        lock.release()
        print("TestCase" + str(name) + " done!")
    except AssertionError as err:
        print("AssertionError occur!""testCase" + str(name))
        print(err)
        # 截取屏幕
        driver.save_screenshot('d:\\screenshoterror' + str(name) + '.png')
        lock.acquire()
        # 向共享列表 failTestCases 中添加执行失败的用例名称
        failTestCases.append("TestCase" + str(name))
        lock.release()
        print("测试用例执行失败")
    except Exception as err:
        print("Exception occur!")
        print(err)
        driver.save_screenshot('d:\\screenshoterror' + str(name) + '.png')
        # 请求获得共享资源操作的锁,操作完后自动释放
        with lock:
            # 向共享列表 failTestCases 中添加执行失败的用例名称
            failTestCases.append("TestCase" + str(name))
        print("测试用例执行失败")
    finally:
        driver.quit()
    end = time.time()
    print('Task %s runs %0.2f seconds.' % (name, (end - start)))

def run(nodeSeq):
    # 创建一个多进程的 Manager 实例
    manager = Manager()
    # 定义一个共享资源列表 successTestCases
    successTestCases = manager.list([])
```

```
    # 定义一个共享资源列表 failTestCases
    failTestCases = manager.list([])
    # 创建一个资源锁
    lock = manager.Lock()
    # 打印主进程的进程 ID
    print('Parent process %s.' % os.getpid())
    # 创建一个容量为 3 的进程池
    p = Pool(processes=3)
    testCaseNumber = len(nodeSeq)
    for i in range(testCaseNumber):
        # 循环创建子进程,并将需要的数据传入子进程
        p.apply_async(node_task, args=(i + 1, lock, nodeSeq[i],
                                       successTestCases, failTestCases))
    print('Waiting for all subprocesses done...')
    # 关闭进程池,不再接受新的请求任务
    p.close()
    # 阻塞主进程直到子进程退出
    p.join()
    return successTestCases, failTestCases

def resultReport(testCaseNumber, successTestCases, failTestCases):
    # 下面代码用于打印本次测试报告
    print("测试报告:\n")
    print("共执行测试用例:" + str(testCaseNumber) + "个\n")
    print("成功的测试用例:", str(len(successTestCases)))
    if len(successTestCases) > 0:
        for t in successTestCases:
            print(t)
    else:
        print("无")
    print("失败的测试用例:", str(len(failTestCases)))
    if len(failTestCases) > 0:
        for t in failTestCases:
            print(t)
    else:
        print("无")

if __name__ == '__main__':
    # 节点列表
    nodeList = [
        {"node":"http://10.0.24.206:6666/wd/hub","browerName":"internet explorer"},
        {"node":"http://10.0.24.206:6666/wd/hub","browerName":"chrome"},
        {"node":"http://10.0.24.206:6666/wd/hub", "browerName":"firefox"}]
    # 获取节点个数
    testCaseNumber = len(nodeList)
    # 开始多进程分布式测试
    successTestCases, failTestCases = run(nodeList)
    print('All subprocesses done.')
    # 在控制台中打印测试报告
    resultReport(testCaseNumber, successTestCases, failTestCases)
```

第15章 自动化测试框架的搭建及实战

本章为高级自动化工程师的必备技能，也是本书最具吸引力的一章。本章将从零开始搭建一个完整的自动化测试框架，请立志成为高级自动化测试专家的读者仔细阅读，建议参照本章的内容在本地计算机环境中进行搭建实战。笔者认为，只有不断地实践，才能真正具备自动化测试框架的搭建能力。

15.1 关于自动化测试框架

大多数的测试从业者都是从手工测试开始自己的职业生涯的，经过多年的手工测试后，开始思考自己的未来发展之路，难道要一辈子靠手工的方式来完成测试吗？一些手工测试工程师开始尝试使用自动化测试工具来替代执行自己每天不断重复的手工测试过程。但是在执行过程中，他们会发现好不容易写好的测试脚本，因为需求变化的原因没过多久就无法执行成功了。这样的情况在软件开发过程中是不可避免的，测试工程师只能去不断修改和维护自动化测试脚本。这样的情况反复出现后，测试工程师发现投入维护脚本的时间和精力比纯手工测试的方式还要多，而且还造成手工测试的时间被明显减少，导致测试的效果大打折扣，软件的质量还不如以前纯手工测试的时候好。有的时候测试脚本刚刚修改一半，软件的需求又发生了变化，这样的情况导致测试工程师只能放弃自动化测试脚本的维护，被迫重新投入到手工测试中。

以上场景大量地出现在各个尝试自动化测试的公司中，大家也在思考和尝试解决这样的问题：如何能够降低测试脚本的维护成本和工作量，提高自动化测试脚本的编写和维护效率，真正让自动化测试能够提高软件测试工程师的工作效率，为企业真正节省测试成本，能够为开发团队快速地反馈当前软件质量状态？在这样的背景下，自动化测试框架应运而生。

1. 自动化测试框架的定义

自动化测试框架是应用于自动化测试的程序框架，它提供了可重用的自动化测试模块，提供最基础的自动化测试功能(例如，打开浏览器、单击链接等功能)，或提供自动化测试执行和管理功能的架构模块(例如，TestNG)。它是由一个或多个自动化测试基础模块、自动化测试管理模块、自动化测试统计模块等组成的工具集合。

2. 自动化测试框架常见的4种模式

(1) 数据驱动测试框架

使用数据数组、测试数据文件或者数据库等方式作为测试过程输入的自动化测试框架，此框架可以将所有测试数据在自动化测试执行的过程中进行自动加载，动态判断测试结果是否符合预期，并自动输出测试报告。

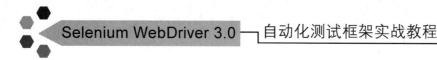

此框架一般用于在一个测试流程中使用多组不同的测试数据,以此来验证被测试系统是否能够正常工作。

(2) 关键字驱动测试框架

关键字驱动测试框架可以理解为高级的数据驱动测试框架,使用被操作的元素对象、操作的方法和操作的数据值作为测试过程输入的自动化测试框架,简单表示为 item.operation(value)。被操作的元素对象、操作的方法和操作的数据值可以保存在数据数组、数据文件、数据库中作为关键字驱动测试框架的输入。例如在页面上的用户名输入框中输入用户名,则可以在数据文件中进行如下定义:

```
用户名输入框,输入,testman
```

关键字驱动测试框架属于更加高级的自动化测试框架,可以兼容更多的自动化测试操作类型,大大提高了自动化测试框架的使用灵活性。

(3) 混合型测试框架

在关键字驱动测试框架中加入了数据驱动的功能,则框架被定义为混合型测试。

(4) 行为驱动测试框架

支持自然语言作为测试用例描述的自动化测试框架,例如前面章节讲到的 lettuce 框架。

3. 自动化测试框架的作用

(1) 能够有效组织和管理测试脚本。

(2) 进行数据驱动或者关键字驱动的测试。

(3) 将基础的测试代码进行封装,降低测试脚本编写的复杂性和重复性。

(4) 提高测试脚本维护和修改的效率。

(5) 自动执行测试脚本,并自动发布测试报告,为持续集成的开发方式提供脚本支持。

(6) 让不具备编程能力的测试工程师开展自动化测试工作。

4. 自动化测试框架的设计核心思想

世上没有最好的自动化测试框架,也没有万能的自动化测试测试框架,各种自动化测试框架都有自身的优点和缺点,所以我们在设计自动化测试框架的时候要考虑到实现一套自动化测试框架到底能够为测试工作本身解决什么样的具体问题,不能为了自动化而自动化,我们要以解决测试中的问题和提高测试工作的效率为主要导向来进行自动化测试框架的设计。

自动化测试框架的核心思想是将常用的脚本代码或者测试逻辑进行抽象和总结,然后将这些代码进行面向对象设计,将需要复用的代码封装到可公用的类方法中。通过调用公用的类方法,测试类中的脚本复杂度会被大大降低,让更多脚本能力不强的测试人员来实施自动化测试。

创建和实施 Web 自动化测试框架的步骤如下:

(1) 根据测试业务的手工测试用例,选出需要可自动化执行的测试用例。

(2) 根据可自动化执行的测试用例,分析出测试框架需要模拟的手工操作和重复高的测试流程或逻辑。

(3) 将手工操作和重复高的测试逻辑用在代码中实现,并在类中进行封装方法的编写。

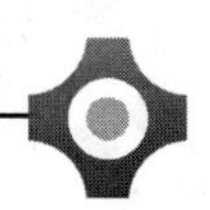

(4) 根据测试业务的类型和本身技术能力，选择数据驱动框架、关键字驱动框架、混合型框架还是行为驱动框架。

(5) 确定框架模型后，将框架中常用的浏览器选择、测试数据处理、文件操作、数据库操作、页面元素的原始操作、日志和报告等功能进行类方法的封装实现。

(6) 对框架代码进行集成测试和系统测试，采用 PageObject 模式和 TestNG 框架(或 JUnit)编写测试脚本，使用框架进行自动化测试，验证框架的功能可以满足自动化测试的需求。

(7) 编写自动化测试框架的常用 API 文档，以供他人参阅。

(8) 在测试组内部进行培训和推广。

(9) 不断收集测试过程中的框架使用问题和反馈意见，不断增加和优化自动化框架的功能，不断增强框架中复杂操作的封装效果，尽量降低测试脚本的编写复杂性。

(10) 定期评估测试框架的使用效果，评估自动化测试的投入产出比，再逐步推广自动化框架的应用范围。

15.2　数据驱动框架及实战

本小节主要讲解数据驱动测试框架的搭建，并且使用此框架来测试 126 邮箱登录和地址薄的相关功能。框架用到的基础知识均在前面的章节中做了详细介绍，本章节重点讲解框架搭建的详细过程。

被测试功能的相关页面描述：

登录页面如图 15-1 所示。

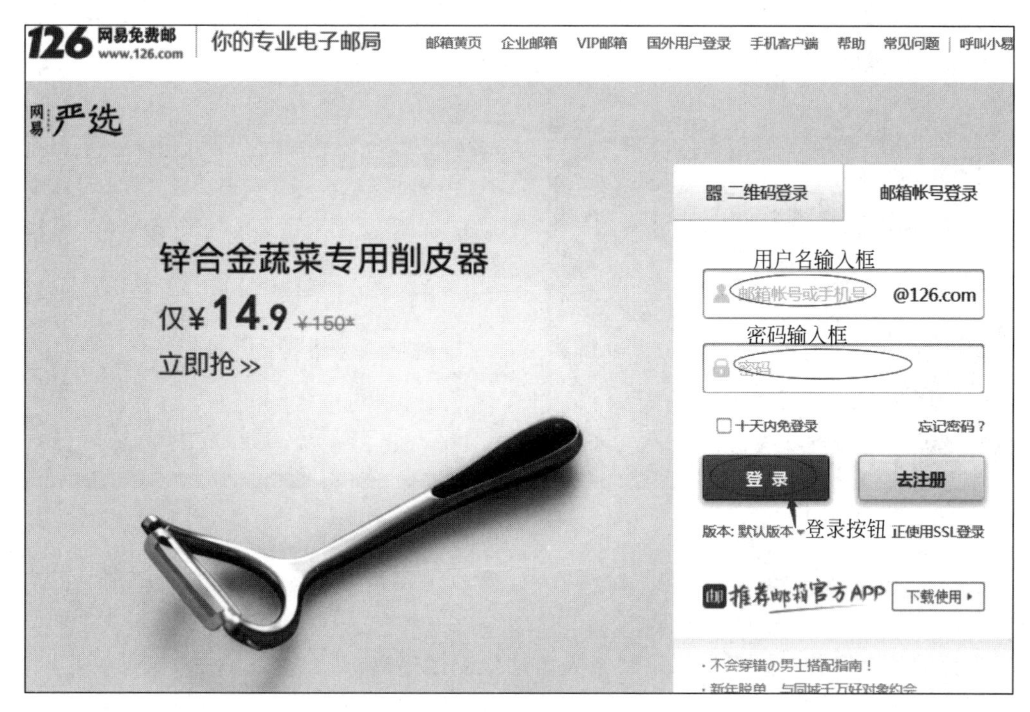

图　15-1

登录后的页面如图 15-2 所示。

图 15-2

单击“通讯录”链接后，进入通讯录主页，如图 15-3 所示。

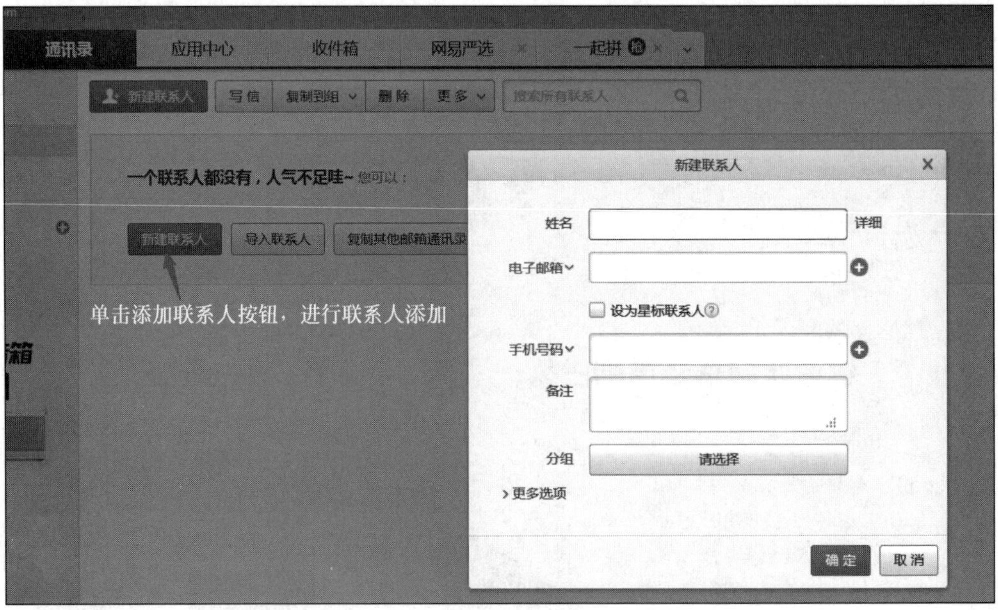

图 15-3

在图 15-3 所示的“新建联系人”对话框中，输入联系人的基本信息，然后单击“确定”按钮保存，显示的页面如图 15-4 所示。

非数据驱动框架时的添加邮箱联系人自动化测试代码：

```
# encoding = utf - 8
```

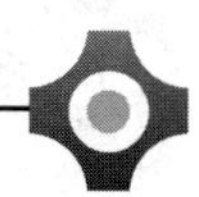

图　15-4

```
from selenium import webdriver
from selenium.webdriver.common.keys import Keys
import time

# 创建 Firefox 浏览器的实例
driver = webdriver.Firefox(executable_path = "c:\\geckodriver")
# 设置隐式等待时间为 10 秒
driver.implicitly_wait(10)
# 最大化浏览器窗口
driver.maximize_window()

# 访问 126 邮箱登录页面
driver.get("http://mail.126.com")
# 暂停 5 秒,以便邮箱登录页面加载完成
time.sleep(5)
iframe = driver.find_element_by_xpath("//iframe[contains(@id,'x-URS-iframe')]")
# 切换进 frame 控件
driver.switch_to.frame(iframe)
# 获取用户名输入框
userName = driver.find_element_by_xpath('//input[@name = "email"]')
# 输入用户名
userName.send_keys("xxx")
# 获取密码输入框
pwd = driver.find_element_by_xpath("//input[@name = 'password']")
# 输入密码
pwd.send_keys("xxx")
# 发送一个回车键
pwd.send_keys(Keys.RETURN)
# 切换至默认句柄,以规避 can't access dead object 异常
driver.switch_to.default_content()

# 等待 10 秒,以便登录成功后的页面加载完成
time.sleep(10)
# 单击"通讯录"按钮
driver.find_element_by_xpath("//div[text() = '通讯录']").click()
time.sleep(2)
# 单击"新建联系人"按钮
driver.find_element_by_xpath("//span[text() = '新建联系人']").click()
```

```
time.sleep(2)
# 输入联系人姓名
driver.find_element_by_xpath(\
    "//a[@title = '编辑详细姓名']/preceding - sibling::div/input").send_keys("lucy")
# 输入联系人电子邮箱
driver.find_element_by_xpath\
    ("//*[@id = 'iaddress_MAIL_wrap']//input").send_keys("xxx@qq.com")
driver.find_element_by_xpath("//span[text() = '设为星标联系人']/preceding - sibling::span/
b").click()
time.sleep(2)
# 输入联系人手机号
driver.find_element_by_xpath\
    ("//*[@id = 'iaddress_TEL_wrap']//dd//input").send_keys("135xxxxxxx")
time.sleep(2)
# 输入备注信息
driver.find_element_by_xpath("//textarea").send_keys("朋友")
time.sleep(2)
# 单击"确定"按钮
driver.find_element_by_xpath("//span[text() = '确 定']").click()
time.sleep(5)
driver.quit()
```

注意在执行上述代码前，请将代码里面的“xxx”换成有效的登录用户名及密码，以后章节也是如此，兹不赘述。

本小节要做的事就是将上面的程序一步步地改成数据驱动测试框架。

数据驱动框架搭建过程：

(1) 在 PyCharm 工具中，新建一个名为 DataDrivenFrameWork 的 Python 工程。

(2) 在工程中新建 4 个 Python Package，分别命名为：

- appModules，用于实现可复用的业务逻辑封装方法。
- pageObjects，用于实现被测试对象的页面对象。
- testScripts，用于实现具体的测试脚本逻辑。
- util，用于实现测试过程中调用的工具类方法，例如读取配置文件、MapObject、页面元素的操作方法、解析 Excel 文件等。

详细结构图如图 15-5 所示。

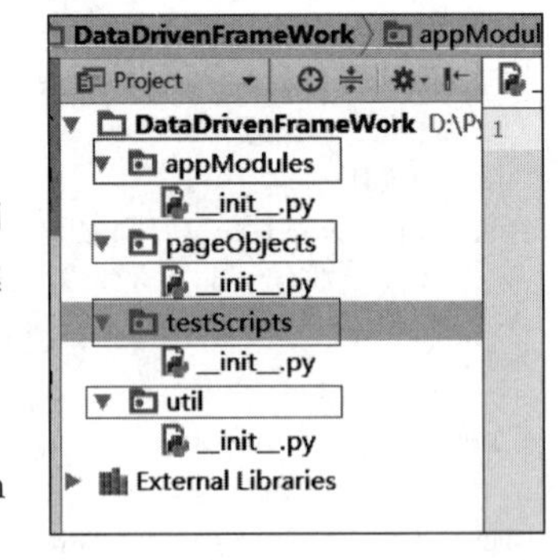

图 15-5

(3) 在 util 包中新建一名叫 ObjectMap.py 的 Python 文件，用于实现定位页面元素的公共方法，具体代码如下：

```
#encoding = utf - 8
from selenium.webdriver.support.ui import WebDriverWait

# 获取单个页面元素对象
def getElement(driver, locateType, locatorExpression):
    try:
        element = WebDriverWait(driver, 30).until\
            (lambda x: x.find_element(by = locateType, value = locatorExpression))
        return element
```

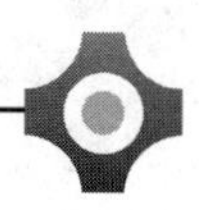

```
        except Exception as err:
            raise err

    # 获取多个相同页面元素对象,以 list 返回
    def getElements(driver, locateType, locatorExpression):
        try:
            elements = WebDriverWait(driver, 30).until\
                (lambda x:x.find_elements(by = locateType, value = locatorExpression))
            return elements
        except Exception as err:
            raise err

    if __name__ == '__main__':
        from selenium import webdriver
        # 进行单元测试
        driver = webdriver.Firefox(executable_path="c:\geckodriver")
        driver.get("http://www.baidu.com")
        searchBox = getElement(driver, "id", "kw")
        # 打印页面对象的标签名
        print(searchBox.tag_name)
        aList = getElements(driver, "tag name", "a")
        print(len(aList))
        driver.quit()
```

以上代码的含义请参阅之前章节的讲解。

(4) 在 pageObjects 包中新建一名为 LoginPage.py 的 Python 文件,用于编写 126 邮箱登录页面的页面元素对象,具体代码如下。

```
#encoding=utf-8
from util.ObjectMap import *

class LoginPage(object):

    def __init__(self, driver):
        self.driver = driver

    def switchToFrame(self):
        locateType, locatorExpression = self.loginOptions\
            ["loginPage.frame".lower()].split(">")
        iframe = getElement(self.driver, locateType, locatorExpression)
        self.driver.switch_to.frame(iframe)

    def switchToDefaultFrame(self):
        self.driver.switch_to.default_content()

    def userNameObj(self):
        try:
            # 获取登录页面的用户名输入框页面对象,并返回给调用者
            elementObj = getElement(self.driver,
                "xpath", '//input[@name="email"]')
            return elementObj
```

```
        except Exception as err:
            raise err

    def passwordObj(self):
        try:
            # 获取登录页面的密码输入框页面对象,并返回给调用者
            elementObj = getElement(self.driver,
                "xpath", "//input[@name = 'password']")
            return elementObj
        except Exception as err:
            raise  err

    def loginButton(self):
        try:
            # 获取登录页面的登录按钮页面对象,并返回给调用者
            elementObj = getElement(self.driver, "id", "dologin")
            return elementObj
        except Exception as err:
            raise err

if __name__ == '__main__':
    # 测试代码
    from selenium import webdriver
    driver = webdriver.Firefox(executable_path = "c:\geckodriver.exe")
    driver.get("http://mail.126.com")
    import time
    time.sleep(5)
    login = LoginPage(driver)
    login.switchToFrame()
    # 输入登录用户名
    login.userNameObj().send_keys("xxx")
    # 输入登录密码
    login.passwordObj().send_keys("xxx")
    login.loginButton().click()
    login.switchToDefaultFrame()
    driver.quit()
```

(5) 在 testScripts 包中新建一名为 TestMail126AddContacts.py 的 Python 文件,用于编写具体的测试操作代码,具体代码如下:

```
#encoding = utf - 8
from selenium import webdriver
from pageObjects.LoginPage import LoginPage
import time

def testMailLogin():
    # 启动 Firefox 浏览器
    driver = webdriver.Firefox(executable_path = "c:\geckodriver.exe")
    try:
        # 访问 126 邮箱首页
        driver.get("http://mail.126.com")
```

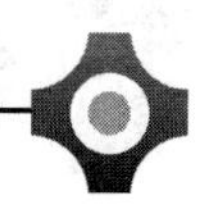

```
        driver.implicitly_wait(30)
        driver.maximize_window()
        loginPage = LoginPage(driver)
        # 将当前焦点切换到登录模块的 frame 中,以便能进行后续登录操作
        loginPage.switchToFrame()
        # 输入登录用户名
        loginPage.userNameObj().send_keys("xxx")
        # 输入登录密码
        loginPage.passwordObj().send_keys("xxx")
        # 单击登录按钮
        loginPage.loginButton().click()
        time.sleep(5)
        # 切换到默认窗体,以兼容 Firefox 浏览器
        loginPage.switchToDefaultFrame()
        assert "未读邮件" in driver.page_source
    except Exception as err:
        raise err
    finally:
        # 退出浏览器
        driver.quit()

if __name__ == '__main__':
    testMailLogin()
    print("登录 126 邮箱成功!")
```

上述代码虽然分了模块,但是仍未做到程序与数据的分离,不能实现框架的通用,所以我们需要继续进行改造。

(6) 在 appModules 包中新建一名叫 LoginAction.py 的 Python 文件,实现登录模块的封装方法,具体代码如下。

```
#encoding = utf-8
from pageObjects.LoginPage import  LoginPage

class LoginAction(object):
    def __init__(self):
        print("login...")

    @staticmethod
    def login(driver, username, password):
        try:
            login = LoginPage(driver)
            # 将当前焦点切换到登录模块的 frame 中,以便能进行后续登录操作
            login.switchToFrame()
            # 输入登录用户名
            login.userNameObj().send_keys(username)
            # 输入登录密码
            login.passwordObj().send_keys(password)
            # 单击登录按钮
            login.loginButton().click()
            # 切回到默认窗体
```

```
            login.switchToDefaultFrame()
        except Exception as err:
            raise err

if __name__ == '__main__':
    from selenium import webdriver
    import time
    # 启动 Firefox 浏览器
    driver = webdriver.Firefox(executable_path = "c:\\geckodriver")
    # 访问 126 邮箱首页
    driver.get("http://mail.126.com")
    time.sleep(5)
    LoginAction.login(driver, username = "xxx", password = "xxx")
    time.sleep(5)
    driver.quit()
```

(7) 修改 testScripts 包中的 TestMail126AddContacts.py 内容如下。

```
#encoding = utf - 8
from selenium import webdriver
from appModules.LoginAction import LoginAction
import time

def testMailLogin():
    try:
        # 启动 Firefox 浏览器
        driver = webdriver.Firefox(executable_path = "c:\\geckodriver")
        # 访问 126 邮箱首页
        driver.get("http://mail.126.com")
        driver.implicitly_wait(30)
        driver.maximize_window()
        time.sleep(5)
        # 登录 126 邮箱
        LoginAction.login(driver, "xxx", "xxx")
        time.sleep(8)
        assert "未读邮件" in driver.page_source
    except Exception as err:
        raise err
    finally:
        driver.quit()

if __name__ == '__main__':
    testMailLogin()
    print("登录 126 邮箱成功!")
```

比较 TestMail126AddContacts.py 修改前后的代码,我们可以发现登录操作的多个步骤被一个函数调用替代了,函数为 LoginAction.login(driver, username = "xxx", password = "xxx")。此种方式实现了业务逻辑的封装,只要调用一个函数就可以实现登录的操作,大大减少了测试脚本的重复编写,从而也实现了代码的封装。

(8) 在 DataDrivenFrameWork 工程中新建一个名叫 config 的 Python Package,并在

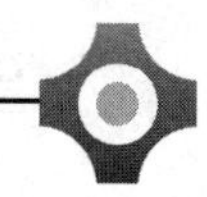

config 包里新建一名为 PageElementLocator. ini 的 File，用于配置定位页面元素的定位表达式，具体内容如下：

```
[126mail_login]
loginPage.frame = xpath>//iframe[contains(@id,'x - URS - iframe')]
loginPage.username = xpath>//input[@name = 'email']
loginPage.password = xpath>//input[@name = 'password']
loginPage.loginbutton = id>dologin
```

(9) 在 config 包中新建一个名为 VarConfig. py 的 Python 文件，此文件主要用于定义一些全局变量，用于存储一些文件路径等，具体内容如下。

```
#encoding = utf - 8
import os

# 获取当前文件所在目录的绝对路径
parentDirPath = os.path.dirname(os.path.dirname(os.path.abspath(__file__)))

# 获取存放页面元素定位表达式文件的绝对路径
pageElementLocatorPath = parentDirPath + "\\config\\PageElementLocator.ini"
```

(10) 在 util 包中新建一个名为 ParseConfigurationFile. py 的 Python 文件，用于解析存储定位页面元素的定位表达式文件，以便获取定位表达式，具体代码如下。

```
#encoding = utf - 8
from configparser import ConfigParser
from config.VarConfig import pageElementLocatorPath

class ParseCofigFile(object):

    def __init__(self):
        self.cf = ConfigParser()
        self.cf.read(pageElementLocatorPath)

    def getItemsSection(self, sectionName):
        # 获取配置文件中指定 section 下的所有 option 键值对
        # 并以字典类型返回给调用者
        """注意:
        使用 self.cf.items(sectionName)此种方法获取到的
        配置文件中的 options 内容均被转换成小写,
        比如:loginPage.frame 被转换成了 loginpage.frame
        """
        optionsDict = dict(self.cf.items(sectionName))
        return optionsDict

    def getOptionValue(self, sectionName, optionName):
        # 获取指定 section 下的指定 option 的值
        value = self.cf.get(sectionName, optionName)
        return value
```

```
if __name__ == '__main__':
    pc = ParseCofigFile()
    print(pc.getItemsSection("126mail_login"))
    print(pc.getOptionValue("126mail_login", "loginPage.frame"))
```

(11) 修改 pageObjects 包中的 LoginPage.py 文件内容如下。

```
#encoding = utf - 8
from util.ObjectMap import *
from util.ParseConfigurationFile import ParseCofigFile

class LoginPage(object):

    def __init__(self, driver):
        self.driver = driver
        self.parseCF = ParseCofigFile()
        self.loginOptions = self.parseCF.getItemsSection("126mail_login")
        print self.loginOptions

    def switchToFrame(self):
        try:
            # 从定位表达式配置文件中读取 frame 的定位表达式
            locateType, locatorExpression = self.loginOptions\
            ["loginPage.frame".lower()].split(">")
            iframe = getElement(self.driver, locateType, locatorExpression)
            self.driver.switch_to.frame(iframe)
        except Exception as err:
            raise err

    def switchToDefaultFrame(self):
        try:
            self.driver.switch_to.default_content()
        except Exception as err:
            raise err

    def userNameObj(self):
        try:
            # 从定位表达式配置文件中读取定位用户名输入框的定位方式和表达式
            locateType, locatorExpression = self.loginOptions\
                ["loginPage.username".lower()].split(">")
            # 获取登录页面的用户名输入框页面对象,并返回给调用者
            elementObj = getElement(self.driver, locateType, locatorExpression)
            return elementObj
        except Exception as err:
            raise err

    def passwordObj(self):
        try:
            # 从定位表达式配置文件中读取定位密码输入框的定位方式和表达式
            locateType, locatorExpression = self.loginOptions\
                ["loginPage.password".lower()].split(">")
```

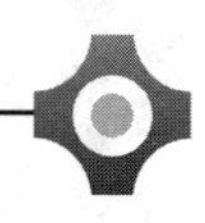

```
            # 获取登录页面的密码输入框页面对象,并返回给调用者
            elementObj = getElement(self.driver, locateType, locatorExpression)
            return elementObj
        except Exception as err:
            raise  err

    def loginButton(self):
        try:
            # 从定位表达式配置文件中读取定位登录按钮的定位方式和表达式
            locateType, locatorExpression = self.loginOptions\
                ["loginPage.loginbutton".lower()].split(">")
            # 获取登录页面的登录按钮页面对象,并返回给调用者
            elementObj = getElement(self.driver, locateType, locatorExpression)
            return elementObj
        except Exception as err:
            raise err

if __name__ == '__main__':
    # 测试代码
    from selenium import webdriver
    driver = webdriver.Firefox(executable_path = "c:\\geckodriver")
    driver.get("http://mail.126.com")
    import time
    time.sleep(5)
    login = LoginPage(driver)
    login.switchToFrame()
    # 输入登录用户名
    login.userNameObj().send_keys("xxx")
    # 输入登录密码
    login.passwordObj().send_keys("xxx")
    login.loginButton().click()
    time.sleep(10)
    login.switchToDefaultFrame()
    assert "未读邮件" in driver.page_source
    driver.quit()
```

如此我们就将定位页面元素的定位表达式跟程序分离出来,并存储在一个专用的配置文件 PageElementLocator.ini 中进行集中管理,后续如果页面结构变了,自动化测试人员只需要修改此文件中的等号(=)后面的定位方式和定位表达式即可,由此提高了脚本维护效率。

(12) 在 pageObjects 包中新建一个名叫 HomePage.py 的 Python 文件和 AddressBookPage.py 文件,并在配置文件 PageElementLocator.ini 中补充两个页面的页面元素的定位表达式。

PageElementLocator.ini 配置文件更新后的内容如下:

```
[126mail_login]
loginPage.frame = xpath>//iframe[contains(@id,'x-URS-iframe')]
loginPage.username = xpath>//input[@name = 'email']
loginPage.password = xpath>//input[@name = 'password']
```

```
loginPage.loginbutton=id>dologin

[126mail_homePage]
homePage.addressbook=xpath>//div[text()='通讯录']

[126mail_addContactsPage]
addContactsPage.createContactsBtn=xpath>//span[text()='新建联系人']
addContactsPage.contactPersonName = xpath >//a[@title = '编辑详细姓名']/preceding-
sibling::div/input
addContactsPage.contactPersonEmail=xpath>//*[@id='iaddress_MAIL_wrap']//input
addContactsPage.starContacts=xpath>//span[text()='设为星标联系人']/preceding-sibling::
span/b
addContactsPage.contactPersonMobile=xpath>//*[@id='iaddress_TEL_wrap']//dd//input
addContactsPage.contactPersonComment=xpath>//textarea
addContactsPage.savecontacePerson=xpath>//span[.='确 定']
```

HomePage.py 文件内容如下：

```python
#encoding=utf-8
from util.ObjectMap import *
from util.ParseConfigurationFile import ParseCofigFile

class HomePage(object):

    def __init__(self, driver):
        self.driver = driver
        self.parseCF = ParseCofigFile()

    def addressLink(self):
        try:
            # 从定位表达式配置文件中读取定位通讯录按钮的定位方式和表达式
            locateType, locatorExpression = self.parseCF.getOptionValue\
                ("126mail_homePage", "homePage.addressbook").split(">")
            # 获取登录成功页面的通讯录页面元素,并返回给调用者
            elementObj = getElement(self.driver, locateType, locatorExpression)
            return elementObj
        except Exception as err:
            raise err
```

AddressBookPage.py 文件内容如下：

```python
#encoding=utf-8
from util.ObjectMap import *
from util.ParseConfigurationFile import ParseCofigFile

class AddressBookPage(object):

    def __init__(self, driver):
        self.driver = driver
        self.parseCF = ParseCofigFile()
        self.addContactsOptions =
        self.parseCF.getItemsSection("126mail_addContactsPage")
```

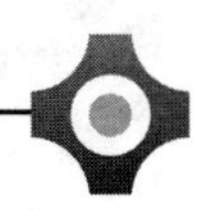

```
        print(self.addContactsOptions)

    def createContactPersonButton(self):
        # 获取新建联系人按钮
        try:
            # 从定位表达式配置文件中读取定位新建联系人按钮的定位方式和表达式
            locateType, locatorExpression = self.addContactsOptions\
                ["addContactsPage.createContactsBtn".lower()].split(">")
            # 获取新建联系人按钮页面元素,并返回给调用者
            elementObj = getElement(self.driver, locateType, locatorExpression)
            return elementObj
        except Exception as err:
            raise err

    def contactPersonName(self):
        # 获取新建联系人界面中的姓名输入框
        try:
            # 从定位表达式配置文件中读取联系人姓名输入框的定位方式和表达式
            locateType, locatorExpression = self.addContactsOptions\
                ["addContactsPage.contactPersonName".lower()].split(">")
            # 获取新建联系人界面的姓名输入框页面元素,并返回给调用者
            elementObj = getElement(self.driver, locateType, locatorExpression)
            return elementObj
        except Exception as err:
            raise err

    def contactPersonEmail(self):
        # 获取新建联系人界面中的电子邮件输入框
        try:
            # 从定位表达式配置文件中读取联系人邮箱输入框的定位方式和表达式
            locateType, locatorExpression = self.addContactsOptions \
                ["addContactsPage.contactPersonEmail".lower()].split(">")
            # 获取新建联系人界面的邮箱输入框页面元素,并返回给调用者
            elementObj = getElement(self.driver, locateType, locatorExpression)
            return elementObj
        except Exception as err:
            raise err

    def starContacts(self):
        # 获取新建联系人界面中的星标联系人选择框
        try:
            # 从定位表达式配置文件中读取星标联系人复选框的定位方式和表达式
            locateType, locatorExpression = self.addContactsOptions \
                ["addContactsPage.starContacts".lower()].split(">")
            # 获取新建联系人界面的星标联系人复选框页面元素,并返回给调用者
            elementObj = getElement(self.driver, locateType, locatorExpression)
            return elementObj
        except Exception as err:
            raise err

    def contactPersonMobile(self):
        # 获取新建联系人界面中的联系人手机号输入框
        try:
            # 从定位表达式配置文件中读取联系人手机号输入框的定位方式和表达式
```

```
            locateType, locatorExpression = self.addContactsOptions \
                ["addContactsPage.contactPersonMobile".lower()].split(">")
            # 获取新建联系人界面的联系人手机号输入框页面元素,并返回给调用者
            elementObj = getElement(self.driver, locateType, locatorExpression)
            return elementObj
        except Exception as err:
            raise err

    def contactPersonComment(self):
        # 获取新建联系人界面中的联系人备注信息输入框
        try:
            # 从定位表达式配置文件中读取联系人备注信息输入框的定位方式和表达式
            locateType, locatorExpression = self.addContactsOptions \
                ["addContactsPage.contactPersonComment".lower()].split(">")
            # 获取新建联系人界面的联系人备注信息输入框页面元素,并返回给调用者
            elementObj = getElement(self.driver, locateType, locatorExpression)
            return elementObj
        except Exception as err:
            raise err

    def saveContacePerson(self):
        # 获取新建联系人界面中的保存联系人按钮
        try:
            # 从定位表达式配置文件中读取保存联系人按钮的定位方式和表达式
            locateType, locatorExpression = self.addContactsOptions \
                ["addContactsPage.saveContacePerson".lower()].split(">")
            # 获取新建联系人界面的保存保存联系人按钮页面元素,并返回给调用者
            elementObj = getElement(self.driver, locateType, locatorExpression)
            return elementObj
        except Exception as err:
            raise err
```

(13) 在 DataDrivenFrameWork 工程下新建一名为 testData 的 Directory(目录),并在该目录中新建一名为"126 邮箱联系人.xlsx"Excel 文件,并在 Excel 文件中创建两个工作表,分别为"126 账号""联系人",其内容分别如表 15-1 和表 15-2 所示。

表 15-1

序号	用户名	密码	数据表	是否执行	测试结果
1	xxx	xxxx	联系人	y	
2	yyy	yyyy	联系人	y	

表 15-2

序号	联系人姓名	联系人邮箱	是否设为星标联系人	联系人手机号	联系人备注信息	验证页面包含的关键字	是否执行	执行时间	测试结果
1	lily	lily@qq.com	是	135xxxxxxx1	常联系人	lily@qq.com	y		
2	张三	zhangsan@qq.com	否	158xxxxxxx3	不常联系人	zhangsan@qq.com	y		
3	amy	amy@qq.com	是	139xxxxx8		amy	n		
4	李四	lisi@qq.com	否	157xxxxxx9		李四	y		

测试过程中需要的数据单独存放到数据文件中，不仅做到了数据与程序的分离，而且方便管理与维护，自动化测试人员只需要修改数据文件中的数据而不需要改动代码，就可以完成针对不同测试数据的测试工作，提高了测试效率和成本。

(14) 在 util 包中新建一个名为 ParseExcel.py 的 Python 文件，用于实现解析 Excel 文件的方法封装，具体内容如下。

```
#encoding = utf - 8
import openpyxl
from openpyxl.styles import Border, Side, Font
import time

class ParseExcel(object):

    def __init__(self):
        self.workbook = None
        self.excelFile = None
        self.font = Font(color = None)                      # 设置字体的颜色
        # 颜色对应的 RGB 值
        self.RGBDict = {'red': 'FFFF3030', 'green': 'FF008B00'}

    def loadWorkBook(self, excelPathAndName):
        # 将 Excel 文件加载到内存,并获取其 workbook 对象
        try:
            self.workbook = openpyxl.load_workbook(excelPathAndName)
        except Exception as err:
            raise err
        self.excelFile = excelPathAndName
        return self.workbook

    def getSheetByName(self, sheetName):
        # 根据 sheet 名获取该 sheet 对象
        try:
            sheet = self.workbook[sheetName]
            return sheet
        except Exception as err:
            raise err

    def getSheetByIndex(self, sheetIndex):
        # 根据 sheet 的索引号获取该 sheet 对象
        try:
            sheetname = self.workbook.sheetnames[sheetIndex]
        except Exception as err:
            raise err
        sheet = self.workbook[sheetname]
        return sheet

    def getRowsNumber(self, sheet):
        # 获取 sheet 中有数据区域的结束行号
        return sheet.max_row
```

```
    def getColsNumber(self, sheet):
        # 获取 sheet 中有数据区域的结束列号
        return sheet.max_column

    def getStartRowNumber(self, sheet):
        # 获取 sheet 中有数据区域的开始的行号
        return sheet.min_row

    def getStartColNumber(self, sheet):
        # 获取 sheet 中有数据区域的开始的列号
        return sheet.min_column

    def getRow(self, sheet, rowNo):
        # 获取 sheet 中某一行,返回的是这一行所有的数据内容组成的 tuple,
        # 下标从 1 开始,sheet.rows[1]表示第一行
        try:
            rows_data = list(sheet.rows)
            return rows_data[rowNo - 1]
        except Exception as err:
            raise err

    def getColumn(self, sheet, colNo):
        # 获取 sheet 中某一列,返回的是这一列所有的数据内容组成 tuple,
        # 下标从 1 开始,sheet.columns[1]表示第一列
        try:
            columns_s_data = list(sheet.columns)
            return columns_s_data[colNo - 1]
        except Exception as err:
            raise err

    def getCellOfValue(self, sheet, rowNo= None, colsNo = None):
        # 根据单元格所在的位置索引获取该单元格中的值,下标从 1 开始,
        # sheet.cell(row = 1, column = 1).value,表示 Excel 中第一行第一列的值
        if rowNo is not None and colsNo is not None:
            try:
                return sheet.cell(row = rowNo, column = colsNo).value
            except Exception as err:
                raise err
        else:
            raise Exception("Insufficient Coordinates of cell !")

    def getCellOfObject(self, sheet, rowNo = None,colsNo = None):
        # 获取某个单元格的对象,可以根据单元格所在位置的数字索引,
        # 也可以直接根据 Excel 中单元格的编码坐标
        # getCellObject(sheet, rowNo = 1, colsNo = 2)
        if rowNo is not None and colsNo is not None:
            try:
                return sheet.cell(row = rowNo,column = colsNo)
            except Exception as err:
                raise err
        else:
```

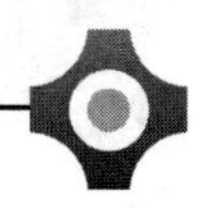

```
            raise Exception("Insufficient Coordinates of cell !")

    def writeCell(self, sheet, content, rowNo = None, colsNo = None, style = None):
        # 根据单元格在 Excel 中的编码坐标或者数字索引坐标向单元格中写入数据，
        # 下标从 1 开始，参 style 表示字体的颜色的名字，比如 red,green
        if rowNo is not None and colsNo is not None:
            try:
                sheet.cell(row = rowNo,column = colsNo).value = content
                if style:
                    sheet.cell(row = rowNo,column = colsNo).\
                        font = Font(color = self.RGBDict[style])
                self.workbook.save(self.excelFile)
            except Exception as err:
                raise err
        else:
            raise Exception("Insufficient Coordinates of cell !")

    def writeCellCurrentTime(self, sheet, rowNo = None, colsNo = None):
        # 写入当前的时间，下标从 1 开始
        now = int(time.time())                          #显示为时间戳
        timeArray = time.localtime(now)
        currentTime = time.strftime("%Y-%m-%d %H:%M:%S", timeArray)
        if rowNo is not None and colsNo is not None:
            try:
                sheet.cell(row = rowNo, column = colsNo).value = currentTime
                self.workbook.save(self.excelFile)
            except Exception as err:
                raise err
        else:
            raise Exception("Insufficient Coordinates of cell !")

if __name__ == '__main__':
    pe = ParseExcel()
    #测试读取 Excel 文件
    pe.loadWorkBook('D:\\126 邮箱联系人.xlsx')
    sheet = pe.getSheetByIndex(0)
    rows = pe.getColumn(sheet, 1)                       #获取第一行
    for i in rows:
        print(i.value)
    # 获取第一行第一列单元格内容
    print(pe.getCellOfValue(sheet, rowNo = 1, colsNo = 1))
    pe.writeCell(sheet, '我爱祖国', rowNo = 10, colsNo = 10)
    pe.writeCellCurrentTime(sheet, rowNo = 10, colsNo = 11)
```

对 Python 解析 Excel 的 openpyxl 模块进行二次封装，以满足我们所需，同时提供给其他模块直接使用，减少重复代码的编写，而且方便维护。

(15) 在 appModules 包中新建一个名为 AddContactPersonAction.py 的 Python 文件，用于实现添加联系人操作，具体内容如下。

```
#encoding = utf-8
```

```
from pageObjects.HomePage import HomePage
from pageObjects.AddressBookPage import AddressBookPage
import traceback
import time
class AddContactPerson(object):

    def __init__(self):
        print("add contact person.")

    @staticmethod
    def add(driver, contactName, contactEmail, isStar, contactPhone, contactComment):
        try:
            # 创建主页实例对象
            hp = HomePage(driver)
            # 单击通讯录链接
            hp.addressLink().click()
            time.sleep(3)
            # 创建添加联系人页实例对象
            apb = AddressBookPage(driver)
            apb.createContactPersonButton().click()
            if contactName:
                # 非必填项
                apb.contactPersonName().send_keys(contactName)
            # 必填项
            apb.contactPersonEmail().send_keys(contactEmail)
            if isStar == "是":
                # 非必填项
                apb.starContacts().click()
            if contactPhone:
                # 非必填项
                apb.contactPersonMobile().send_keys(contactPhone)
            if contactComment:
                apb.contactPersonComment().send_keys(contactComment)
            apb.saveContacePerson().click()
        except Exception as err:
            # 打印堆栈异常信息
            print(traceback.print_exc())
            raise err

if __name__ == '__main__':
    from appModules.LoginAction import LoginAction
    from selenium import webdriver
    # 启动 Firefox 浏览器
    driver = webdriver.Firefox(executable_path = r"c:\geckodriver")
    # 访问 126 邮箱首页
    driver.get("http://mail.126.com")
    # driver.maximize_window()
    time.sleep(5)
    LoginAction.login(driver, "xxx", "xxx")
    time.sleep(5)
    AddContactPerson.add(driver, "张三", "zs@qq.com", "是", "", "")
```

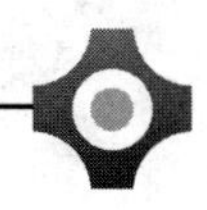

```
    time.sleep(3)
    assert "张三" in driver.page_source
    driver.quit()
```

(16) 修改 config 包下的 VarConfig.py 文件内容如下。

```
#encoding=utf-8
import os

# 获取当前文件所在目录的父目录的绝对路径
parentDirPath = os.path.dirname(os.path.dirname(os.path.abspath(__file__)))

# 获取存放页面元素定位表达式文件的绝对路径
pageElementLocatorPath = parentDirPath + "\\config\\PageElementLocator.ini"

# 获取数据文件存放绝对路径
dataFilePath = parentDirPath + "\\testData\\126 邮箱联系人.xlsx"

# 126 账号工作表中,每列对应的数字序号
account_username = 2
account_password = 3
account_dataBook = 4
account_isExecute = 5
account_testResult = 6

# 联系人工作表中,每列对应的数字序号
contacts_contactPersonName = 2
contacts_contactPersonEmail = 3
contacts_isStar = 4
contacts_contactPersonMobile = 5
contacts_contactPersonComment = 6
contacts_assertKeyWords = 7
contacts_isExecute = 8
contacts_runTime = 9
contacts_testResult = 10
```

(17) 修改 testScripts 包中的 TestMail126AddContacts.py 文件内容如下。

```
#encoding=utf-8
from selenium import webdriver
from util.ParseExcel import ParseExcel
from config.VarConfig import *
from appModules.LoginAction import LoginAction
from appModules.AddContactPersonAction import AddContactPerson
import traceback
from time import sleep

# 创建解析 Excel 对象
excelObj = ParseExcel()
# 将 Excel 数据文件加载到内存
excelObj.loadWorkBook(dataFilePath)
```

```
def LaunchBrowser():
    # 创建浏览器实例对象
    driver = webdriver.Firefox(executable_path = r"c:\geckodriver")
    # 访问 126 邮箱首页
    driver.get("http://mail.126.com")
    sleep(3)
    return driver

def test126MailAddContacts():
    try:
        # 根据 Excel 文件中 sheet 名称获取此 sheet 对象
        userSheet = excelObj.getSheetByName("126 账号")
        # 获取 126 账号 sheet 中是否执行列
        isExecuteUser = excelObj.getColumn(userSheet, account_isExecute)
        # 获取 126 账号 sheet 中的数据表列
        dataBookColumn = excelObj.getColumn(userSheet, account_dataBook)
        print("测试为 126 邮箱添加联系人执行开始...")

        for idx, i in enumerate(isExecuteUser[1:]):
            # 循环遍历 126 账号表中的账号,为需要执行的账号添加联系人
            if i.value == "y": # 表示要执行
                # 获取第 i 行的数据
                userRow = excelObj.getRow(userSheet, idx + 2)
                # 获取第 i 行中的用户名
                username = userRow[account_username - 1].value
                # 获取第 i 行中的密码
                password = str(userRow[account_password - 1].value)
                print(username, password)

                driver = LaunchBrowser()

                # 登录 126 邮箱
                LoginAction.login(driver, username, password)
                # 等待 3 秒,让浏览器启动完成,以便正常进行后续操作
                sleep(3)
                # 获取为第 i 行中用户添加的联系人数据表 sheet 名
                dataBookName = dataBookColumn[idx + 1].value
                # 获取对应的数据表对象
                dataSheet = excelObj.getSheetByName(dataBookName)
                # 获取联系人数据表中是否执行列对象
                isExecuteData = excelObj.getColumn(dataSheet, contacts_isExecute)
                contactNum = 0 # 记录添加成功联系人个数
                isExecuteNum = 0 # 记录需要执行联系人个数

                for id, data in enumerate(isExecuteData[1:]):
                    # 循环遍历是否执行添加联系人列,
                    # 如果被设置为添加,则进行联系人添加操作
                    if data.value == "y":
                        # 如果第 id 行的联系人被设置为执行,则 isExecuteNum 自增 1
                        isExecuteNum += 1
                        # 获取联系人表第 id + 2 行对象
```

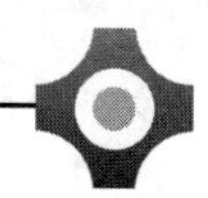

```
                rowContent = excelObj.getRow(dataSheet, id + 2)
                # 获取联系人姓名
                contactPersonName = \
                    rowContent[contacts_contactPersonName - 1].value
                # 获取联系人邮箱
                contactPersonEmail = \
                    rowContent[contacts_contactPersonEmail - 1].value
                # 获取是否设置为星标联系人
                isStar = rowContent[contacts_isStar - 1].value
                # 获取联系人手机号
                contactPersonPhone = \
                    rowContent[contacts_contactPersonMobile - 1].value
                # 获取联系人备注信息
                contactPersonComment = \
                    rowContent[contacts_contactPersonComment - 1].value
                # 添加联系人成功后,断言的关键字
                assertKeyWord = \
                    rowContent[contacts_assertKeyWords - 1].value
                print(contactPersonName, contactPersonEmail, assertKeyWord)
                print(contactPersonPhone, contactPersonComment, isStar)
                # 执行新建联系人操作
                AddContactPerson.add(driver,
                      contactPersonName,
                      contactPersonEmail,
                      isStar,
                      contactPersonPhone,
                      contactPersonComment)
                sleep(1)
                # 在联系人工作表中写入添加联系人执行时间
                excelObj.writeCellCurrentTime(dataSheet,
                    rowNo = id + 2, colsNo = contacts_runTime)
                try:
                    # 断言给定的关键字是否出现在页面中
                    assert assertKeyWord in driver.page_source
                except AssertionError as err:
                    # 断言失败,在联系人工作表中写入添加联系人测试失败信息
                    excelObj.writeCell(dataSheet,"faild",rowNo = id+2,
                        colsNo = contacts_testResult,style = "red")
                else:
                    # 断言成功,写入添加联系人成功信息
                    excelObj.writeCell(dataSheet,"pass",rowNo = id+2,
                        colsNo = contacts_testResult,style = "green")
                    contactNum += 1
        print("contactNum = %s, isExecuteNum = %s"
               % (contactNum, isExecuteNum))
        if contactNum == isExecuteNum:
            # 如果成功添加的联系人数与需要添加的联系人数相等,
            # 说明给第 i 个用户添加联系人测试用例执行成功,
            # 在 126 账号工作表中写入成功信息,否则写入失败信息
            excelObj.writeCell(userSheet, "pass", rowNo = idx + 2,
                colsNo = account_testResult, style = "green")
```

```
                    print("为用户 %s 添加 %d 个联系人,测试通过!"
                          % (username, contactNum))
                else:
                    excelObj.writeCell(userSheet, "faild", rowNo = idx + 2,
                        colsNo = account_testResult, style = "red")
                driver.quit()
            else:
                print("用户 %s 被设置为忽略执行!" % excelObj.getCellOfValue
                    (userSheet, rowNo = idx + 2, colsNo = account_username))
    except Exception as err:
        print("数据驱动框架主程序发生异常,异常信息为:")
        # 打印异常堆栈信息
        print(traceback.print_exc())

if __name__ == '__main__':
    test126MailAddContacts()
    print("登录 126 邮箱成功!")
```

(18) 在 DataDrivenFrameWork 工程根目录中创建一个名叫 RunTest.py 的 Python 文件,用于编写整个数据驱动框架运行的主入口代码,具体代码如下。

```
#encoding = utf-8
from testScripts.TestMail126AddContacts import *

if __name__ == '__main__':
    test126MailAddContacts()
```

在 config 包中的 VarConfig.py 文件中定义了多个常量,在测试脚本文件 TestMail126AddContacts.py 中多行代码调用了这些常量,实现了测试数据在测试方法中的重复使用,如果需要修改数据,只修改 VarConfig.py 文件中的常量值就可以实现在全部测试过程生效,减少了代码的维护成本,同时也增加了测试代码的可读性。

在 TestMail126AddContacts.py 文件中改为从 Excel 数据文件中读取测试数据,作为数据驱动框架测试过程中的数据来源,执行完某条测试用例后,则会在 Excel 数据文件最后两列分别写入"测试执行时间"和"测试结果"。

经过以上步骤,一个简单的数据驱动框架雏形就完成了,但我们还需要加入打印日志的功能,让它看起来更完善一点。

(19) 通过 logging 模块,为数据驱动框架加入打印日志功能。在 config 包中新建一个名叫 Logger.conf 的文件,用于配置日志基本信息,具体内容如下。

```
#logger.conf
###############################################
[loggers]
keys = root,example01,example02
[logger_root]
level = DEBUG
handlers = hand01,hand02

[logger_example01]
```

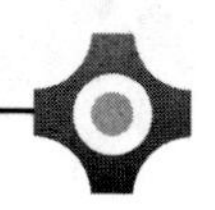

```
handlers = hand01,hand02
qualname = example01
propagate = 0

[logger_example02]
handlers = hand01,hand03
qualname = example02
propagate = 0

###############################################
[handlers]
keys = hand01,hand02,hand03

[handler_hand01]
class = StreamHandler
level = INFO
formatter = form01
args = (sys.stderr,)

[handler_hand02]
class = FileHandler
level = DEBUG
formatter = form01
args = ('log\\DataDrivenFrameWork.log', 'a')

[handler_hand03]
class = handlers.RotatingFileHandler
level = INFO
formatter = form01
args = ('log\\DataDrivenFrameWork.log', 'a', 10 * 1024 * 1024, 5)

###############################################
[formatters]
keys = form01,form02

[formatter_form01]
format = %(asctime)s %(filename)s[line: %(lineno)d] %(levelname)s %(message)s
datefmt = %Y-%m-%d %H:%M:%S

[formatter_form02]
format = %(name)-12s: %(levelname)-8s %(message)s
datefmt = %Y-%m-%d %H:%M:%S
```

(20) 在 util 包中新建一个名为 Log.py 的 Python 文件，用于初始化日志对象，具体内容如下。

```
# -*- coding: UTF-8 -*-
import logging
import logging.config
from config.VarConfig import parentDirPath
```

```
# 读取日志配置文件
logging.config.fileConfig(parentDirPath + "\config\Logger.conf")
# 选择一个日志格式
logger = logging.getLogger("example02")                    #或者 example01

def debug(message):
    # 定义 dubug 级别日志打印方法
    logger.debug(message)

def info(message):
    # 定义 info 级别日志打印方法
    logger.info(message)

def warning(message):
    # 定义 warning 级别日志打印方法
    logger.warning(message)
```

(21) 在 DataDrivenFrameWork 工程根目录下创建一个名为 log 的目录,然后修改 testScripts 包中的 TestMail126AddContacts.py 文件内容如下。

```
#encoding = utf - 8
from selenium import webdriver
from util.ParseExcel import ParseExcel
from config.VarConfig import *
from appModules.LoginAction import LoginAction
from appModules.AddContactPersonAction import AddContactPerson
import traceback
from time import sleep
from util.Log import *

# 创建解析 Excel 对象
excelObj = ParseExcel()
# 将 Excel 数据文件加载到内存
excelObj.loadWorkBook(dataFilePath)

def LaunchBrowser():
    # 创建浏览器实例对象
    driver = webdriver.Firefox(executable_path = r"c:\geckodriver")
    # 访问 126 邮箱首页
    driver.get("http://mail.126.com")
    sleep(3)
    return driver

def test126MailAddContacts():
    logging.info("126 邮箱添加联系人数据驱动测试开始...")
    try:
        # 根据 Excel 文件中 sheet 名称获取此 sheet 对象
        userSheet = excelObj.getSheetByName("126 账号")
        # 获取 126 账号 sheet 中是否执行列
        isExecuteUser = excelObj.getColumn(userSheet, account_isExecute)
        # 获取 126 账号 sheet 中的数据表列
```

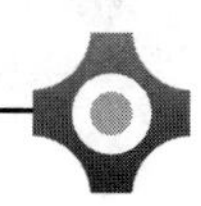

```
dataBookColumn = excelObj.getColumn(userSheet, account_dataBook)

for idx, i in enumerate(isExecuteUser[1:]):
    # 循环遍历 126 账号表中的账号,为需要执行的账号添加联系人
    if i.value == "y":                              # 表示要执行
        # 获取第 i 行的数据
        userRow = excelObj.getRow(userSheet, idx + 2)
        # 获取第 i 行中的用户名
        username = userRow[account_username - 1].value
        # 获取第 i 行中的密码
        password = str(userRow[account_password - 1].value)

        driver = LaunchBrowser()
        logging.info("启动浏览器,访问 126 邮箱主页")

        # 登录 126 邮箱
        LoginAction.login(driver, username, password)
        # 等待 3 秒,让浏览器启动完成,以便正常进行后续操作
        sleep(3)
        try:
            # 断言登录后跳转页面的标题是否包含"网易邮箱"
            assert "收 信" in driver.page_source
            logging.info \
                ("用户 %s 登录后,断言页面关键字"收 信"成功" % username)
        except AssertionError as err:
            logging.debug("用户 %s 登录后,断言页面关键字"收 信"失败,"
                "异常信息:%s" % (username, str(traceback.format_exc())))

        # 获取为第 i 行中用户添加的联系人数据表 sheet 名
        dataBookName = dataBookColumn[idx + 1].value
        # 获取对应的数据表对象
        dataSheet = excelObj.getSheetByName(dataBookName)
        # 获取联系人数据表中是否执行列对象
        isExecuteData = excelObj.getColumn(dataSheet, contacts_isExecute)
        contactNum = 0                              # 记录添加成功联系人个数
        isExecuteNum = 0                            # 记录需要执行联系人个数

        for id, data in enumerate(isExecuteData[1:]):
            # 循环遍历是否执行添加联系人列,
            # 如果被设置为添加,则进行联系人添加操作
            if data.value == "y":
                # 如果第 id 行的联系人被设置为执行,则 isExecuteNum 自增 1
                isExecuteNum += 1
                # 获取联系人表第 id + 2 行对象
                rowContent = excelObj.getRow(dataSheet, id + 2)
                # 获取联系人姓名
                contactPersonName = \
                    rowContent[contacts_contactPersonName - 1].value
                # 获取联系人邮箱
                contactPersonEmail = \
                    rowContent[contacts_contactPersonEmail - 1].value
```

```
                    # 获取是否设置为星标联系人
                    isStar = rowContent[contacts_isStar - 1].value
                    # 获取联系人手机号
                    contactPersonPhone = \
                        rowContent[contacts_contactPersonMobile - 1].value
                    # 获取联系人备注信息
                    contactPersonComment = \
                        rowContent[contacts_contactPersonComment - 1].value
                    # 添加联系人成功后,断言的关键字
                    assertKeyWord = \
                        rowContent[contacts_assertKeyWords - 1].value
                    # 执行新建联系人操作
                    AddContactPerson.add(driver,
                         contactPersonName,
                         contactPersonEmail,
                         isStar,
                         contactPersonPhone,
                         contactPersonComment)
                    sleep(1)
                    logging.info("添加联系人 %s 成功" % contactPersonEmail)
                    # 在联系人工作表中写入添加联系人执行时间
                    excelObj.writeCellCurrentTime(dataSheet,
                        rowNo = id + 2, colsNo = contacts_runTime)
                    try:
                        # 断言给定的关键字是否出现在页面中
                        assert assertKeyWord in driver.page_source
                    except AssertionError as err:
                        # 断言失败,在联系人工作表中写入添加联系人测试失败信息
                        excelObj.writeCell(dataSheet,"faild",rowNo = id+2,
                            colsNo = contacts_testResult,style = "red")
                        logging.info("断言关键字"%s"失败" % assertKeyWord)
                    else:
                        # 断言成功,写入添加联系人成功信息
                        excelObj.writeCell(dataSheet,"pass",rowNo = id+2,
                            colsNo = contacts_testResult,style = "green")
                        contactNum += 1
                        logging.info("断言关键字"%s"成功" % assertKeyWord)
            if contactNum == isExecuteNum:
                # 如果成功添加的联系人数与需要添加的联系人数相等,
                # 说明给第 i 个用户添加联系人测试用例执行成功,
                # 在 126 账号工作表中写入成功信息,否则写入失败信息
                excelObj.writeCell(userSheet, "pass", rowNo = idx + 2,
                    colsNo = account_testResult, style = "green")
                logging.info("为用户 %s 添加 %d 个联系人,测试通过!"
                        % (username, contactNum))
            else:
                excelObj.writeCell(userSheet, "faild", rowNo = idx + 2,
                    colsNo = account_testResult, style = "red")
            driver.quit()
        else:
            logging.info("用户 %s 被设置为忽略执行!" % excelObj.getCellOfValue
```

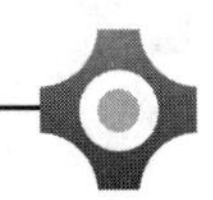

```
                    (userSheet, rowNo = idx + 2, colsNo = account_username))
    except Exception as err:
        logging.debug("数据驱动框架主程序执行过程发生异常,异常信息:%s" \
                      % str(traceback.format_exc()))

if __name__ == '__main__':
    test126MailAddContacts()
    print("登录 126 邮箱成功!")
```

(22) 运行 RunTest.py 文件,执行结束后可以在工程目录下的 Log 目录中看到打印的日志文件 DataDrivenFrameWork.log,文件所包含的内容如图 15-6 所示,这里只粗略地打印了一些执行过程信息,读者可以添加更加详细的打印日志信息,以便查看更详细执行过程逻辑,后续用于测试执行中的问题分析和过程监控。

DataDrivenFrameWork.log - 记事本

文件(F) 编辑(E) 格式(O) 查看(V) 帮助(H)

```
2017-01-11 17:58:51 TestMail126AddContacts.py[line:24] INFO 126邮箱添加联系人数据驱动测试开始...
2017-01-11 17:59:01 TestMail126AddContacts.py[line:46] INFO 启动浏览器, 访问126邮箱主页
2017-01-11 17:59:06 TestMail126AddContacts.py[line:55] INFO 用户AgilityToSR登录后, 断言页面关键字“收 信”成功
2017-01-11 17:59:15 TestMail126AddContacts.py[line:102] INFO 添加联系人lily@qq.com成功
2017-01-11 17:59:15 TestMail126AddContacts.py[line:120] INFO 断言关键字“lily@qq.com”成功
2017-01-11 17:59:20 TestMail126AddContacts.py[line:102] INFO 添加联系人zhangsan@qq.com成功
2017-01-11 17:59:20 TestMail126AddContacts.py[line:120] INFO 断言关键字“zhangsan@qq.com”成功
2017-01-11 17:59:20 TestMail126AddContacts.py[line:122] INFO 联系人zhangsan@qq.com被忽略执行
2017-01-11 17:59:24 TestMail126AddContacts.py[line:102] INFO 添加联系人lisi@qq.com成功
2017-01-11 17:59:25 TestMail126AddContacts.py[line:120] INFO 断言关键字“李四”成功
2017-01-11 17:59:25 TestMail126AddContacts.py[line:133] INFO 为用户AgilityToSR添加3个联系人, 3个成功

2017-01-11 17:59:28 TestMail126AddContacts.py[line:138] INFO 用户chenliangjuen被忽略执行
```

图　15-6

打开 Excel 数据文件,可以看到两个表中都写入了本次执行成功或失败的信息,如图 15-7 和图 15-8 所示。

A	B	C	D	E	F
序号	用户名	密码	数据表	是否执行	测试结果
1	AgilityToSR	[illegible]	联系人	y	pass
2	chenliangjuen	[illegible]	联系人	n	

126账号　联系人

图　15-7

A	B	C	D	E	F	G	H	I	J
序号	联系人姓名	联系人邮箱	是否	联系人手机号	联系人备注信	验证页面包含的关	是否执行	执行时间	测试结果
1	lily	lily@qq.com	是	135xxxxxxx1	常联系人	lily@qq.com	y	2017-01-11 17:59:15	pass
2	张三	zhangsan@qq.com	否	158xxxxxxx3	不常联系人	zhangsan@qq.com	y	2017-01-11 17:59:20	pass
3	amy	amy@qq.com	是	139xxxxxxx8		amy	n		
4	李四	lisi@qq.com	否	157xxxxxxx9		李四	y	2017-01-11 17:59:25	pass

126账号　联系人

图　15-8

至此,数据驱动测试框架全部搭建完成,在 PyCharm 工具中,整个工程的结构如图 15-9 所示。

数据测试驱动框架的优点分析：

(1) 通过配置文件，实现页面元素定位表达式和测试代码的分离。

(2) 使用 ObjectMap 方式，简化页面元素定位相关的代码工作量。

(3) 使用 PageObject(页面对象)模式，封装了网页中的页面元素，方便测试代码调用，也实现了一处维护全局生效的目标。

(4) 在 appModules 的 Package 中封装了常用的页面对象操作方法，简化了测试脚本编写的工作量。

(5) 在 Excel 文件中定义多个测试数据，每个 126 用户都一一对应存放联系人数据的工作表，测试框架可自动调用测试数据完成数据驱动测试。

(6) 实现了测试执行过程中的日志记录功能，可以通过日志文件分析测试脚本执行的情况。

(7) 在 Excel 数据文件测试数据行中，通过设定“测试数据是否执行”列的内容为 y 或者 n，可自定义选择测试数据，测试执行结束后会在“测试结果”列中显示测试执行的时间和结果，方便测试人员查看。

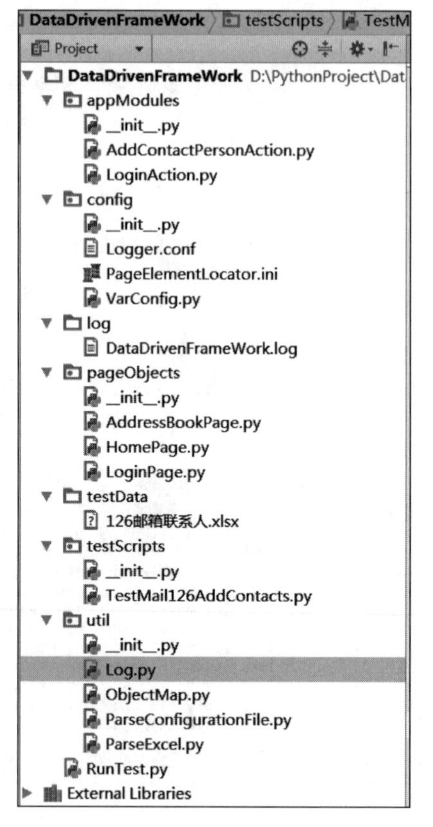

图 15-9

本例中使用操作 Excel 文件的方式定义和维护测试数据及测试结果，如果读者擅长数据库和网页开发技术，可以借鉴此框架的思想实现基于数据库和网页架构的数据驱动框架。借助 Web 方式，测试人员可以通过浏览器来进行数据驱动测试，完成测试数据的定义、测试用例的执行和测试结果的查看。

15.3 关键字驱动框架及实战

本节主要讲解关键字驱动测试框架的搭建过程，并且使用此框架来测试 126 邮箱登录和发送邮件等相关功能。关键字框架用到的基础知识均在前面的章节做了详细介绍，本节重点讲解框架搭建的详细过程。

被测试功能的相关页面描述：

登录页面如图 15-10 所示。

登录后的页面如图 15-11 所示。

单击“写信”链接后，进入写信页面，如图 15-12 所示。

发送邮件成功后显示的页面如图 15-13 所示。

非关键字驱动框架时的发送邮件自动化测试代码：

```
# encoding = utf - 8
from selenium import webdriver
from selenium.webdriver.common.keys import Keys
```

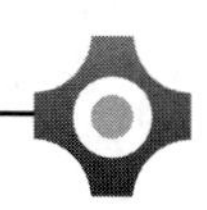

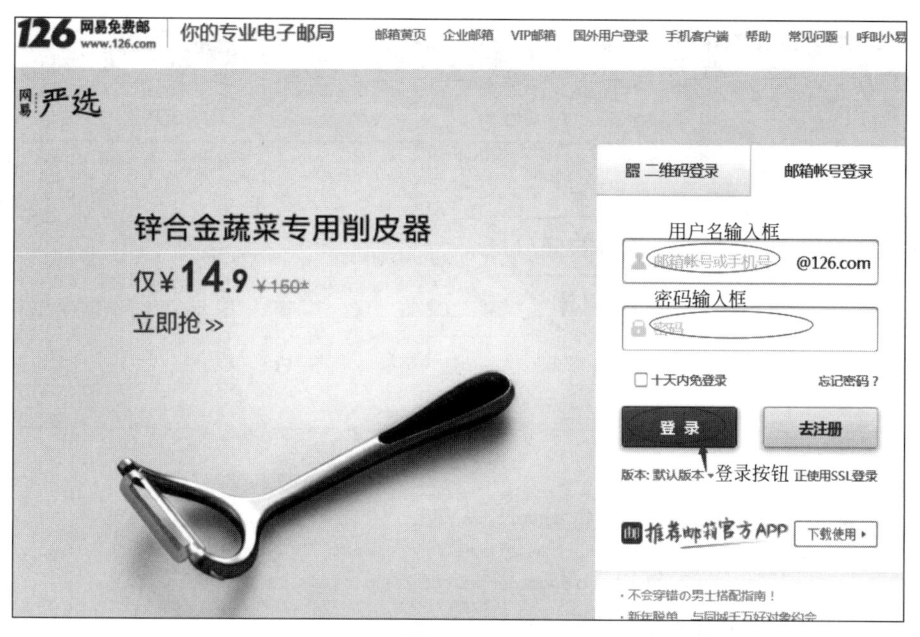

图　15-10

图　15-11

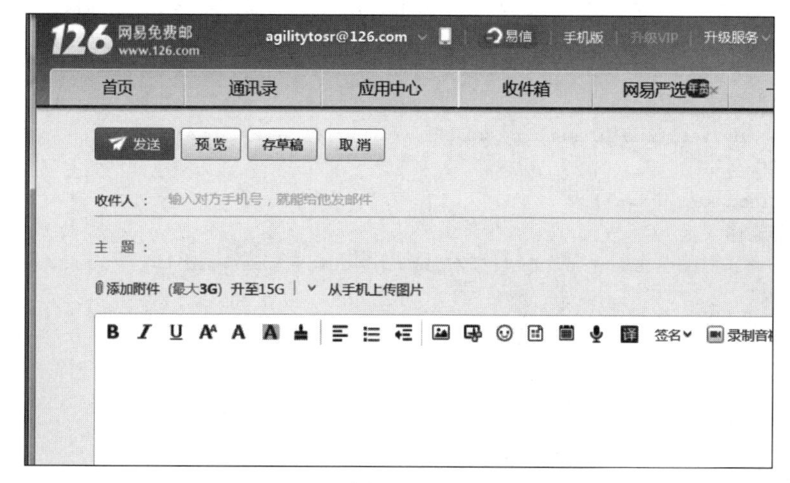

图　15-12

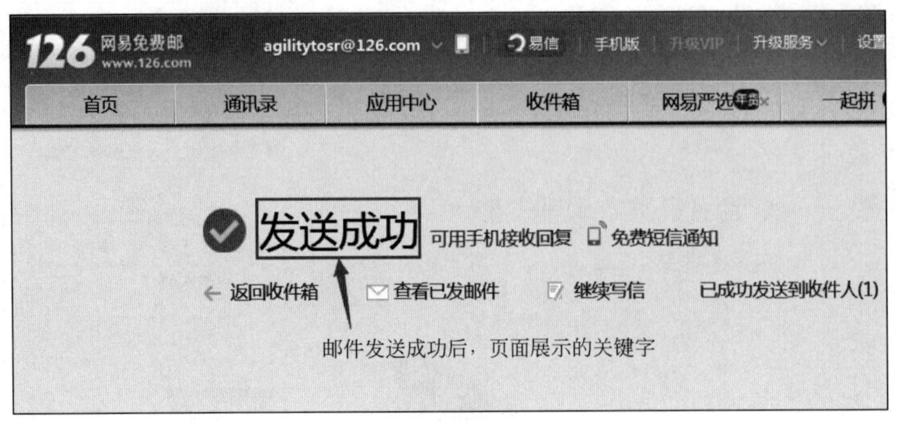

图　15-13

```
from selenium.webdriver.common.by import By
from selenium.webdriver.support.ui import WebDriverWait
from selenium.webdriver.support import expected_conditions as EC
import time

print("启动浏览器...")
# 创建 Firefox 浏览器的实例
driver = webdriver.Firefox(executable_path = "c:\\geckodriver")
# 最大化浏览器窗口
driver.maximize_window()
print("启动浏览器成功")
print("访问 126 邮箱登录页...")
driver.get("http://mail.126.com")
# 暂停 5 秒,以便邮箱登录页面加载完成
time.sleep(5)
assert "126 网易免费邮 -- 你的专业电子邮局" in driver.title
print("访问 126 邮箱登录页成功")
# 创建显示等待
wait = WebDriverWait(driver, 30)
# 检查 id 为 x - URS - iframe 的 frame 是否存在,存在则切换进 frame 控件
wait.until(EC.frame_to_be_available_and_switch_to_it((By.XPATH,
            "//iframe[contains(@id,'x - URS - iframe')]")))
# 获取用户名输入框
userName = driver.find_element_by_xpath('//input[@name = "email"]')
# 输入用户名
userName.send_keys("xxx")
# 获取密码输入框
pwd = driver.find_element_by_xpath("//input[@name = 'password']")
# 输入密码
pwd.send_keys("xxx")
# 发送一个回车键
pwd.send_keys(Keys.RETURN)
print("用户登录...")
# 等待 5 秒,以便登录成功后的页面加载完成
time.sleep(5)
```

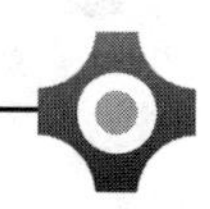

```
assert "网易邮箱" in driver.title
print("登录成功")
# 切换至默认句柄,以规避 can't access dead object 异常
driver.switch_to.default_content()
print("写信...")
# 显示等待写信链接页面元素的出现
element = wait.until\
    (EC.visibility_of_element_located((By.XPATH, "//span[text() = '写 信']")))
# 单击写信链接
element.click()
# 写入收件人地址
driver.find_element_by_xpath\
    ("//div[contains(@id,'_mail_emailinput')]/input").send_keys("xxx")
# 写入邮件主题
driver.find_element_by_xpath\
    ("//div[@aria-label='邮件主题输入框,请输入邮件主题']/input").send_keys("新邮件")
# 切换进 frame 控件
driver.switch_to.frame(driver.find_element_by_xpath("//iframe[@tabindex=1]"))
editBox = driver.find_element_by_xpath("/html/body")
editBox.send_keys("发给光荣之路的一封信")
driver.switch_to.default_content()
print("写信完成")
driver.find_element_by_xpath("//header//span[text()='发送']").click()
print("开始发送邮件...")
time.sleep(3)
assert "发送成功" in driver.page_source
print("邮件发送成功")
driver.quit()
```

本小节要做的事就是将上面的程序一步步地改成关键字驱动测试框架,同时为发送的邮件增加附件。

关键字驱动测试框架搭建步骤:

(1) 在 PyCharm 工具中新建一个名叫 KeyWordsFrameWork 的 Python 工程。

(2) 在工程中新建 3 个 Python Package,1 个目录,分别命名为:

- config 包,主要用于实现框架中各种配置。
- util 包,主要用于实现测试过程中调用的工具类方法,例如读取配置文件、MapObject、页面元素的操作方法、解析 Excel 文件等。
- testData 目录,主要用于存放框架所需要的测试数据文件。
- testScripts 包,用于实现具有测试逻辑的测试脚本。

(3) 在 util 包中新建 ObjectMap.py 模块,用于实现定位页面元素,具体代码如下。

```
#encoding=utf-8
from selenium.webdriver.support.ui import WebDriverWait

# 获取单个页面元素对象
def getElement(driver, locationType, locatorExpression):
    try:
        element = WebDriverWait(driver, 30).until\
```

```
            (lambda x: x.find_element(by = locationType, value = locatorExpression))
        return element
    except Exception, e:
        raise e

# 获取多个相同页面元素对象,以 list 返回
def getElements(driver, locationType, locatorExpression):
    try:
        elements = WebDriverWait(driver, 30).until\
            (lambda x:x.find_elements(by = locationType, value = locatorExpression))
        return elements
    except Exception as err:
        raise err

if __name__ == '__main__':
    from selenium import webdriver
    # 进行单元测试
    driver = webdriver.Firefox(executable_path = "c:\geckodriver.exe")
    driver.get("http://www.baidu.com")
    searchBox = getElement(driver, "id", "kw")
    # 打印页面对象的标签名
    print(searchBox.tag_name)
    aList = getElements(driver, "tag name", "a")
    print(len(aList))
    driver.quit()
```

(4) 在 util 包中新建一名为 WaitUtil.py 文件,用于实现智能等待页面元素的出现,具体代码如下。

```
#encoding = utf - 8
from selenium.webdriver.common.by import By
from selenium.webdriver.support.ui import WebDriverWait
from selenium.webdriver.support import expected_conditions as EC

class WaitUtil(object):

    def __init__(self, driver):
        self.locationTypeDict = {
            "xpath": By.XPATH,
            "id": By.ID,
            "name": By.NAME,
            "class_name": By.CLASS_NAME,
            "tag_name": By.TAG_NAME,
            "link_text": By.LINK_TEXT,
            "partial_link_text": By.PARTIAL_LINK_TEXT
        }
        self.driver = driver
        self.wait = WebDriverWait(self.driver, 30)

    def frame_available_and_switch_to_it(self, locationType, locatorExpression):
        '''检查 frame 是否存在,存在则切换进 frame 控件中
```

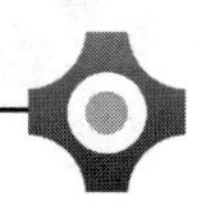

```
            '''
            try:
                self.wait.until(EC.frame_to_be_available_and_switch_to_it
                    ((self.locationTypeDict[locationType.lower()], locatorExpression)))
            except Exception as err:
                # 抛出异常信息给上层调用者
                raise err

        def visibility_element_located(self, locationType, locatorExpression):
            '''显示等待页面元素的出现'''
            try:
                element = self.wait.until(EC.visibility_of_element_located
                    ((self.locationTypeDict[locationType.lower()], locatorExpression)))
                return element
            except Exception as err:
                raise err

    if __name__ == '__main__':
        from selenium import webdriver
        driver = webdriver.Firefox(executable_path = "c:\\geckodriver")
        driver.get("http://mail.126.com")
        waitUtil = WaitUtil(driver)
        waitUtil.frame_available_and_switch_to_it("xpath", "//iframe[contains(@id,'x-URS-
    iframe')]")
        e = waitUtil.visibility_element_located("xpath", "//input[@name='email']")
        e.send_keys("success")
        driver.quit()
```

(5) 在 util 包中新建一名叫 KeyBoardUtil.py 的 Python 文件，用于实现模拟键盘单个或组合按键，具体内容如下。

```
# encoding = utf-8
import win32api
import win32con

class KeyboardKeys(object):
    '''
    模拟键盘按键类
    '''
    VK_CODE = {
        'enter': 0x0D,
        'ctrl': 0x11,
        'v': 0x56}

    @staticmethod
    def keyDown(keyName):
        # 按下按键
        win32api.keybd_event(KeyboardKeys.VK_CODE[keyName], 0, 0, 0)

    @staticmethod
    def keyUp(keyName):
```

```
        # 释放按键
        win32api.keybd_event(KeyboardKeys.VK_CODE[keyName],
                                      0, win32con.KEYEVENTF_KEYUP, 0)

    @staticmethod
    def oneKey(key):
        # 模拟单个按键
        KeyboardKeys.keyDown(key)
        KeyboardKeys.keyUp(key)

    @staticmethod
    def twoKeys(key1, key2):
        # 模拟两个组合键
        KeyboardKeys.keyDown(key1)
        KeyboardKeys.keyDown(key2)
        KeyboardKeys.keyUp(key2)
        KeyboardKeys.keyUp(key1)
```

(6) 在 util 包中新建一个名为 ClipboardUtil.py 的 Python 文件,用于实现将数据设置到剪切板中,具体内容如下。

```
#encoding = utf - 8
import win32clipboard as w
import win32con

class Clipboard(object):
    '''
    模拟 Windows 设置剪切板
    '''
    #读取剪切板
    @staticmethod
    def getText():
        # 打开剪切板
        w.OpenClipboard()
        # 获取剪切板中的数据
        d = w.GetClipboardData(win32con.CF_TEXT)
        # 关闭剪切板
        w.CloseClipboard()
        # 返回剪切板数据给调用者
        return d

    #设置剪切板内容
    @staticmethod
    def setText(aString):
        # 打开剪切板
        w.OpenClipboard()
        # 清空剪切板
        w.EmptyClipboard()
        # 将数据 aString 写入剪切板
        w.SetClipboardData(win32con.CF_UNICODETEXT, aString)
        # 关闭剪切板
        w.CloseClipboard()
```

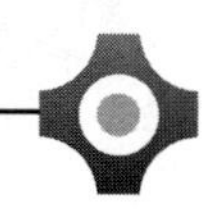

(7) 在 testScripts 包中新建一个名为 TestSendMailWithAttachment.py 的 Python 文件,用于编写具体的测试逻辑代码,具体内容如下。

```
#encoding = utf - 8
from util.ObjectMap import *
from util.KeyBoardUtil import KeyboardKeys
from util.ClipboardUtil import Clipboard
from util.WaitUtil import WaitUtil
from selenium import webdriver
from selenium.webdriver.common.keys import Keys
import time

def TestSendMailWithAttachment():
    # 创建 Firefox 浏览器的实例
    driver = webdriver.Firefox(executable_path = "c:\\geckodriver")
    # 最大化浏览器窗口
    driver.maximize_window()
    print("启动浏览器成功")
    print("访问 126 邮箱登录页...")
    driver.get("http://mail.126.com")
    # 暂停 5 秒,以便邮箱登录页面加载完成
    time.sleep(5)
    assert "126 网易免费邮 -- 你的专业电子邮局" in driver.title
    print("访问 126 邮箱登录页成功")

    wait = WaitUtil(driver)
    wait.frame_available_and_switch_to_it("xpath",
            "//iframe[contains(@id,'x - URS - iframe')]")
    print("输入登录用户名")
    username = getElement(driver, "xpath", "//input[@name = 'email']")
    username.send_keys("xxx")
    print("输入登录密码")
    passwd = getElement(driver, "xpath", "//input[@name = 'password']")
    passwd.send_keys("xxx")
    print("登录...")
    passwd.send_keys(Keys.ENTER)
    # 等待 5 秒,以便登录成功后的页面加载完成
    time.sleep(5)
    assert "网易邮箱" in driver.title
    print("登录成功")
    # 切换至默认句柄,以规避 can't access dead object 异常
    driver.switch_to.default_content()

    element = wait.visibility_element_located("xpath", "//span[text() = '写 信']")
    element.click()
    print("写信...")
    receiver = getElement(driver, "xpath",
                "//div[contains(@id,'_mail_emailinput')]/input")
    # 输入收信人地址
    receiver.send_keys("38617389@qq.com")
```

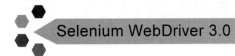

```
    subject = getElement(driver, "xpath",
        "//div[@aria-label='邮件主题输入框,请输入邮件主题']/input")
    # 输入邮件主题
    subject.send_keys("新邮件")
    # 设置剪切板内容
    Clipboard.setText("d:\\aa.txt")
    # 获取剪切板内容
    Clipboard.getText()
    attachment = getElement(driver, "xpath", "//div[contains(@title, '600 首 MP3')]")
    # 单击上传附件链接
    attachment.click()
    time.sleep(3)
    # 在上传附件 Windows 弹窗中粘贴剪切板中的内容
    KeyboardKeys.twoKeys("ctrl", "v")
    time.sleep(3)
    # 模拟回车键,以便加载要上传的附件
    KeyboardKeys.oneKey("enter")
    # 切换进邮件正文的 frame
    wait.frame_available_and_switch_to_it("xpath", "//iframe[@tabindex=1]")
    body = getElement(driver, "xpath", "/html/body")
    # 输入邮件正文
    body.send_keys("发给光荣之路的一封信")
    # 切出邮件正文的 frame 框
    driver.switch_to.default_content()
    print("写信完成")
    getElement(driver, "xpath", "//header//span[text()='发送']").click()
    print("开始发送邮件...")
    time.sleep(3)
    assert "发送成功" in driver.page_source
    print("邮件发送成功")
    driver.quit()

if __name__ == '__main__':
    TestSendMailWithAttachment()
```

代码解释:

将 TestSendMailWithAttachment.py 文件中的"xxx"改成有效的邮箱地址及密码,然后执行该文件,就可以看到浏览器自动登录 126 邮箱,并且自动发了一封带有附件的邮件。在上面的代码中可以看到我们将定位页面元素的方法、模拟键盘按键等封装成了公共方法,方便复用,同时还加入了智能等待,让 UI 自动化测试执行更稳定。但上面的代码仍未做到数据与程序的分离,并且也不能实现高度复用,因此需要继续改造。

(8) 在 config 包中新建一名为 VarConfig.py 的 Python 文件,用于定义整个框架中所需要的一些全局常量值,方便维护,具体代码如下。

```
#encoding=utf-8
import os

firefoxDriverFilePath = "c:\geckodriver"
```

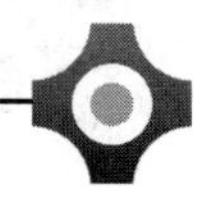

```
# 获取当前文件所在目录的父目录的绝对路径
parentDirPath = os.path.dirname(os.path.dirname(os.path.abspath(__file__)))
screenPicturesDir = parentDirPath + "\\exceptionpictures\\"
```

(9) 在 util 包中新建一名为 DirAndTime.py 的 Python 文件，用于获取当前日期及时间，以及创建异常截图存放目录，具体内容如下。

```
#encoding = utf - 8
import time, os
from datetime import datetime
from config.VarConfig import screenPicturesDir

# 获取当前的日期
def getCurrentDate():
    timeTup = time.localtime()
    currentDate = str(timeTup.tm_year) + "-" + \
        str(timeTup.tm_mon) + "-" + str(timeTup.tm_mday)
    return currentDate

# 获取当前的时间
def getCurrentTime():
    timeStr = datetime.now()
    nowTime = timeStr.strftime('%H-%M-%S-%f')
    return nowTime

# 创建截图存放的目录
def createCurrentDateDir():
    dirName = os.path.join(screenPicturesDir, getCurrentDate())
    if not os.path.exists(dirName):
        os.makedirs(dirName)
    return dirName

if __name__ == '__main__':
    print(getCurrentDate())
    print(createCurrentDateDir())
    print(getCurrentTime())
```

(10) 修改 util 包中的 WaitUtil.py 文件内容如下。

```
#encoding = utf - 8
from selenium.webdriver.common.by import By
from selenium.webdriver.support.ui import WebDriverWait
from selenium.webdriver.support import expected_conditions as EC

class WaitUtil(object):

    def __init__(self, driver):
        self.locationTypeDict = {
            "xpath": By.XPATH,
            "id": By.ID,
            "name": By.NAME,
```

```
            "css_selector": By.CSS_SELECTOR,
            "class_name": By.CLASS_NAME,
            "tag_name": By.TAG_NAME,
            "link_text": By.LINK_TEXT,
            "partial_link_text": By.PARTIAL_LINK_TEXT
        }
        self.driver = driver
        self.wait = WebDriverWait(self.driver, 30)

    def presenceOfElementLocated(self, locatorMethod, locatorExpression, *arg):
        '''显示等待页面元素出现在 DOM 中,并一定可以见,
        存在则返回该页面元素对象'''
        try:
            element = self.wait.until(EC.presence_of_element_located((
                self.locationTypeDict[locatorMethod.lower()],
                locatorExpression)))
            return element
        except Exception as err:
            raise err

    def frameToBeAvailableAndSwitchToIt(self,locationType,locatorExpression, *arg):
        '''检查 frame 是否存在,存在则切换进 frame 控件中
        '''
        try:
            self.wait.until(
                EC.frame_to_be_available_and_switch_to_it((
                    self.locationTypeDict[locationType.lower()],
                    locatorExpression)))
        except Exception as err:
            # 抛出异常信息给上层调用者
            raise err

    def visibilityOfElementLocated(self, locationType, locatorExpression, *arg):
        '''显示等待页面元素出现在 DOM 中,并且可见,存在则返回该页面元素对象'''
        try:
            element = self.wait.until(
                EC.visibility_of_element_located((
                    self.locationTypeDict[locationType.lower()],
                    locatorExpression)))
            return element
        except Exception as err:
            raise err

if __name__ == '__main__':
    from selenium import webdriver
    driver = webdriver.Firefox(executable_path = "c:\\geckodriver")
    driver.get("http://mail.126.com")
    waitUtil = WaitUtil(driver)
    waitUtil.frameToBeAvailableAndSwitchToIt("xpath",
                "//iframe[contains(@id,'x-URS-iframe')]")
    waitUtil.visibilityOfElementLocated("xpath", "//input[@name='email']")
```

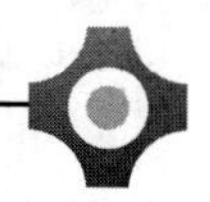

```
    waitUtil.presenceOfElementLocated("xpath", "//input[@name = 'email']")
    driver.quit()
```

(11) 在 KeyWordsFrameWork 工程中新建一个名为 action 的 Python package，并在此包中新建一个名为 PageAction.py 的 Python 文件，用于实现具体的页面动作，比如在输入框中输入数据，单击页面按钮等，具体内容如下。

```
#encoding = utf-8
from selenium import webdriver
from config.VarConfig import firefoxDriverFilePath
from util.ObjectMap import getElement
from util.ClipboardUtil import Clipboard
from util.KeyBoardUtil import KeyboardKeys
from util.DirAndTime import *
from util.WaitUtil import WaitUtil
import time

# 定义全局 driver 变量
driver = None
# 全局的等待类实例对象
waitUtil = None

def open_browser(*arg):
    # 打开浏览器
    global driver, waitUtil
    try:
        driver = webdriver.Firefox(executable_path = firefoxDriverFilePath)
        # driver 对象创建成功后,创建等待类实例对象
        waitUtil = WaitUtil(driver)
    except Exception as err:
        raise err

def visit_url(url, *arg):
    # 访问某个网址
    global driver
    try:
        driver.get(url)
    except Exception as err:
        raise err

def close_browser(*arg):
    # 关闭浏览器
    global driver
    try:
        driver.quit()
    except Exception as err:
        raise err

def sleep(sleepSeconds, *arg):
    # 强制等待
    try:
```

```
            time.sleep(int(sleepSeconds))
        except Exception as err:
            raise err

def clear(locationType, locatorExpression, *arg):
    # 清除输入框默认内容
    global driver
    try:
        getElement(driver, locationType, locatorExpression).clear()
    except Exception as err:
        raise err

def input_string(locationType, locatorExpression, inputContent):
    # 在页面输入框中输入数据
    global driver
    try:
        getElement(driver, locationType,
            locatorExpression).send_keys(inputContent)
    except Exception as err:
        raise err

def click(locationType, locatorExpression, *arg):
    # 单击页面元素
    global driver
    try:
        getElement(driver, locationType, locatorExpression).click()
    except Exception as err:
        raise err

def assert_string_in_pagesource(assertString, *arg):
    # 断言页面源码是否存在某关键字或关键字符串
    global driver
    try:
        assert assertString in driver.page_source, \
            "%s not found in page source!" % assertString
    except AssertionError as err:
        raise AssertionError(err)
    except Exception as e:
        raise e

def assert_title(titleStr, *args):
    # 断言页面标题是否存在给定的关键字符串
    global driver
    try:
        assert titleStr in driver.title, \
            "%s not found in title!" % titleStr
    except AssertionError as err:
        raise AssertionError(err)
    except Exception as e:
        raise e
```

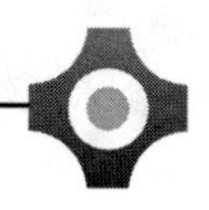

```
def getTitle( * arg):
    # 获取页面标题
    global driver
    try:
        return driver.title
    except Exception as err:
        raise err

def getPageSource( * arg):
    # 获取页面源码
    global driver
    try:
        return driver.page_source
    except Exception as err:
        raise err

def switch_to_frame(locationType, frameLocatorExpression,  * arg):
    # 切换进入 frame
    global driver
    try:
        driver.switch_to.frame(getElement
            (driver, locationType, frameLocatorExpression))
    except Exception as err:
        raise err

def switch_to_default_content( * arg):
    # 切出 frame
    global driver
    try:
        driver.switch_to.default_content()
    except Exception as err:
        raise err

def paste_string(pasteString,  * arg):
    # 模拟 Ctrl + v 操作
    try:
        Clipboard.setText(pasteString)
        # 等待 2 秒,防止代码执行得太快,而未成功粘贴内容
        time.sleep(2)
        KeyboardKeys.twoKeys("ctrl", "v")
    except Exception as err:
        raise err

def press_tab_key( * arg):
    # 模拟 Tab 键
    try:
        KeyboardKeys.oneKey("tab")
    except Exception as err:
        raise err

def press_enter_key( * arg):
```

```
    # 模拟回车键
    try:
        KeyboardKeys.oneKey("enter")
    except Exception as err:
        raise err

def maximize_browser():
    # 窗口最大化
    global driver
    try:
        driver.maximize_window()
    except Exception as err:
        raise err

def capture_screen( * args):
    # 截取屏幕图片
    global driver
    currTime = getCurrentTime()
    picNameAndPath = str(createCurrentDateDir()) + "\\" + str(currTime) + ".png"
    try:
        driver.get_screenshot_as_file(picNameAndPath.replace('\\', r'\\'))
    except Exception as err:
        raise err
    else:
        return picNameAndPath

def waitPresenceOfElementLocated(locationType, locatorExpression, * arg):
    '''显示等待页面元素出现在 DOM 中,并一定可以见,
            存在则返回该页面元素对象'''
    global waitUtil
    try:
        waitUtil.presenceOfElementLocated(locationType, locatorExpression)
    except Exception as err:
        raise err

def waitFrameToBeAvailableAndSwitchToIt(locationType, locatorExpression, * args):
    '''检查 frame 是否存在,存在则切换进 frame 控件中'''
    global waitUtil
    try:
        waitUtil.frameToBeAvailableAndSwitchToIt(locationType, locatorExpression)
    except Exception as err:
        raise err

def waitVisibilityOfElementLocated(locationType, locatorExpression, * args):
    '''显示等待页面元素出现在 DOM 中,并且可见,存在返回该页面元素对象'''
    global waitUtil
    try:
        waitUtil.visibilityOfElementLocated(locationType, locatorExpression)
    except Exception as err:
        raise err
```

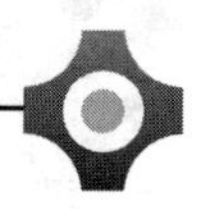

（12）修改 testScripts 包中的 TestSendMailWithAttachment.py 文件内容如下。

```
#encoding = utf-8
from action.PageAction import *
import time

def TestSendMailWithAttachment():
    print("启动浏览器")
    open_browser()
    maximize_browser()
    print("访问 126 邮箱登录页")
    visit_url("http://mail.126.com")
    sleep(5)
    assert_string_in_pagesource("126 网易免费邮 -- 你的专业电子邮局")
    print("访问 126 邮箱登录页成功")
    waitFrameToBeAvailableAndSwitchToIt("xpath",
                "//iframe[contains(@id,'x-URS-iframe')]")
    print("输入登录用户名")
    input_string("xpath", "//input[@name = 'email']", "xxx")
    print("输入登录密码")
    input_string("xpath", "//input[@name = 'password']", "xxx")
    click("id", "dologin")
    sleep(5)
    assert_title(u"网易邮箱")
    print("登录成功")
    switch_to_default_content()
    waitVisibilityOfElementLocated("xpath", "//span[text() = '写 信']")
    click("xpath", "//span[text() = '写 信']")
    print("开始写信")
    print("输入收件人地址")
    input_string("xpath",
        "//div[contains(@id,'_mail_emailinput')]/input",
        "xxx@qq.com")
    print("输入邮件主题")
    input_string("xpath",
        "//div[@aria-label = '邮件主题输入框,请输入邮件主题']/input",
        "新邮件")
    print("单击上传附件按钮")
    click("xpath", "//div[contains(@title, '600 首 MP3')]")
    sleep(3)
    print("上传附件")
    paste_string("d:\\aa.txt")
    press_enter_key()
    sleep(2)
    waitFrameToBeAvailableAndSwitchToIt("xpath", "//iframe[@tabindex = 1]")
    print("写入邮件正文")
    input_string("xpath", "/html/body", "发给光荣之路的一封信")
    switch_to_default_content()
    print("写信完成")
    print("开始发送邮件...")
    click("xpath", "//header//span[text() = '发送']")
```

```
    time.sleep(3)
    assert_string_in_pagesource("发送成功")
    print("邮件发送成功")
    close_browser()

if __name__ == '__main__':
    TestSendMailWithAttachment()
```

替换 TestSendMailWithAttachment.py 文件中的“xxx”为有效的 126 邮箱登录账号及密码后执行该文件，可以看到程序会自动启动 Firefox 浏览器，然后访问 126 邮箱并发送一封带附件的邮件。

(13) 在 util 包中新建一个名为 ParseExcel.py 的 Python 文件，用于实现读取 Excel 数据文件代码封装，具体内容如下。

```
#encoding = utf-8
import openpyxl
from openpyxl.styles import Border, Side, Font
import time

class ParseExcel(object):

    def __init__(self):
        self.workbook = None
        self.excelFile = None
        self.font = Font(color = None)                    # 设置字体的颜色
        # 颜色对应的 RGB 值
        self.RGBDict = {'red': 'FFFF3030', 'green': 'FF008B00'}

    def loadWorkBook(self, excelPathAndName):
        # 将 Excel 文件加载到内存，并获取其 workbook 对象
        try:
            self.workbook = openpyxl.load_workbook(excelPathAndName)
        except Exception, e:
            raise e
        self.excelFile = excelPathAndName
        return self.workbook

    def getSheetByName(self, sheetName):
        # 根据 sheet 名获取该 sheet 对象
        try:
            sheet = self.workbook.get_sheet_by_name(sheetName)
            return sheet
        except Exception, e:
            raise e

    def getSheetByIndex(self, sheetIndex):
        # 根据 sheet 的索引号获取该 sheet 对象
        try:
            sheetname = self.workbook.get_sheet_names()[sheetIndex]
        except Exception, e:
```

```
            raise e
        sheet = self.workbook.get_sheet_by_name(sheetname)
        return sheet

    def getRowsNumber(self, sheet):
        # 获取 sheet 中有数据区域的结束行号
        return sheet.max_row

    def getColsNumber(self, sheet):
        # 获取 sheet 中有数据区域的结束列号
        return sheet.max_column

    def getStartRowNumber(self, sheet):
        # 获取 sheet 中有数据区域的开始的行号
        return sheet.min_row

    def getStartColNumber(self, sheet):
        # 获取 sheet 中有数据区域的开始的列号
        return sheet.min_column

    def getRow(self, sheet, rowNo):
        # 获取 sheet 中某一行,返回的是这一行所有的数据内容组成的 tuple,
        # 下标从 1 开始,sheet.rows[1]表示第一行
        try:
            return sheet.rows[rowNo - 1]
        except Exception, e:
            raise e

    def getColumn(self, sheet, colNo):
        # 获取 sheet 中某一列,返回的是这一列所有的数据内容组成 tuple,
        # 下标从 1 开始,sheet.columns[1]表示第一列
        try:
            return sheet.columns[colNo - 1]
        except Exception, e:
            raise e

    def getCellOfValue(self, sheet, coordinate = None,
                       rowNo = None, colsNo = None):
        # 根据单元格所在的位置索引获取该单元格中的值,下标从 1 开始,
        # sheet.cell(row = 1, column = 1).value,表示 Excel 中第一行第一列的值
        if coordinate != None:
            try:
                return sheet.cell(coordinate = coordinate).value
            except Exception, e:
                raise e
        elif coordinate is None and rowNo is not None and colsNo is not None:
            try:
                return sheet.cell(row = rowNo, column = colsNo).value
            except Exception, e:
                raise e
        else:
```

```
            raise Exception("Insufficient Coordinates of cell !")

    def getCellOfObject(self, sheet, coordinate = None,
                        rowNo = None, colsNo = None):
        # 获取某个单元格的对象,可以根据单元格所在位置的数字索引,
        # 也可以直接根据 Excel 中单元格的编码坐标
        # 如 getCellObject(sheet, coordinate = 'A1') or
        # getCellObject(sheet, rowNo = 1, colsNo = 2)
        if coordinate != None:
            try:
                return sheet.cell(coordinate = coordinate)
            except Exception, e:
                raise e
        elif coordinate == None and rowNo is not None and colsNo is not None:
            try:
                return sheet.cell(row = rowNo,column = colsNo)
            except Exception, e:
                raise e
        else:
            raise Exception("Insufficient Coordinates of cell !")

    def writeCell(self, sheet, content, coordinate = None,
        rowNo = None, colsNo = None, style = None):
        # 根据单元格在 Excel 中的编码坐标或者数字索引坐标向单元格中写入数据,
        # 下标从 1 开始,参 style 表示字体的颜色的名字,比如 red,green
        if coordinate is not None:
            try:
                sheet.cell(coordinate = coordinate).value = content
                if style is not None:
                    sheet.cell(coordinate = coordinate).\
                        font = Font(color = self.RGBDict[style])
                self.workbook.save(self.excelFile)
            except Exception, e:
                raise e
        elif coordinate == None and rowNo is not None and colsNo is not None:
            try:
                sheet.cell(row = rowNo,column = colsNo).value = content
                if style:
                    sheet.cell(row = rowNo,column = colsNo).\
                        font = Font(color = self.RGBDict[style])
                self.workbook.save(self.excelFile)
            except Exception, e:
                raise e
        else:
            raise Exception("Insufficient Coordinates of cell !")

    def writeCellCurrentTime(self, sheet, coordinate = None,
                rowNo = None, colsNo = None):
        # 写入当前的时间,下标从 1 开始
        now = int(time.time())                         #显示为时间戳
        timeArray = time.localtime(now)
```

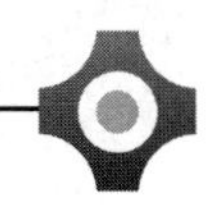

```
        currentTime = time.strftime("%Y-%m-%d %H:%M:%S", timeArray)
        if coordinate is not None:
            try:
                sheet.cell(coordinate = coordinate).value = currentTime
                self.workbook.save(self.excelFile)
            except Exception, e:
                raise e
        elif coordinate == None and rowNo is not None and colsNo is not None:
            try:
                sheet.cell(row = rowNo, column = colsNo).value = currentTime
                self.workbook.save(self.excelFile)
            except Exception, e:
                raise e
        else:
            raise Exception("Insufficient Coordinates of cell !")

if __name__ == '__main__':
    pe = ParseExcel()
    #测试所用的 Excel 文件"126 邮箱联系人.xlsx"请自行创建
    pe.loadWorkBook(u'D:\\PythonProject\\126 邮箱联系人.xlsx')
    print "通过名称获取 sheet 对象的名字:", pe.getSheetByName(u"联系人").title
    print "通过 index 序号获取 sheet 对象的名字:", pe.getSheetByIndex(0).title
    sheet = pe.getSheetByIndex(0)
    print type(sheet)
    print pe.getRowsNumber(sheet)                    #获取最大行号
    print pe.getColsNumber(sheet)                    #获取最大列号
    rows = pe.getRow(sheet, 1)                       #获取第一行
    for i in rows:
        print i.value
    # 获取第一行第一列单元格内容
    print pe.getCellOfValue(sheet, rowNo = 1, colsNo = 1)
    pe.writeCell(sheet, u'我爱祖国', rowNo = 10, colsNo = 10)
    pe.writeCellCurrentTime(sheet, rowNo = 10, colsNo = 11)
```

（14）在 testData 目录中新建一个名为“126 邮箱发送邮件.xlsx”的 Excel 文件，并在此 Excel 文件中新建三个工作表，分别命名为“测试用例”“登录”及“发邮件”。

“测试用例”工作表用于存放测试用例，内容如表 15-3 所示。

表　15-3

序号	用例名称	用例描述	步骤 sheet 名	是否执行	执行结束时间	结果
1	登录 126 邮箱	使用有效的账号登录 126 邮箱	登录	y		
2	发送带附件的邮件	登录 126 邮箱后，发送一封带附件的邮件	发邮件	y		

表格中“是否执行”列中的 y 表示本条用例需要执行，n 表示不执行。

“登录”工作表用于存放登录 126 邮箱的所有步骤信息，内容如表 15-4 所示。

表 15-4

序号	测试步骤描述	关键字	操作元素的定位方式	操作元素的定位表达式	操作值	测试执行时间	测试结果	错误信息	错误截图
1	打开浏览器	open_browser			chrome				
2	访问被测试网址 http://www.126.com	visit_url			http://www.126.com				
3	最大化窗口	maximize_browser							
4	等待 126 邮箱登录主页加载完成	sleep			5				
5	断言当前活动页面源码中是否包含"126 网易免费邮--你的专业电子邮局"	assert_string_in_pagesource			126 网易免费邮--你的专业电子邮局				
6	显示等待 id 属性值为 x-URS-iframe 的 frame 框的出现,然后切换进入该 frame 框中	waitFrameToBeAvailableAndSwitchToIt	xpath	//iframe[contains(@ id, ' x-URS-iframe')]					
7	输入登录用户名	input_string	xpath	//input[@name='email']	xxx				
8	输入登录密码	input_string	xpath	//input[@name='password']	xxx				
9	单击登录按钮	click	id	dologin					
10	等待	sleep			5				
11	切回默认会话窗体	switch_to_default_content							
12	断言登录成功后的页面标题是否包含"网易邮箱 6.0 版"关键内容	assert_title			网易邮箱 6.0 版				

"发邮件"工作表用于存放登录成功后,发送邮件所有步骤信息,内容如表 15-5 所示。

表 15-5

序号	测试步骤描述	关键字	操作元素的定位方式	操作元素的定位表达式	操作值	测试执行时间	测试结果	错误信息	错误截图
1	判断"写信"按钮是否在页面上可见	waitVisibilityOfElementLocated	xpath	//span[text()='写 信']					
2	单击"写信"按钮	click	xpath	//span[text()='写 信']					

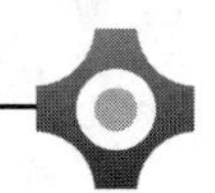

续表

序号	测试步骤描述	关键字	操作元素的定位方式	操作元素的定位表达式	操作值	测试执行时间	测试结果	错误信息	错误截图
3	输入收件人地址	input_string	xpath	//div[contains(@ id, '_ mail _ emailinput')]/input	xxx				
4	输入邮件主题	input_string	xpath	//div[@aria-label='邮件主题输入框，请输入邮件主题']/input	带附件的邮件				
5	单击"上传附件"链接	click	xpath	//div[contains(@ title, '600 首 MP3')]					
6	输入附件所在绝对路径	paste_string			d:\\a. txt				
7	模拟键盘回车键	press_enter_key							
8	显示等待附件上传完毕	waitVisibilityOfElementLocated	xpath	//span[text()="上传完成"]					
9	如果邮件正文的 frame 框是否可见，切换进该 frame 中	waitFrameToBeAvailableAndSwitchToIt	xpath	//iframe[@tabindex=1]					
10	输入邮件正文	input_string	xpath	/html/body	发给光荣之路的一封信				
11	退出邮件正文的 frame	switch_to_default_content							
12	单击邮件发送按钮	click	xpath	//header//span[text()='发送']					
13	等待邮件发送成功，返回结果	sleep			3				
14	断言页面源码中是否出现"发送成功"关键内容	assert_string_in_pagesource			发送成功				
15	关闭浏览器	close_browser							

说明：

"登录"和"发邮件"工作表中的"关键字"列内容对应 action 包中 PageAction. py 文件中的函数名；"操作元素定位方式列"表示定位页面元素所使用的定位方式，比如 xpath、id 等；"操作元素定位表达式"列，表示定位页面元素所使用的定位方式对应的定位表达式；"操作值"列，表示页面输入框需要输入的内容、断言函数需要的关键内容等。

"测试用例"工作表中的"序号"和"用例描述"以及用例步骤表中的"序号"不作为关键字

驱动框架使用的列，因此可以不填，但这些列必须存在，并且表格的顺序不能改变，如果不想要这两列，可以将其删除，或者修改了上面几张表格的顺序，需要在 config 包中的 VarConfig. py 文件中，同步更新一下框架中用到的各个工作表中的列的数字编号。

从数据表的设计可以看出，程序中使用的定位表达式及操作值不再与代码混合在一起，如果定位方式、定位表达式或者操作值有变化，只需要修改数据文件中相关内容即可，不仅方便维护，还可以让不懂代码的测试人员实现自动化测试，提高了自动化测试的效率。

(15) 修改 config 包中的 VarConfig. py 文件内容如下。

```
#encoding = utf - 8
import os

firefoxDriverFilePath = "c:\geckodriver"

# 当前文件所在目录的父目录的绝对路径
parentDirPath = os.path.dirname(os.path.dirname(os.path.abspath(__file__)))

# 异常截图存放目录绝对路径
screenPicturesDir = parentDirPath + "\\exceptionpictures\\"

# 测试数据文件存放绝对路径
dataFilePath = parentDirPath + "\\testData\\126邮箱发送邮件.xlsx"

# 测试数据文件中,测试用例表中部分列对应的数字序号
testCase_testCaseName = 2
testCase_testStepSheetName = 4
testCase_isExecute = 5
testCase_runTime = 6
testCase_testResult = 7

# 用例步骤表中,部分列对应的数字序号
testStep_testStepDescribe = 2
testStep_keyWords = 3
testStep_locationType = 4
testStep_locatorExpression = 5
testStep_operateValue = 6
testStep_runTime = 7
testStep_testResult = 8
testStep_errorInfo = 9
testStep_errorPic = 10
```

(16) 修改 testScripts 包中 TestSendMailWithAttachment. py 文件内容如下。

```
#encoding = utf - 8
from action.PageAction import *
from util.ParseExcel import ParseExcel
from config.VarConfig import *
import time
import traceback
```

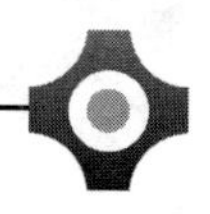

```
# 创建解析 Excel 对象
excelObj = ParseExcel()
# 将 Excel 数据文件加载到内存
excelObj.loadWorkBook(dataFilePath)

# 用例或用例步骤执行结束后,向 Excel 中写执行结果信息
def writeTestResult(sheetObj, rowNo, colsNo, testResult,
                    errorInfo = None, picPath = None):
    # 测试通过结果信息为绿色,失败为红色
    colorDict = {"pass":"green", "faild":"red"}

    # 因为"测试用例"工作表和"用例步骤 sheet 表"中都有测试执行时间和
    # 测试结果列,定义此字典对象是为了区分具体应该写哪个工作表
    colsDict = {
        "testCase":[testCase_runTime, testCase_testResult],
        "caseStep":[testStep_runTime, testStep_testResult]}
    try:
        # 在测试步骤 sheet 中,写入测试时间
        excelObj.writeCellCurrentTime(sheetObj,
                rowNo = rowNo, colsNo = colsDict[colsNo][0])
        # 在测试步骤 sheet 中,写入测试结果
        excelObj.writeCell(sheetObj, content = testResult,
                rowNo = rowNo, colsNo = colsDict[colsNo][1],
                style = colorDict[testResult])
        if errorInfo and picPath:
            # 在测试步骤 sheet 中,写入异常信息
            excelObj.writeCell(sheetObj, content = errorInfo,
                    rowNo = rowNo, colsNo = testStep_errorInfo)
            # 在测试步骤 sheet 中,写入异常截图路径
            excelObj.writeCell(sheetObj, content = picPath,
                    rowNo = rowNo, colsNo = testStep_errorPic)
        else:
            # 在测试步骤 sheet 中,清空异常信息单元格
            excelObj.writeCell(sheetObj, content = "",
                    rowNo = rowNo, colsNo = testStep_errorInfo)
            # 在测试步骤 sheet 中,清空异常信息单元格
            excelObj.writeCell(sheetObj, content = "",
                    rowNo = rowNo, colsNo = testStep_errorPic)
    except Exception as err:
        print("写 excel 出错,", traceback.print_exc())


def TestSendMailWithAttachment():
    try:
        # 根据 Excel 文件中的 sheet 名获取 sheet 对象
        caseSheet = excelObj.getSheetByName("测试用例")
        # 获取测试用例 sheet 中是否执行列对象
        isExecuteColumn = excelObj.getColumn(caseSheet, testCase_isExecute)
        # 记录执行成功的测试用例个数
        successfulCase = 0
        # 记录需要执行的用例个数
```

```
requiredCase = 0
for idx, i in enumerate(isExecuteColumn[1:]):
    # 因为用例 sheet 中第一行为标题行,无须执行
    # print (i.value)
    # 循环遍历"测试用例"表中的测试用例,执行被设置为执行的用例
    if i.value.lower() == "y":
        requiredCase += 1
        # 获取"测试用例"表中第 idx + 2 行数据
        caseRow = excelObj.getRow(caseSheet, idx + 2)
        # 获取第 idx + 2 行的"步骤 sheet"单元格内容
        caseStepSheetName = caseRow[testCase_testStepSheetName - 1].value
        # print (caseStepSheetName)

        # 根据用例步骤名获取步骤 sheet 对象
        stepSheet = excelObj.getSheetByName(caseStepSheetName)
        # 获取步骤 sheet 中步骤数
        stepNum = excelObj.getRowsNumber(stepSheet)
        # print (stepNum)
        # 记录测试用例 i 的步骤成功数
        successfulSteps = 0
        print("开始执行用例" % s""
              % caseRow[testCase_testCaseName - 1].value)
        for step in range(2, stepNum + 1):
            # 因为步骤 sheet 中的第一行为标题行,无须执行
            # 获取步骤 sheet 中第 step 行对象
            stepRow = excelObj.getRow(stepSheet, step)
            # 获取关键字作为调用的函数名
            keyWord = stepRow[testStep_keyWords - 1].value
            # 获取操作元素定位方式作为调用函数的参数
            locationType = stepRow[testStep_locationType - 1].value
            # 获取操作元素的定位表达式作为调用函数的参数
            locatorExpression = stepRow[testStep_locatorExpression - 1].value
            # 获取操作值作为调用函数的参数
            operateValue = stepRow[testStep_operateValue - 1].value

            # 将操作值为数字类型的数据转成字符串类型,方便字符串拼接
            if isinstance(operateValue, int):
                operateValue = str(operateValue)
            # print(keyWord, locationType, locatorExpression, operateValue)

            expressionStr = ""
            # 构造需要执行的 Python 语句,
            # 对应的是 PageAction.py 文件中的页面动作函数调用的字符串表示
            if keyWord and operateValue and \
                    locationType is None and locatorExpression is None:
                expressionStr = keyWord.strip() + "('" + operateValue + "')"
            elif keyWord and operateValue is None and \
                locationType is None and locatorExpression is None:
                expressionStr = keyWord.strip() + "()"
            elif keyWord and locationType and operateValue and \
                            locatorExpression is None:
```

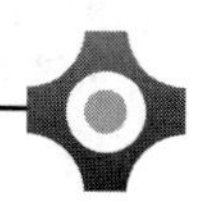

```
            expressionStr = keyWord.strip() + \
                "('" + locationType.strip() + "', '" + operateValue + "')"
        elif keyWord and locationType and locatorExpression \
                and operateValue:
            expressionStr = keyWord.strip() + \
                "('" + locationType.strip() + "', '" + \
                locatorExpression.replace("'", '"').strip() + \
                "', '" + operateValue + "')"
        elif keyWord and locationType and locatorExpression \
                and operateValue is None:
            expressionStr = keyWord.strip() + \
                "('" + locationType.strip() + "', '" + \
                locatorExpression.replace("'", '"').strip() + "')"
        # print expressionStr
        try:
            # 通过 eval 函数,将拼接的页面动作函数调用的字符串表示
            # 当成有效的 Python 表达式执行,从而执行测试步骤的 sheet 中
            # 关键字在 ageAction.py 文件中对应的映射方法,
            # 来完成对页面元素的操作
            eval(expressionStr)
            # 在测试执行时间列写入执行时间
            excelObj.writeCellCurrentTime(
                stepSheet, rowNo = step,
                colsNo = testStep_runTime)
        except Exception as err:
            # 截取异常屏幕图片
            capturePic = capture_screen()
            # 获取详细的异常堆栈信息
            errorInfo = traceback.format_exc()
            # 在测试步骤 Sheet 中写入失败信息
            writeTestResult(
                stepSheet, step, "caseStep",
                "faild", errorInfo, capturePic)
            print("步骤" %s"执行失败!"
                % stepRow[testStep_testStepDescribe - 1].value)
        else:
            # 在测试步骤 Sheet 中写入成功信息
            writeTestResult(stepSheet, step, "caseStep", "pass")
            # 每成功一步,successfulSteps 变量自增 1
            successfulSteps += 1
            print("步骤" %s"执行通过!"
                % stepRow[testStep_testStepDescribe - 1].value)
    if successfulSteps == stepNum - 1:
        # 当测试用例步骤 sheet 中所有的步骤都执行成功,
        # 方认为此测试用例执行通过,然后将成功信息写入
        # 测试用例工作表中,否则写入失败信息
        writeTestResult(caseSheet, idx + 2, "testCase", "pass")
        successfulCase += 1
    else:
        writeTestResult(caseSheet, idx + 2, "testCase", "faild")
print("共%d 条用例,%d 条需要被执行,本次执行通过%d 条。"
```

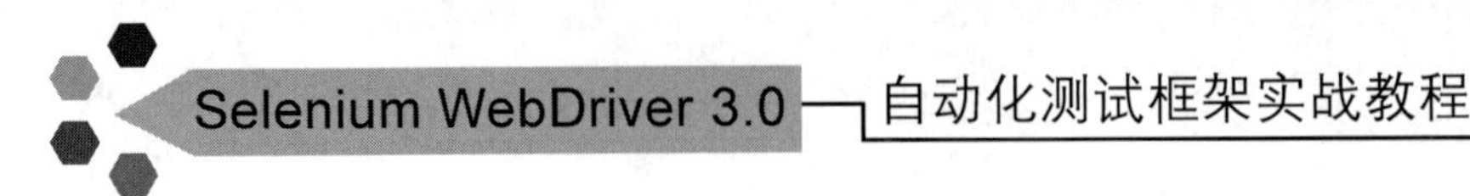


```
                % (len(isExecuteColumn) - 1, requiredCase, successfulCase))
        except Exception as err:
            # 打印详细的异常堆栈信息
            print(traceback.print_exc())

if __name__ == '__main__':
    TestSendMailWithAttachment()
```

(17) 在 KeyWordsFrameWork 工程根目录中创建一名为 RunTest.py 的 Python 文件,用于实现关键字框架运行的入口代码,具体内容如下。

```
#encoding = utf - 8
from testScripts.TestSendMailWithAttachment import TestSendMailWithAttachment

if __name__ == '__main__':
    TestSendMailWithAttachment()
```

(18) 将数据文件"126 邮箱发送邮件.xlsx"中的各表格"操作值"列中的"xxx"改成有效的邮箱账号及密码,然后在 Pycharm 中执行 RunTest.py 文件,可以看到程序自动打开了浏览器,并发送了一封带附件的邮件,执行结果如图 15-14 所示。

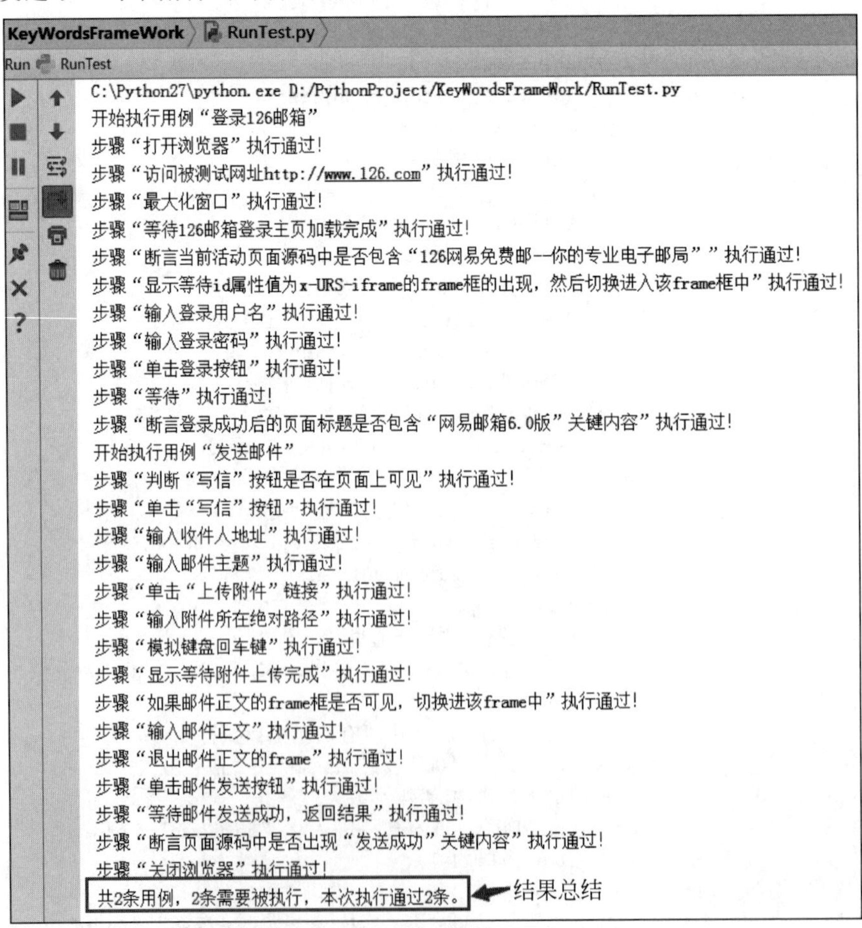

图 15-14

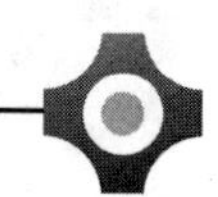

上面我们只是把测试过程及结果总结信息输出到控制台，但不方便错误的排查以及后期结果信息的查看，为此我们加入打印日志功能。

(19) 通过 logging 模块，为关键字驱动框架加入打印日志功能。在 config 包中新建一个名叫 Logger.conf 的文件，用于配置日志基本信息，具体内容如下。

```
#logger.conf
###############################################
[loggers]
keys = root,example01,example02
[logger_root]
level = DEBUG
handlers = hand01,hand02

[logger_example01]
handlers = hand01,hand02
qualname = example01
propagate = 0

[logger_example02]
handlers = hand01,hand03
qualname = example02
propagate = 0

###############################################
[handlers]
keys = hand01,hand02,hand03

[handler_hand01]
class = StreamHandler
level = INFO
formatter = form01
args = (sys.stderr,)

[handler_hand02]
class = FileHandler
level = DEBUG
formatter = form01
args = ('log\\Mail126TestLogfile.log', 'a')

[handler_hand03]
class = handlers.RotatingFileHandler
level = INFO
formatter = form01
args = ('log\\Mail126TestLogfile.log', 'a', 10 * 1024 * 1024, 5)

###############################################
[formatters]
keys = form01,form02

[formatter_form01]
```

```
format = %(asctime)s %(filename)s[line: %(lineno)d] %(levelname)s %(message)s
datefmt = %Y-%m-%d %H:%M:%S

[formatter_form02]
format = %(name)-12s: %(levelname)-8s %(message)s
datefmt = %Y-%m-%d %H:%M:%S
```

(20) 在 util 包中新建一个名为 Log.py 的 Python 文件,用于初始化日志对象,具体内容如下。

```
#encoding = utf-8
import logging
import logging.config
from config.VarConfig import parentDirPath

# 读取日志配置文件
logging.config.fileConfig(parentDirPath + "\\config\\Logger.conf")
# 选择一个日志格式
logger = logging.getLogger("example02")                    #或者 example01

def debug(message):
    # 定义 dubug 级别日志打印方法
    logger.debug(message)

def info(message):
    # 定义 info 级别日志打印方法
    logger.info(message)

def warning(message):
    # 定义 warning 级别日志打印方法
    logger.warning(message)
```

(21) 在 DataDrivenFrameWork 工程根目录下创建一名为 log 的目录,然后修改 testScripts 包中的 TestSendMailWithAttachment.py 文件的内容如下。

```
#encoding = utf-8
from action.PageAction import *
from util.ParseExcel import ParseExcel
from config.VarConfig import *
from util.Log import *
import traceback

# 创建解析 Excel 对象
excelObj = ParseExcel()
# 将 Excel 数据文件加载到内存
excelObj.loadWorkBook(dataFilePath)

# 用例或用例步骤执行结束后,向 Excel 中写入执行结果信息
def writeTestResult(sheetObj, rowNo, colsNo, testResult,
                    errorInfo = None, picPath = None):
    # 测试通过结果信息为绿色,失败为红色
```

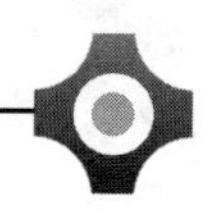

```
    colorDict = {"pass":"green", "faild":"red"}

    # 因为"测试用例"工作表和"用例步骤 sheet 表"中都有测试执行时间和
    # 测试结果列,定义此字典对象是为了区分具体应该写哪个工作表
    colsDict = {
        "testCase":[testCase_runTime, testCase_testResult],
        "caseStep":[testStep_runTime, testStep_testResult]}
    try:
        # 在测试步骤 sheet 中,写入测试时间
        excelObj.writeCellCurrentTime(sheetObj,
                rowNo = rowNo, colsNo = colsDict[colsNo][0])
        # 在测试步骤 sheet 中,写入测试结果
        excelObj.writeCell(sheetObj, content = testResult,
                rowNo = rowNo, colsNo = colsDict[colsNo][1],
                style = colorDict[testResult])
        if errorInfo and picPath:
            # 在测试步骤 sheet 中,写入异常信息
            excelObj.writeCell(sheetObj, content = errorInfo,
                    rowNo = rowNo, colsNo = testStep_errorInfo)
            # 在测试步骤 sheet 中,写入异常截图路径
            excelObj.writeCell(sheetObj, content = picPath,
                    rowNo = rowNo, colsNo = testStep_errorPic)
        else:
            # 在测试步骤 sheet 中,清空异常信息单元格
            excelObj.writeCell(sheetObj, content = "",
                    rowNo = rowNo, colsNo = testStep_errorInfo)
            # 在测试步骤 sheet 中,清空异常信息单元格
            excelObj.writeCell(sheetObj, content = "",
                    rowNo = rowNo, colsNo = testStep_errorPic)
    except Exception as err:
        logging.debug("写 Excel 出错, %s" %traceback.print_exc())


def TestSendMailWithAttachment():
    try:
        # 根据 Excel 文件中的 sheet 名获取 sheet 对象
        caseSheet = excelObj.getSheetByName("测试用例")
        # 获取测试用例 sheet 中是否执行列对象
        isExecuteColumn = excelObj.getColumn(caseSheet, testCase_isExecute)
        # 记录执行成功的测试用例个数
        successfulCase = 0
        # 记录需要执行的用例个数
        requiredCase = 0
        for idx, i in enumerate(isExecuteColumn[1:]):
            # 因为用例 sheet 中第一行为标题行,无须执行
            # print (i.value)
            # 循环遍历"测试用例"表中的测试用例,执行被设置为执行的用例
            if i.value.lower() == "y":
                requiredCase += 1
                # 获取"测试用例"表中第 idx + 2 行数据
                caseRow = excelObj.getRow(caseSheet, idx + 2)
```

```
# 获取第 idx + 2 行的"步骤 sheet"单元格内容
caseStepSheetName = caseRow[testCase_testStepSheetName - 1].value
# print (caseStepSheetName)

# 根据用例步骤名获取步骤 sheet 对象
stepSheet = excelObj.getSheetByName(caseStepSheetName)
# 获取步骤 sheet 中步骤数
stepNum = excelObj.getRowsNumber(stepSheet)
# print (stepNum)
# 记录测试用例 i 的步骤成功数
successfulSteps = 0
logging.info("开始执行用例" % s""
       % caseRow[testCase_testCaseName - 1].value)
for step in range(2, stepNum + 1):
    # 因为步骤 sheet 中的第一行为标题行,无须执行
    # 获取步骤 sheet 中第 step 行对象
    stepRow = excelObj.getRow(stepSheet, step)
    # 获取关键字作为调用的函数名
    keyWord = stepRow[testStep_keyWords - 1].value
    # 获取操作元素定位方式作为调用的函数的参数
    locationType = stepRow[testStep_locationType - 1].value
    # 获取操作元素的定位表达式作为调用函数的参数
    locatorExpression = stepRow[testStep_locatorExpression - 1].value
    # 获取操作值作为调用函数的参数
    operateValue = stepRow[testStep_operateValue - 1].value

    # 将操作值为数字类型的数据转成字符串类型,方便字符串拼接
    if isinstance(operateValue, int):
        operateValue = str(operateValue)
    # print(keyWord, locationType, locatorExpression, operateValue)

    expressionStr = ""
    # 构造需要执行的 python 语句,
    # 对应的是 PageAction.py 文件中的页面动作函数调用的字符串表示
    if keyWord and operateValue and \
            locationType is None and locatorExpression is None:
        expressionStr = keyWord.strip() + "('" + operateValue + "')"
    elif keyWord and operateValue is None and \
        locationType is None and locatorExpression is None:
        expressionStr = keyWord.strip() + "()"
    elif keyWord and locationType and operateValue and \
                    locatorExpression is None:
        expressionStr = keyWord.strip() + \
            "('" + locationType.strip() + "', '" + operateValue + "')"
    elif keyWord and locationType and locatorExpression \
            and operateValue:
        expressionStr = keyWord.strip() + \
            "('" + locationType.strip() + "', '" + \
            locatorExpression.replace("'", '"').strip() + \
            "', '" + operateValue + "')"
    elif keyWord and locationType and locatorExpression \
```

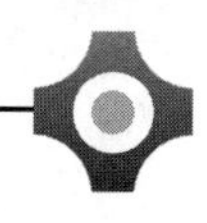

```
                        and operateValue is None:
                    expressionStr = keyWord.strip() + \
                        "('" + locationType.strip() + "', '" + \
                        locatorExpression.replace("'", '"').strip() + "')"
                # print expressionStr
                try:
                    # 通过 eval 函数,将拼接的页面动作函数调用的字符串表示
                    # 当成有效的 Python 表达式执行,从而执行测试步骤的 sheet 中
                    # 关键字在 ageAction.py 文件中对应的映射方法,
                    # 来完成对页面元素的操作
                    eval(expressionStr)
                    # 在测试执行时间列写入执行时间
                    excelObj.writeCellCurrentTime(
                        stepSheet, rowNo = step,
                        colsNo = testStep_runTime)
                except Exception as err:
                    # 截取异常屏幕图片
                    capturePic = capture_screen()
                    # 获取详细的异常堆栈信息
                    errorInfo = traceback.format_exc()
                    # 在测试步骤 Sheet 中写入失败信息
                    writeTestResult(
                        stepSheet, step, "caseStep",
                        "faild", errorInfo, capturePic)
                    logging.info("步骤" %s"执行失败!"
                         % stepRow[testStep_testStepDescribe - 1].value)
                else:
                    # 在测试步骤 Sheet 中写入成功信息
                    writeTestResult(stepSheet, step, "caseStep", "pass")
                    # 每成功一步,successfulSteps 变量自增 1
                    successfulSteps += 1
                    logging.info("步骤" %s"执行通过!"
                         % stepRow[testStep_testStepDescribe - 1].value)
            if successfulSteps == stepNum - 1:
                # 当测试用例步骤 sheet 中所有的步骤都执行成功,
                # 方认为此测试用例执行通过,然后将成功信息写入
                # 测试用例工作表中,否则写入失败信息
                writeTestResult(caseSheet, idx + 2, "testCase", "pass")
                successfulCase += 1
            else:
                writeTestResult(caseSheet, idx + 2, "testCase", "faild")
        logging.info("共%d条用例,%d条需要被执行,本次执行通过%d条。"
             % (len(isExecuteColumn) - 1, requiredCase, successfulCase))
    except Exception as err:
        # 打印详细的异常堆栈信息
        logging.debug(traceback.print_exc())

if __name__ == '__main__':
    TestSendMailWithAttachment()
```

(22) 运行 RunTest.py 文件,执行结束后可以在工程目录下的 log 目录中看到打印的日志文件 Mail126TestLogfile.log,文件所包含的内容如图 15-15 所示,这里只粗略地打印

了一些执行过程信息，读者可以添加更加详细的日志信息，以便查看更详细的执行过程逻辑，便于测试执行中过程的问题分析及过程监控。

```
2017-01-17 16:16:43 TestSendMailWithAttachment.py[line:73] INFO 开始执行用例“登录126邮箱”
2017-01-17 16:16:47 TestSendMailWithAttachment.py[line:145] INFO 步骤“打开浏览器”执行通过!
2017-01-17 16:16:48 TestSendMailWithAttachment.py[line:145] INFO 步骤“访问被测试网址http://www.126.com”执行通过!
2017-01-17 16:16:49 TestSendMailWithAttachment.py[line:145] INFO 步骤“最大化窗口”执行通过!
2017-01-17 16:16:54 TestSendMailWithAttachment.py[line:145] INFO 步骤“等待126邮箱登录主页加载完成”执行通过!
2017-01-17 16:16:54 TestSendMailWithAttachment.py[line:145] INFO 步骤“断言当前活动页面源码中是否包含“126网易免费邮—你的专业电子邮局””执行通过!
2017-01-17 16:16:54 TestSendMailWithAttachment.py[line:145] INFO 步骤“显示等待id属性值为x-URS-iframe的frame框的出现，然后切换进入该frame框中”执行通过!
2017-01-17 16:16:54 TestSendMailWithAttachment.py[line:145] INFO 步骤“输入登录用户名”执行通过!
2017-01-17 16:16:55 TestSendMailWithAttachment.py[line:145] INFO 步骤“输入登录密码”执行通过!
2017-01-17 16:16:55 TestSendMailWithAttachment.py[line:145] INFO 步骤“单击登录按钮”执行通过!
2017-01-17 16:17:00 TestSendMailWithAttachment.py[line:145] INFO 步骤“等待”执行通过!
2017-01-17 16:17:01 TestSendMailWithAttachment.py[line:145] INFO 步骤“断言登录成功后的页面标题是否包含“网易邮箱6.0版”关键内容”执行通过!
2017-01-17 16:17:01 TestSendMailWithAttachment.py[line:73] INFO 开始执行用例“发送邮件”
2017-01-17 16:17:04 TestSendMailWithAttachment.py[line:145] INFO 步骤“判断“写信”按钮是否在页面上可见”执行通过!
2017-01-17 16:17:04 TestSendMailWithAttachment.py[line:145] INFO 步骤“单击“写信”按钮”执行通过!
2017-01-17 16:17:05 TestSendMailWithAttachment.py[line:145] INFO 步骤“输入收件人地址”执行通过!
2017-01-17 16:17:07 TestSendMailWithAttachment.py[line:145] INFO 步骤“输入邮件主题”执行通过!
2017-01-17 16:17:08 TestSendMailWithAttachment.py[line:145] INFO 步骤“单击“上传附件”链接”执行通过!
2017-01-17 16:17:10 TestSendMailWithAttachment.py[line:145] INFO 步骤“输入附件所在绝对路径”执行通过!
2017-01-17 16:17:11 TestSendMailWithAttachment.py[line:145] INFO 步骤“模拟键盘回车键”执行通过!
2017-01-17 16:17:11 TestSendMailWithAttachment.py[line:145] INFO 步骤“显示等待附件上传完成”执行通过!
2017-01-17 16:17:12 TestSendMailWithAttachment.py[line:145] INFO 步骤“如果邮件正文的frame框是否可见，切换进该frame中”执行通过!
2017-01-17 16:17:12 TestSendMailWithAttachment.py[line:145] INFO 步骤“输入邮件正文”执行通过!
2017-01-17 16:17:12 TestSendMailWithAttachment.py[line:145] INFO 步骤“退出邮件正文的frame”执行通过!
2017-01-17 16:17:13 TestSendMailWithAttachment.py[line:145] INFO 步骤“单击邮件发送按钮”执行通过!
2017-01-17 16:17:16 TestSendMailWithAttachment.py[line:145] INFO 步骤“等待邮件发送成功，返回结果”执行通过!
2017-01-17 16:17:16 TestSendMailWithAttachment.py[line:145] INFO 步骤“断言页面源码中是否出现“发送成功”关键内容”执行通过!
2017-01-17 16:17:19 TestSendMailWithAttachment.py[line:145] INFO 步骤“关闭浏览器”执行通过!
2017-01-17 16:17:20 TestSendMailWithAttachment.py[line:155] INFO 共2条用例，2条需要被执行，本次执行通过2条。
```

图 15-15

更多说明：

在 TestSendMailWithAttachment.py 文件中每个 try 和 except 的语句块中均添加了打印日志的语句，这样可以实现在测试执行结束后，通过查看日志信息来查看测试执行的过程信息及异常详细信息，以便定位错误。同时测试数据“126 邮箱发送邮件.xlsx”文件中，分别在每个测试用例后面及用例步骤后面写入了本次执行成功与否的信息、异常信息及发生异常页面的截图路径及名称。由于篇幅所限，这里就不再贴数据文件的截图。

至此，关键字驱动测试框架全部搭建完成，在 PyCharm 工具中，整个工程的结构如图 15-16 所示。

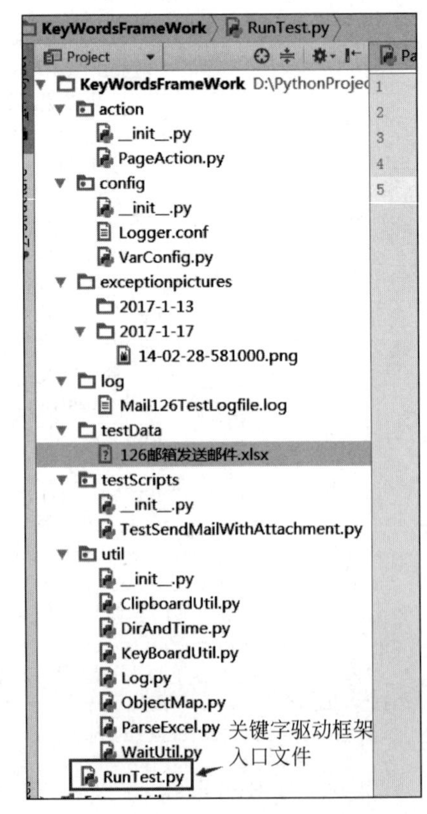

图 15-16

关键字驱动测试框架比数据驱动框架更加高级，可以进一步提高自动化测试工作实施的效率，其具体特点如下：

(1) 使用外部测试数据文件，使用 Excel 管理测试用例的集合和每个测试用例的所有执行测试步骤，实现在一个文件中完成测试用例的维护工作。

(2) 每个测试用例的测试结果可以在一个文件中查看和统计。

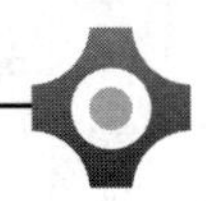

(3) 通过定义关键字、操作元素的定位方式及定位表达式和操作值就可以实现每个测试步骤的执行,可以更加灵活地实现自动化测试的需求。

(4) 实现定位表达式和测试代码的分离,实现定位表达式直接在数据文件中进行维护。

(5) 框架提供日志功能,方便调试和监控自动化测试程序的执行。

(6) 基于关键字测试框架,即使不懂开发技术的测试人员也可以实施自动化测试,便于在整个测试团队中推广和使用自动化测试技术,降低自动化测试实施的技术门槛。

(7) 基于关键字的方式,可以进行任意关键字的扩展,以满足更加复杂项目的自动化测试需求。

15.4 关键字&数据混合驱动框架及实战

在前面两小节我们完成了数据驱动框架及关键字驱动框架的搭建,但在实际工作中,仍可能出现不满足测试需求的情况,因此本小节我们将实践把数据驱动和关键字驱动框架整合到一起,搭建一个关键字 & 数据混合驱动框架。

本小节实战的被测试功能是登录 126 邮箱、批量添加联系人及发送一封带附件的邮件。

非混合框架时向 126 邮箱添加联系人及发邮件测试代码:

```
# encoding = utf - 8
from selenium import webdriver
from selenium.webdriver.common.keys import Keys
from selenium.webdriver.common.by import By
from selenium.webdriver.support.ui import WebDriverWait
from selenium.webdriver.support import expected_conditions as EC
import time

print("启动浏览器...")
# 创建 Chrome 浏览器的实例
driver = webdriver.Firefox(executable_path = "d:\\driver\\geckodriver")
# 最大化浏览器窗口
driver.maximize_window()
print("启动浏览器成功")
print("访问 126 邮箱登录页...")
driver.get("http://mail.126.com")
# 暂停 5 秒,以便邮箱登录页面加载完成
time.sleep(5)
assert u"126 网易免费邮 -- 你的专业电子邮局" in driver.title
print("访问 126 邮箱登录页成功")
# 创建显式等待
wait = WebDriverWait(driver, 30)
# 检查 id 为 x - URS - iframe 的 frame 是否存在,存在则切换进 frame 控件
wait.until(EC.frame_to_be_available_and_switch_to_it((By.XPATH, "//iframe[contains(@id,'x
- URS - iframe')]")))
# 获取用户名输入框
userName = driver.find_element_by_xpath('//input[@name = "email"]')
# 输入用户名
userName.send_keys("xxx")
```

```
# 获取密码输入框
pwd = driver.find_element_by_xpath("//input[@name = 'password']")
# 输入密码
pwd.send_keys("xxx")
# 发送一个回车键
pwd.send_keys(Keys.RETURN)
print("用户登录...")
# 等待 5 秒,以便登录成功后的页面加载完成
time.sleep(5)
assert "网易邮箱" in driver.title
print("登录成功")
# 切换至默认句柄,以规避 can't access dead object 异常
driver.switch_to.default_content()
print("添加联系人...")
# 显示等待通讯录链接页面元素的出现
addressBook = wait.until(EC.visibility_of_element_located
    ((By.XPATH, "//div[text() = '通讯录']")))
# 单击"通讯录"按钮
addressBook.click()
print("单击通讯录按钮")
# 显示等待新建联系人按钮出现
newContact = wait.until(EC.visibility_of_element_located
    ((By.XPATH, "//span[text() = '新建联系人']")))
# 单击新建联系人按钮
newContact.click()
print("单击新建联系人按钮")
# 显示等待输入联系人输入框的出现
contactName = wait.until(EC.visibility_of_element_located
    ((By.XPATH,
    "//a[@title = '编辑详细姓名']/preceding - sibling::div/input")))
contactName.send_keys("lily")
# 输入联系人电子邮箱
driver.find_element_by_xpath(
    "// * [@id = 'iaddress_MAIL_wrap']//input"
    ).send_keys("lily@qq.com")
driver.find_element_by_xpath(
    "//span[text() = '设为星标联系人']/preceding - sibling::span/b"
    ).click()
# 输入联系人手机号
driver.find_element_by_xpath(
    "// * [@id = 'iaddress_TEL_wrap']//dd//input").send_keys("185xxxxxx")
# 输入备注信息
driver.find_element_by_xpath("//textarea").send_keys("朋友")
# 单击"确定"按钮
driver.find_element_by_xpath("//span[text() = '确 定']").click()
time.sleep(2)
assert "lily@qq.com" in driver.page_source
print("添加联系人成功")

print("进入首页")
driver.find_element_by_xpath('//div[. = "首页"]').click()
```

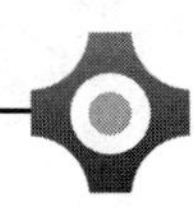

```
print("写信...")
# 显示等待写信链接页面元素的出现
element = wait.until(EC.visibility_of_element_located
    ((By.XPATH, "//span[text() = '写 信']")))
# 单击写信链接
element.click()
# 写入收件人地址
driver.find_element_by_xpath(
    "//div[contains(@id,'_mail_emailinput')]/input"
    ).send_keys("xxx@qq.com")
# 写入邮件主题
driver.find_element_by_xpath(
    "//div[@aria - label = '邮件主题输入框,请输入邮件主题']/input"
    ).send_keys("光荣之路")
# 切换进 frame 控件
driver.switch_to.frame(driver.find_element_by_xpath("//iframe[@tabindex = 1]"))
editBox = driver.find_element_by_xpath("/html/body")
editBox.send_keys("发给光荣之路的一封信")
driver.switch_to.default_content()
print("写信完成")
driver.find_element_by_xpath("//header//span[text() = '发送']").click()
print("开始发送邮件...")
time.sleep(3)
assert "发送成功" in driver.page_source
print("邮件发送成功")
driver.quit()
```

本小节将把上面的测试程序，一步步地改造成关键字 & 数据驱动混合框架模式，同时为发送的邮件增加附件。

关键字 & 数据混合驱动测试框架搭建步骤：

(1) 在 PyCharm 工具中新建一个名叫 KeyWordAndDataDrivenFrameWork 的 Python 工程。

(2) 在工程中新建 3 个 Python Package，1 个目录，分别命名为：

- config 包，用于实现框架中各种配置信息。
- util 包，用于实现测试过程中调用的工具类或方法，例如读取配置文件、MapObject、页面元素的操作方法、解析 Excel 文件等。
- testData 目录，用于存放框架所需要的测试数据文件。
- testScripts 包，用于实现具有测试逻辑的测试脚本。

(3) 在 util 包中新建 ObjectMap.py 模块，用于实现定位页面元素，具体代码如下。

```
#encoding = utf - 8
from selenium.webdriver.support.ui import WebDriverWait

# 获取单个页面元素对象
def getElement(driver, locationType, locatorExpression):
    try:
        element = WebDriverWait(driver, 30).until(
            lambda x: x.find_element(by = locationType, value = locatorExpression))
```

```
            return element
        except Exception as err:
            raise err

# 获取多个相同页面元素对象,以 list 返回
def getElements(driver, locationType, locatorExpression):
    try:
        elements = WebDriverWait(driver, 30).until(
            lambda x:x.find_elements(by = locationType, value = locatorExpression))
        return elements
    except Exception as err:
        raise err

if __name__ == '__main__':
    from selenium import webdriver
    # 进行单元测试
    driver = webdriver.Firefox(executable_path = "c:\geckodriver.exe")
    driver.get("http://www.baidu.com")
    searchBox = getElement(driver, "id", "kw")
    # 打印页面对象的标签名
    print(searchBox.tag_name)
    aList = getElements(driver, "tag name", "a")
    print(len(aList))
    driver.quit()
```

(4) 在 util 包中新建一名为 WaitUtil.py 文件,用于实现智能等待页面元素的出现,具体代码如下。

```
#encoding = utf - 8
from selenium.webdriver.common.by import By
from selenium.webdriver.support.ui import WebDriverWait
from selenium.webdriver.support import expected_conditions as EC

class WaitUtil(object):

    def __init__(self, driver):
        self.locationTypeDict = {
            "xpath": By.XPATH,
            "id": By.ID,
            "name": By.NAME,
            "css_selector": By.CSS_SELECTOR,
            "class_name": By.CLASS_NAME,
            "tag_name": By.TAG_NAME,
            "link_text": By.LINK_TEXT,
            "partial_link_text": By.PARTIAL_LINK_TEXT
        }
        self.driver = driver
        self.wait = WebDriverWait(self.driver, 30)

    def presenceOfElementLocated(self, locatorMethod, locatorExpression, *arg):
        '''显示等待页面元素出现在 DOM 中,并一定可以见,
```

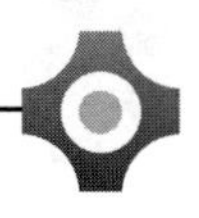

```
        存在则返回该页面元素对象'''
        try:
            element = self.wait.until(EC.presence_of_element_located((
                self.locationTypeDict[locatorMethod.lower()],
                locatorExpression)))
            return element
        except Exception as err:
            raise err

    def frameToBeAvailableAndSwitchToIt(self,locationType,locatorExpression, *arg):
        '''检查 frame 是否存在,存在则切换进 frame 控件中
        '''
        try:
            self.wait.until(
                EC.frame_to_be_available_and_switch_to_it((
                    self.locationTypeDict[locationType.lower()],
                    locatorExpression)))
        except Exception as err:
            # 抛出异常信息给上层调用者
            raise err

    def visibilityOfElementLocated(self, locationType, locatorExpression, *arg):
        '''显示等待页面元素出现在 DOM 中,并且可见,存在则返回该页面元素对象'''
        try:
            element = self.wait.until(
                EC.visibility_of_element_located((
                    self.locationTypeDict[locationType.lower()],
                    locatorExpression)))
            return element
        except Exception as err:
            raise err

if __name__ == '__main__':
    from selenium import webdriver
    driver = webdriver.Firefox(executable_path = "c:\\geckodriver")
    driver.get("http://mail.126.com")
    waitUtil = WaitUtil(driver)
    waitUtil.frameToBeAvailableAndSwitchToIt("xpath",
                "//iframe[contains(@id,'x-URS-iframe')]")
    waitUtil.visibilityOfElementLocated("xpath", "//input[@name='email']")
    waitUtil.presenceOfElementLocated("xpath", "//input[@name='email']")
    driver.quit()
```

(5) 在 util 包中新建一个名为 ClipboardUtil.py 的 Python 文件,用于实现剪切板功能,具体内容如下。

```
#encoding=utf-8
import win32clipboard as w
import win32con

class Clipboard(object):
```

```
    '''
    模拟 Windows 设置剪切板
    '''
    #读取剪切板
    @staticmethod
    def getText():
        # 打开剪切板
        w.OpenClipboard()
        # 获取剪切板中的数据
        d = w.GetClipboardData(win32con.CF_TEXT)
        # 关闭剪切板
        w.CloseClipboard()
        # 返回剪切板数据给调用者
        return d

    #设置剪切板内容
    @staticmethod
    def setText(aString):
        # 打开剪切板
        w.OpenClipboard()
        # 清空剪切板
        w.EmptyClipboard()
        # 将数据 aString 写入剪切板
        w.SetClipboardData(win32con.CF_UNICODETEXT, aString)
        # 关闭剪切板
        w.CloseClipboard()
```

(6) 在 util 包中新建一个名为 KeyBoardUtil.py 的 Python 文件,用于实现模拟键盘按键功能代码,具体内容如下。

```
#encoding = utf - 8
import win32api
import win32con

class KeyboardKeys(object):
    '''
    模拟键盘按键类
    '''
    VK_CODE = {
        'enter': 0x0D,
        'ctrl': 0x11,
        'v': 0x56}

    @staticmethod
    def keyDown(keyName):
        # 按下按键
        win32api.keybd_event(KeyboardKeys.VK_CODE[keyName], 0, 0, 0)

    @staticmethod
    def keyUp(keyName):
        # 释放按键
```

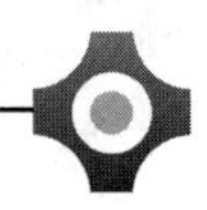

```
        win32api.keybd_event(KeyboardKeys.VK_CODE[keyName],
                             0, win32con.KEYEVENTF_KEYUP, 0)

    @staticmethod
    def oneKey(key):
        # 模拟单个按键
        KeyboardKeys.keyDown(key)
        KeyboardKeys.keyUp(key)

    @staticmethod
    def twoKeys(key1, key2):
        # 模拟两个组合键
        KeyboardKeys.keyDown(key1)
        KeyboardKeys.keyDown(key2)
        KeyboardKeys.keyUp(key2)
        KeyboardKeys.keyUp(key1)
```

(7) 在 testScripts 包中新建一个名为 TestSendMailWithAttachment.py 的 Python 文件,用于实现具体的测试逻辑代码,具体代码如下。

```
#encoding = utf-8
from util.ObjectMap import *
from util.KeyBoardUtil import KeyboardKeys
from util.ClipboardUtil import Clipboard
from util.WaitUtil import WaitUtil
from selenium import webdriver
from selenium.webdriver.common.keys import Keys
import time

def TestSendMailWithAttachment():
    # 创建 Firefox 浏览器的实例
    driver = webdriver.Firefox(executable_path = "d:\\driver\\geckodriver")
    # 最大化浏览器窗口
    driver.maximize_window()
    print("启动浏览器成功")
    print("访问 126 邮箱登录页...")
    driver.get("http://mail.126.com")
    # 暂停 5 秒,以便邮箱登录页面加载完成
    time.sleep(5)
    assert "126 网易免费邮--你的专业电子邮局" in driver.title
    print("访问 126 邮箱登录页成功")

    wait = WaitUtil(driver)
    # 显示等待 id="x-URS-iframe"的框架出现,并切换进去
    wait.frameToBeAvailableAndSwitchToIt("xpath",
            "//iframe[contains(@id,'x-URS-iframe')]")
    print("输入登录用户名")
    username = getElement(driver, "xpath", "//input[@name='email']")
    username.send_keys("xxx")
    print("输入登录密码")
    passwd = getElement(driver, "xpath", "//input[@name='password']")
```

```
passwd.send_keys("xxx")
print("登录...")
passwd.send_keys(Keys.ENTER)
# 等待5秒,以便登录成功后的页面加载完成
time.sleep(5)
assert "网易邮箱" in driver.title
print("登录成功")
# 切换至默认句柄,以规避 can't access dead object 异常
driver.switch_to.default_content()
print("添加联系人...")
# 显示等待通讯录链接页面元素的出现
addressBook = wait.visibilityOfElementLocated("xpath", "//div[text() = '通讯录']")
# 单击通讯录链接
addressBook.click()
print("单击通讯录按钮")
# 显示等待新建联系人按钮的出现
newContact = wait.visibilityOfElementLocated("xpath",
                "//span[text() = '新建联系人']")
# 单击新建联系人按钮
newContact.click()
print("单击新建联系人按钮")
# 显示等待输入联系人姓名输入框出现
contactName = wait.visibilityOfElementLocated(
    "xpath", "//a[@title = '编辑详细姓名']/preceding - sibling::div/input")
contactName.send_keys("lily")
email = getElement(driver, "xpath", "//*[@id = 'iaddress_MAIL_wrap']//input")
email.send_keys("lily@qq.com")
# 设定为星标联系人
getElement(driver, "xpath",
    "//span[text() = '设为星标联系人']/preceding - sibling::span/b").click()
mobile = getElement(driver, "xpath","//*[@id = 'iaddress_TEL_wrap']//dd//input")
# 输入联系人手机号
mobile.send_keys("183xxxxxxx")
# 输入备注信息
getElement(driver, "xpath", "//textarea").send_keys("朋友")
# 单击确定按钮,保存新联系人
getElement(driver, "xpath","//span[text() = '确 定']").click()
time.sleep(1)
assert "lily@qq.com" in driver.page_source
print("添加联系人成功")
time.sleep(2)
getElement(driver, "xpath", '//div[. = "首页"]').click()
element = wait.visibilityOfElementLocated("xpath", "//span[text() = '写 信']")
element.click()
print("写信...")
receiver = getElement(driver, "xpath",
            "//div[contains(@id,'_mail_emailinput')]/input")
# 输入收信人地址
receiver.send_keys("xxx")
subject = getElement(driver, "xpath",
    "//div[@aria - label = '邮件主题输入框,请输入邮件主题']/input")
```

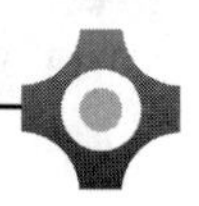

```
        # 输入邮件主题
        subject.send_keys("新邮件")
        # 设置剪切板内容
        Clipboard.setText("d:\\aa.txt")
        # 获取剪切板内容
        Clipboard.getText()
        attachment = getElement(driver, "xpath",
                    "//div[contains(@title, '600 首 MP3')]")
        # 单击上传附件链接
        attachment.click()
        time.sleep(3)
        # 在上传附件 Windows 弹窗中粘贴剪切板中的内容
        KeyboardKeys.twoKeys("ctrl", "v")
        # 模拟回车键,以便加载要上传的附件
        KeyboardKeys.oneKey("enter")
        # 切换进邮件正文的 frame
        wait.frameToBeAvailableAndSwitchToIt("xpath", "//iframe[@tabindex = 1]")
        body = getElement(driver, "xpath", "/html/body")
        # 输入邮件正文
        body.send_keys("发给光荣之路的一封信")
        # 切出邮件正文的 frame 框
        driver.switch_to.default_content()
        print("写信完成")
        getElement(driver, "xpath", "//header//span[text() = '发送']").click()
        print("开始发送邮件...")
        time.sleep(3)
        assert "发送成功" in driver.page_source
        print("邮件发送成功")
        driver.quit()

if __name__ == '__main__':
    TestSendMailWithAttachment()
```

代码解释：

将 TestSendMailWithAttachment.py 文件中的“xxx”改成有效的邮箱地址及密码，然后执行该文件，就可以看到浏览器自动登录 126 邮箱，然后添加一个联系人，并且自动发了一封带有附件的邮件。在上面的代码中可以看到我们将定位页面元素的方法、模拟键盘按键等封装成了公共方法，方便复用，同时还加入了智能等待，让 UI 自动化测试执行更稳定。但上面的代码还并未实现数据驱动、关键字驱动，为此我们需要继续改造。

(8) 在 config 包中新建一名为 VarConfig.py 的 Python 文件，用于定义整个框架中所需要的一些全局常量值，方便维护，具体内容如下。

```
#encoding = utf - 8
import os

firefoxDriverFilePath = "c:\geckodriver"

# 获取当前文件所在目录的父目录的绝对路径
parentDirPath = os.path.dirname(os.path.dirname(os.path.abspath(__file__)))
```

```
# 异常图片存放目录
screenPicturesDir = parentDirPath + "\\exceptionpictures\\"
```

(9) 在 util 包中新建一名为 DirAndTime.py 的 Python 文件,用于获取当前日期及时间,以及创建异常截图存放目录,具体内容如下。

```
#encoding = utf - 8
import time, os
from datetime import datetime
from config.VarConfig import screenPicturesDir

# 获取当前的日期
def getCurrentDate():
    timeTup = time.localtime()
    currentDate = str(timeTup.tm_year) + "-" + \
        str(timeTup.tm_mon) + "-" + str(timeTup.tm_mday)
    return currentDate

# 获取当前的时间
def getCurrentTime():
    timeStr = datetime.now()
    nowTime = timeStr.strftime('%H-%M-%S-%f')
    return nowTime

# 创建截图存放的目录
def createCurrentDateDir():
    dirName = os.path.join(screenPicturesDir, getCurrentDate())
    if not os.path.exists(dirName):
        os.makedirs(dirName)
    return dirName

if __name__ == '__main__':
    print(getCurrentDate())
    print(createCurrentDateDir())
    print(getCurrentTime())
```

(10) 在 KeyWordsFrameWork 工程中新建一名为 action 的 Python package,并在此包中新建一名为 PageAction.py 的 Python 文件,用于实现具体页面动作的封装,比如在输入框中输入数据,单击页面按钮等,具体内容如下。

```
#encoding = utf - 8
from selenium import webdriver
from config.VarConfig import firefoxDriverFilePath
from util.ObjectMap import getElement
from util.ClipboardUtil import Clipboard
from util.KeyBoardUtil import KeyboardKeys
from util.DirAndTime import *
from util.WaitUtil import WaitUtil
import time

# 定义全局 driver 变量
```

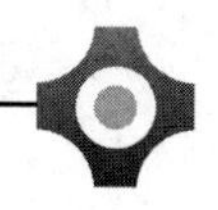

```
driver = None
# 全局的等待类实例对象
waitUtil = None

def open_browser( * arg):
    # 打开浏览器
    global driver, waitUtil
    try:
        driver = webdriver.Firefox(executable_path = firefoxDriverFilePath)
        # driver 对象创建成功后,创建等待类实例对象
        waitUtil = WaitUtil(driver)
    except Exception as err:
        raise err

def visit_url(url, * arg):
    # 访问某个网址
    global driver
    try:
        driver.get(url)
    except Exception as err:
        raise err

def close_browser( * arg):
    # 关闭浏览器
    global driver
    try:
        driver.quit()
    except Exception as err:
        raise err

def sleep(sleepSeconds, * arg):
    # 强制等待
    try:
        time.sleep(int(sleepSeconds))
    except Exception as err:
        raise err

def clear(locationType, locatorExpression, * arg):
    # 清除输入框默认内容
    global driver
    try:
        getElement(driver, locationType, locatorExpression).clear()
    except Exception as err:
        raise err

def input_string(locationType, locatorExpression, inputContent):
    # 在页面输入框中输入数据
    global driver
    try:
        getElement(driver, locationType,
            locatorExpression).send_keys(inputContent)
```

```
        except Exception as err:
            raise err

def click(locationType, locatorExpression, *arg):
    # 单击页面元素
    global driver
    try:
        getElement(driver, locationType, locatorExpression).click()
    except Exception as err:
        raise err

def assert_string_in_pagesource(assertString, *arg):
    # 断言页面源码是否存在某关键字或关键字符串
    global driver
    try:
        assert assertString in driver.page_source, \
            "%s not found in page source!" % assertString
    except AssertionError as err:
        raise AssertionError(err)
    except Exception as e:
        raise e

def assert_title(titleStr, *args):
    # 断言页面标题是否存在给定的关键字符串
    global driver
    try:
        assert titleStr in driver.title, \
            "%s not found in title!" % titleStr
    except AssertionError as err:
        raise AssertionError(err)
    except Exception as e:
        raise e

def getTitle(*arg):
    # 获取页面标题
    global driver
    try:
        return driver.title
    except Exception as err:
        raise err

def getPageSource(*arg):
    # 获取页面源码
    global driver
    try:
        return driver.page_source
    except Exception as err:
        raise err

def switch_to_frame(locationType, frameLocatorExpression, *arg):
    # 切换进入 frame
```

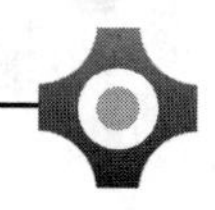

```
    global driver
    try:
        driver.switch_to.frame(getElement
            (driver, locationType, frameLocatorExpression))
    except Exception as err:
        raise err

def switch_to_default_content( * arg):
    # 切出 frame
    global driver
    try:
        driver.switch_to.default_content()
    except Exception as err:
        raise err

def paste_string(pasteString, * arg):
    # 模拟 Ctrl + v 操作
    try:
        Clipboard.setText(pasteString)
        # 等待 2 秒,防止代码执行得太快,而未成功粘贴内容
        time.sleep(2)
        KeyboardKeys.twoKeys("ctrl", "v")
    except Exception as err:
        raise err

def press_tab_key( * arg):
    # 模拟 Tab 键
    try:
        KeyboardKeys.oneKey("tab")
    except Exception as err:
        raise err

def press_enter_key( * arg):
    # 模拟回车键
    try:
        KeyboardKeys.oneKey("enter")
    except Exception as err:
        raise err

def maximize_browser():
    # 窗口最大化
    global driver
    try:
        driver.maximize_window()
    except Exception as err:
        raise err

def capture_screen( * args):
    # 截取屏幕图片
    global driver
    currTime = getCurrentTime()
```

```
    picNameAndPath = str(createCurrentDateDir()) + "\\" + str(currTime) + ".png"
    try:
        driver.get_screenshot_as_file(picNameAndPath.replace('\\', r'\\'))
    except Exception as err:
        raise err
    else:
        return picNameAndPath

def waitPresenceOfElementLocated(locationType, locatorExpression, *arg):
    '''显示等待页面元素出现在 DOM 中,但并不一定可见,
            存在则返回该页面元素对象'''
    global waitUtil
    try:
        element = waitUtil.presenceOfElementLocated(locationType, locatorExpression)
    return element
    except Exception as err:
        raise err

def waitFrameToBeAvailableAndSwitchToIt(locationType, locatorExpression, *args):
    '''检查 frame 是否存在,存在则切换进 frame 控件中'''
    global waitUtil
    try:
        waitUtil.frameToBeAvailableAndSwitchToIt(locationType, locatorExpression)
    except Exception as err:
        raise err

def waitVisibilityOfElementLocated(locationType, locatorExpression, *args):
    '''显示等待页面元素出现在 DOM 中,并且可见,存在返回该页面元素对象'''
    global waitUtil
    try:
        element = waitUtil.visibilityOfElementLocated(locationType, locatorExpression)
            return element
    except Exception as err:
        raise err
```

(11) 修改 testScripts 包中的 TestSendMailWithAttachment.py 文件内容如下。

```
#encoding = utf - 8
from action.PageAction import *
import time

def TestSendMailWithAttachment():
    print("启动 Firefox 浏览器")
    open_browser()
    maximize_browser()
    print("访问 126 邮箱登录页...")
    visit_url("http://mail.126.com")
    sleep(5)
    # 断言页面出现的关键内容
    assert_string_in_pagesource("126 网易免费邮 -- 你的专业电子邮局")
    print("访问 126 邮箱登录页成功")
```

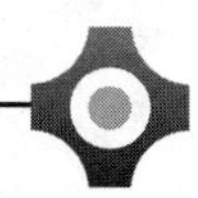

```
waitFrameToBeAvailableAndSwitchToIt("xpath", "//iframe[contains(@id,'x-URS-iframe')]")
print("输入登录用户名")
input_string("xpath", "//input[@name='email']", "AgilityToSR")
print("输入登录密码")
input_string("xpath", "//input[@name='password']", "AutoTest123")
# 单击登录按钮
click("id", "dologin")
sleep(5)
assert_title("网易邮箱")
print("登录成功")
switch_to_default_content()
print("添加联系人")
# 显示等待通讯录链接在页面上可见
waitVisibilityOfElementLocated("xpath", "//div[text()='通讯录']")
# 单击通讯录链接
click("xpath", "//div[text()='通讯录']")
# 单击新建联系人按钮
click("xpath", "//span[text()='新建联系人']")
# 输入联系人姓名
input_string("xpath",
    "//a[@title='编辑详细姓名']/preceding-sibling::div/input",
    "lily")
# 输入联系人邮箱
input_string("xpath",
     "//*[@id='iaddress_MAIL_wrap']//input",
     "lily@qq.com")
# 单击星标联系人复选框
click("xpath",
    "//span[text()='设为星标联系人']/preceding-sibling::span/b")
# 输入联系人手机号
input_string("xpath", "//*[@id='iaddress_TEL_wrap']//dd//input",
     "185xxxxx")
# 输入联系人备注
input_string("xpath", "//textarea", "朋友")
# 单击确定按钮,保存新联系人
click("xpath","//span[text()='确 定']")
time.sleep(1)
# 断言页面是否出现关键内容
assert_string_in_pagesource("lily@qq.com")
print("添加联系人成功")
# 单击首页链接,进入首页界面
click("xpath", "//div[.='首页']")
# 显示等待写信链接出现在页面上
waitVisibilityOfElementLocated("xpath", "//span[text()='写 信']")
# 单击写信链接按钮,进入写信页面
click("xpath", "//span[text()='写 信']")
print("开始写信")
print("输入收件人地址")
input_string("xpath",
     "//div[contains(@id,'_mail_emailinput')]/input",
     "38617389@qq.com")
```

```
    print("输入邮件主题")
    input_string("xpath",
        "//div[@aria-label='邮件主题输入框,请输入邮件主题']/input",
        "新邮件")
    print("单击上传附件按钮")
    click("xpath", "//div[contains(@title, '600 首 MP3')]")
    print("上传附件")
    paste_string("d:\\aa.txt")
    sleep(2)
    press_enter_key()
    sleep(2)
    waitVisibilityOfElementLocated("xpath", '//span[.="上传完成"]')
    print("上传附件成功")
    # 进入邮件正文的 frame 框体中
    waitFrameToBeAvailableAndSwitchToIt("xpath", "//iframe[@tabindex=1]")
    print("写入邮件正文")
    input_string("xpath", "/html/body", "发给光荣之路的一封信")
    # 退出邮件正文的 frame 框体,进入默认会话框体
    switch_to_default_content()
    print("写信完成")
    print("开始发送邮件...")
    click("xpath", "//header//span[text()='发送']")
    time.sleep(3)
    assert_string_in_pagesource("发送成功")
    print("邮件发送成功")
    close_browser()

if __name__ == '__main__':
    TestSendMailWithAttachment()
```

替换 TestSendMailWithAttachment.py 文件中的"xxx"为有效的 126 邮箱登录账号及密码后执行该文件,可以看到程序会自动启动 Firefox 浏览器,然后访问 126 邮箱,然后向邮箱中添加一个联系人,并发送一封带附件的邮件。

(12) 在 util 包中新建一名叫 ParseExcel.py 的 Python 文件,用于实现读取 Excel 数据文件代码封装,具体内容如下。

```
#encoding=utf-8
import openpyxl
from openpyxl.styles import Border, Side, Font
import time

class ParseExcel(object):

    def __init__(self):
        self.workbook = None
        self.excelFile = None
        self.font = Font(color = None)                    # 设置字体的颜色
        # 颜色对应的 RGB 值
        self.RGBDict = {'red': 'FFFF3030', 'green': 'FF008B00'}
```

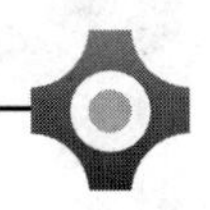

```
def loadWorkBook(self, excelPathAndName):
    # 将 Excel 文件加载到内存,并获取其 workbook 对象
    try:
        self.workbook = openpyxl.load_workbook(excelPathAndName)
    except Exception as err:
        raise err
    self.excelFile = excelPathAndName
    return self.workbook

def getSheetByName(self, sheetName):
    # 根据 sheet 名获取该 sheet 对象
    try:
        sheet = self.workbook[sheetName]
        print(sheet)
        return sheet
    except Exception as err:
        raise err

def getSheetByIndex(self, sheetIndex):
    # 根据 sheet 的索引号获取该 sheet 对象
    try:
        sheetname = self.workbook.sheetnames[sheetIndex]
    except Exception as err:
        raise err
    sheet = self.workbook[sheetname]
    print(sheet)
    return sheet

def getRowsNumber(self, sheet):
    # 获取 sheet 中有数据区域的结束行号
    return sheet.max_row

def getColsNumber(self, sheet):
    # 获取 sheet 中有数据区域的结束列号
    return sheet.max_column

def getStartRowNumber(self, sheet):
    # 获取 sheet 中有数据区域的开始的行号
    return sheet.min_row

def getStartColNumber(self, sheet):
    # 获取 sheet 中有数据区域的开始的列号
    return sheet.min_column

def getRow(self, sheet, rowNo):
    # 获取 sheet 中某一行,返回的是这一行所有的数据内容组成的 tuple,
    # 下标从 1 开始,sheet.rows[1]表示第一行
    try:
        rows_data = list(sheet.rows)
        return rows_data[rowNo - 1]
    except Exception as err:
```

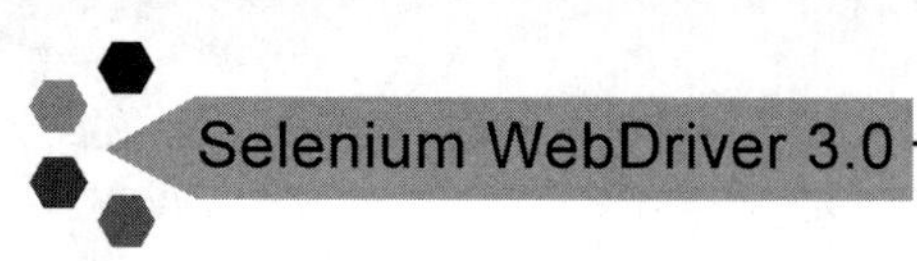


```
            raise err

    def getColumn(self, sheet, colNo):
        # 获取 sheet 中某一列,返回的是这一列所有的数据内容组成 tuple,
        # 下标从 1 开始,sheet.columns[1]表示第一列
        try:
            columns_s_data = list(sheet.columns)
            return columns_s_data[colNo - 1]
        except Exception as err:
            raise err

    def getCellOfValue(self, sheet, rowNo = None, colsNo = None):
        # 根据单元格所在的位置索引获取该单元格中的值,下标从 1 开始,
        # sheet.cell(row = 1, column = 1).value,表示 excel 中第一行第一列的值
        if rowNo is not None and colsNo is not None:
            try:
                return sheet.cell(row = rowNo, column = colsNo).value
            except Exception as err:
                raise err
        else:
            raise Exception("Insufficient Coordinates of cell !")

    def getCellOfObject(self, sheet, rowNo = None,colsNo = None):
        # 获取某个单元格的对象,可以根据单元格所在位置的数字索引,
        # 也可以直接根据 Excel 中单元格的编码坐标
        # getCellObject(sheet, rowNo = 1, colsNo = 2)
        if rowNo is not None and colsNo is not None:
            try:
                return sheet.cell(row = rowNo,column = colsNo)
            except Exception as err:
                raise err
        else:
            raise Exception("Insufficient Coordinates of cell !")

    def writeCell(self, sheet, content, rowNo = None, colsNo = None, style = None):
        # 根据单元格在 Excel 中的编码坐标或者数字索引坐标向单元格中写入数据,
        # 下标从 1 开始,参 style 表示字体的颜色的名字,比如 red,green
        if rowNo is not None and colsNo is not None:
            try:
                sheet.cell(row = rowNo,column = colsNo).value = content
                if style:
                    sheet.cell(row = rowNo,column = colsNo).\
                        font = Font(color = self.RGBDict[style])
                self.workbook.save(self.excelFile)
            except Exception as err:
                raise err
        else:
            raise Exception("Insufficient Coordinates of cell !")

    def writeCellCurrentTime(self, sheet, rowNo = None, colsNo = None):
        # 写入当前的时间,下标从 1 开始
```

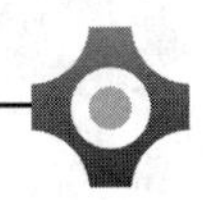

```
        now = int(time.time())   # 显示为时间戳
        timeArray = time.localtime(now)
        currentTime = time.strftime("%Y-%m-%d %H:%M:%S", timeArray)
        if rowNo is not None and colsNo is not None:
            try:
                sheet.cell(row = rowNo, column = colsNo).value = currentTime
                self.workbook.save(self.excelFile)
            except Exception as err:
                raise err
        else:
            raise Exception("Insufficient Coordinates of cell !")

if __name__ == '__main__':
    pe = ParseExcel()
    # 测试所用的 Excel 文件"126 邮箱联系人.xlsx"请自行创建
    pe.loadWorkBook('D:\\126 邮箱联系人.xlsx')
    sheet = pe.getSheetByIndex(0)
    rows = pe.getColumn(sheet, 1)                          # 获取第一行
    for i in rows:
        print(i.value)
    # 获取第一行第一列单元格内容
    print(pe.getCellOfValue(sheet, rowNo = 1, colsNo = 1))
    pe.writeCell(sheet, '我爱祖国', rowNo = 10, colsNo = 10)
    pe.writeCellCurrentTime(sheet, rowNo = 10, colsNo = 11)
```

(13) 在 testData 目录中新建一名为“126 邮箱创建联系人并发邮件.xlsx”的 Excel 文件，在该 Excel 文件中创建 5 张工作表，分别命名为“测试用例”“登录”“联系人”“发邮件”及“创建联系人”。

“测试用例”工作表用于存放测试用例，具体内容如表 15-6 所示。

表　15-6

用例名称	用例描述	调用框架类型	用例步骤 sheet 名	数据驱动的数据源 sheet 名	是否执行	执行结束时间	执行结果
登录 126 邮箱	使用有效的账号登录 126 邮箱	关键字	登录		y		
添加联系人	批量添加联系人	数据	创建联系人	联系人	y		
发送带附件的邮件	登录 126 邮箱后，发送一封带附件的邮件	关键字	发邮件		y		

表 15-6 中“调用框架类型”列，表示执行本条测试用例用到的框架类型，如果使用关键字驱动框架执行，必须填写“关键字”三个关键字串，如果使用的是数据驱动框架执行，必须填写“数据”两个关键字串，测试框架将根据这两关键字来识别具体使用的何种框架执行。数据驱动框架所需要的数据表对应“数据驱动的数据源 sheet 名”列所给的工作表名。

“登录”工作表用于存放登录 126 邮箱的步骤中所用到的关键字及需要的输入数据等，具体内容如表 15-7 所示。

表 15-7

测试步骤描述	关键字	操作元素的定位方式	操作元素的定位表达式	操作值	测试执行时间	测试结果	错位信息	错误截图
打开浏览器	open_browser			firefox				
访问被测试网址 http://www.126.com	visit_url			http://www.126.com				
最大化窗口	maximize_browser							
等待 126 邮箱登录主页加载完成	sleep			5				
断言当前活动页面源码中是否包含"126 网易免费邮--你的专业电子邮局"	assert_string_in_pagesource			126 网易免费邮--你的专业电子邮局				
显示等待 id 属性值为 x-URS-iframe 的 frame 框的出现，然后切换进入该 frame 框中	waitFrameToBeAvailableAndSwitchToIt	xpath	//iframe[contains(@id, 'x-URS-iframe')]					
输入登录用户名	input_string	xpath	//input[@name='email']	xxx				
输入登录密码	input_string	xpath	//input[@name='password']	xxx				
单击登录按钮	click	id	dologin					
等待	sleep			5				
切回默认会话窗体	switch_to_default_content							
断言登录成功后的页面标题是否包含"网易邮箱 6.0 版"关键内容	assert_title			网易邮箱 6.0 版				

表 15-7 中的"关键字"列表示的字符串将与 action 包中 PageAction.py 文件中的函数名一一映射，后面临近的 3 列均是这些函数需要的参数值，其中"操作值"列表示操作页面元素需要的输入数据，或者断言关键数据等。从"测试执行时间列"及以后的列都是在测试过程中写入测试结果信息的列，后续表中这些列的作用均一致，将不再做解释。

"联系人"工作表用于存放添加联系人模块所需要的联系人数据，具体内容如表 15-8 所示。

表 15-8

姓名	电子邮箱	手机号	是否设为星标联系人	备　注	是否执行	执行结束时间	执行结果
sr	sr@qq.com	15xxxxxxxxx	否	别名	y		
wcx	wcx@qq.com	1555xxxxxxx	是	星标联系人	y		
lucy	lucy@126.com	187xxxxxxxx	否	添加新的联系人	y		

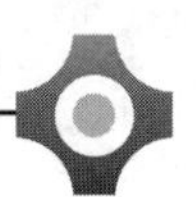

续表

姓名	电子邮箱	手机号	是否设为星标联系人	备　　注	是否执行	执行结束时间	执行结果
lily	lily@qq.com	13778xxxxxx	是	同事	y		
amy	amy@qq.com	158xxxxxxxx	否	一面之缘的人	n		
anne	anne@126.com	1398xxxxxxx	是	邮件往来之人	y		

“发邮件”工作表用于存放发送邮件模块的实现步骤，其表结构同“登录”工作表，具体内容如表 15-9 所示。

表　15-9

测试步骤描述	关键字	操作元素的定位方式	操作元素的定位表达式	操作值	测试执行时间	测试结果	错位信息	错误截图
单击“首页”链接，进入邮箱首页	click	xpath	//div[.='首页']					
判断“写信”按钮是否在页面上可见	waitVisibilityOf-ElementLocated	xpath	//span[text()='写信']					
单击“写信”按钮	click	xpath	//span[text()='写信']					
输入收件人地址	input_string	xpath	//div[contains(@id,'_mail_emailinput')]/input	xxx@qq.com				
输入邮件主题	input_string	xpath	//div[@aria-label='邮件主题输入框，请输入邮件主题']/input	带附件的邮件				
单击“上传附件”链接	click	xpath	//div[contains(@title,'600 首 MP3')]					
输入附件所在绝对路径	paste_string			d:\\a.txt				
模拟键盘回车键	press_enter_key							
显示等待附件上传完成	waitVisibilityOf-ElementLocated	xpath	//span[text()="上传完成"]					
如果邮件正文的 frame 框是否可见，切换进该 frame 中	waitFrameTo-BeAvailable-AndSwitchToIt	xpath	//iframe[@tabindex=1]					
输入邮件正文	input_string	xpath	/html/body	发给光荣之路的一封信				
退出邮件正文的 frame	switch_to_default_content							
单击邮件发送按钮	click	xpath	//header//span[text()='发送']					

续表

测试步骤描述	关键字	操作元素的定位方式	操作元素的定位表达式	操作值	测试执行时间	测试结果	错位信息	错误截图
等待邮件发送成功，返回结果	sleep			3				
断言页面源码中是否出现“发送成功”关键内容	assert_string_in_pagesource			发送成功				
关闭浏览器	close_browser							

“创建联系人”工作表用于存放创建联系人模块的实现步骤，具体内容如表 15-10 所示。

表　15-10

测试步骤描述	关键字	操作元素的定位方式	操作元素的定位表达式	操作值
单击“通讯录”链接，进入通讯录页面	click	xpath	//div[text()="通讯录"]	
显示等待“新建联系人”按钮在页面上可见	waitVisibilityOfElementLocated	xpath	//span[text()="新建联系人"]	
单击“新建联系人”按钮	click	xpath	//span[text()="新建联系人"]	
显示等待输入联系人姓名框是否在页面上可见	waitVisibilityOfElementLocated	xpath	//a[@title='编辑详细姓名']/preceding-sibling::div/input	
输入联系人姓名	input_string	xpath	//a[@title='编辑详细姓名']/preceding-sibling::div/input	A-1
输入电子邮箱	input_string	xpath	//*[@id='iaddress_MAIL_wrap']//input	B-2
设为星标联系人	click	xpath	//span[text()='设为星标联系人']/preceding-sibling::span/b	D-4
输入手机号	input_string	xpath	//*[@id='iaddress_TEL_wrap']//dd//input	C-3
输入联系人备注信息	input_string	xpath	//textarea	E-5
单击“确定”按钮保存新联系人	click	xpath	//span[.='确 定']	
等待 2 秒	sleep			2
断言联系人是否成功添加	assert_string_in_pagesource			B-2

表 15-10 中的“操作值”列的数据对应“联系人”工作表中对应的数据列的字母编号以及对应的数字编号，表示该测试步骤需要的数据从“联系人”工作表中相应的列中取，目的是为了区分其他的操作值，比如等待 2 秒中的数字 2。

(14) 修改 config 包中的 VarConfig.py 文件内容如下。

```
#encoding = utf-8
import os
```

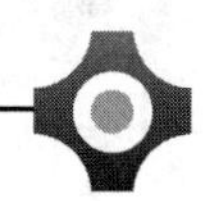

```
firefoxDriverFilePath = "c:\geckodriver"

# 获取当前文件所在目录的父目录的绝对路径
parentDirPath = os.path.dirname(os.path.dirname(os.path.abspath(__file__)))
# 异常图片存放目录
screenPicturesDir = parentDirPath + "\\exceptionpictures\\"

# 测试数据文件存放绝对路径
dataFilePath = parentDirPath + u"\\testData\\126 邮箱创建联系人并发邮件.xlsx"

# 测试数据文件中,测试用例表中部分列对应的数字序号
testCase_testCaseName = 1
testCase_frameWorkName = 3
testCase_testStepSheetName = 4
testCase_dataSourceSheetName = 5
testCase_isExecute = 6
testCase_runTime = 7
testCase_testResult = 8

# 用例步骤表中,部分列对应的数字序号
testStep_testStepDescribe = 1
testStep_keyWords = 2
testStep_locationType = 3
testStep_locatorExpression = 4
testStep_operateValue = 5
testStep_runTime = 6
testStep_testResult = 7
testStep_errorInfo = 8
testStep_errorPic = 9

# 数据源表中,是否执行列对应的数字编号
dataSource_isExecute = 6
dataSource_email = 2
dataSource_runTime = 7
dataSource_result = 8
```

(15) 修改 testScripts 包中的__init__.py 文件内容如下。

```
#encoding = utf-8

from action.PageAction import *
from util.ParseExcel import ParseExcel
from config.VarConfig import *
import time
import traceback

# 设置此次测试的环境编码为 utf8
import sys
reload(sys)
sys.setdefaultencoding("utf-8")
```

```
# 创建解析 Excel 对象
excelObj = ParseExcel()
# 将 Excel 数据文件加载到内存
excelObj.loadWorkBook(dataFilePath)
```

(16) 在 testScripts 包中新建一名为 WriteTestResult.py 的 Python 文件,用于实现向 Excel 中写入测试结果信息的公共方法,具体内容如下。

```
#encoding = utf-8
from . import *

# 用例或用例步骤执行结束后,向 Excel 中写入执行结果信息
def writeTestResult(sheetObj, rowNo, colsNo, testResult,
                errorInfo = None, picPath = None):
    # 测试通过结果信息为绿色,失败为红色
    colorDict = {"pass":"green", "faild":"red", "":None}

    # 因为"测试用例"工作表和"用例步骤 sheet 表"中都有测试执行时间和
    # 测试结果列,定义此字典对象是为了区分具体应该写哪个工作表
    colsDict = {
        "testCase":[testCase_runTime, testCase_testResult],
        "caseStep":[testStep_runTime, testStep_testResult],
        "dataSheet":[dataSource_runTime, dataSource_result]}
    try:
        # 在测试步骤 sheet 中,写入测试结果
        excelObj.writeCell(sheetObj, content = testResult,
                rowNo = rowNo, colsNo = colsDict[colsNo][1],
                style = colorDict[testResult])
        if testResult == "":
            # 清空时间单元格内容
            excelObj.writeCell(sheetObj, content = "",
                rowNo = rowNo, colsNo = colsDict[colsNo][0])
        else:
            # 在测试步骤 sheet 中,写入测试时间
            excelObj.writeCellCurrentTime(sheetObj,
                rowNo = rowNo, colsNo = colsDict[colsNo][0])
        if errorInfo and picPath:
            # 在测试步骤 sheet 中,写入异常信息
            excelObj.writeCell(sheetObj, content = errorInfo,
                    rowNo = rowNo, colsNo = testStep_errorInfo)
            # 在测试步骤 sheet 中,写入异常截图路径
            excelObj.writeCell(sheetObj, content = picPath,
                    rowNo = rowNo, colsNo = testStep_errorPic)
        else:
            if colsNo == "caseStep":
                # 在测试步骤 sheet 中,清空异常信息单元格
                excelObj.writeCell(sheetObj, content = "",
                        rowNo = rowNo, colsNo = testStep_errorInfo)
                # 在测试步骤 sheet 中,清空异常信息单元格
                excelObj.writeCell(sheetObj, content = "",
                        rowNo = rowNo, colsNo = testStep_errorPic)
```

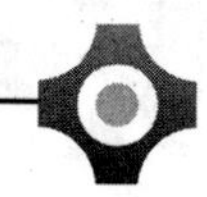

```
    except Exception as err:
        print("写 Excel 时发生异常")
        print(traceback.print_exc())
```

(17) 在 testScripts 包中新建一名为 CreateContacts.py 的 Python 文件,用于实现向 126 邮箱添加联系人的数据驱动框架部分,具体内容如下。

```
# encoding = utf - 8
from . import *
from .WriteTestResult import writeTestResult

def dataDriverFun(dataSourceSheetObj, stepSheetObj):
    try:
        # 获取数据源表中是否执行列对象
        dataIsExecuteColumn = excelObj.getColumn(
            dataSourceSheetObj,
            dataSource_isExecute)
        # 获取数据源表中"电子邮箱"列对象
        emailColumn = excelObj.getColumn(
            dataSourceSheetObj, dataSource_email)
        # 获取测试步骤表中存在数据区域的行数
        stepRowNums = excelObj.getRowsNumber(stepSheetObj)
        # 记录成功执行的数据条数
        successDatas = 0
        # 记录被设置为执行的数据条数
        requiredDatas = 0
        for idx, data in enumerate(dataIsExecuteColumn[1:]):
            # 遍历数据源表,准备进行数据驱动测试
            # 因为第一行是标题行,所以从第二行开始遍历
            if data.value == "y":
                print("开始添加联系人" % s"" % emailColumn[idx + 1].value)
                requiredDatas += 1
                # 定义记录执行成功步骤数变量
                successStep = 0
                for index in range(2, stepRowNums + 1):
                    # 获取数据驱动测试步骤表中
                    # 第 index 行对象
                    rowObj = excelObj.getRow(stepSheetObj, index)
                    # 获取关键字作为调用的函数名
                    keyWord = rowObj[testStep_keyWords - 1].value
                    # 获取操作元素定位方式作为调用函数的参数
                    locationType = rowObj[testStep_locationType - 1].value
                    # 获取操作元素的定位表达式作为调用函数的参数
                    locatorExpression = rowObj[
                        testStep_locatorExpression - 1].value
                    # 获取操作值作为调用函数的参数
                    operateValue = rowObj[testStep_operateValue - 1].value
                    if isinstance(operateValue, int):
                        operateValue = str(operateValue)
                    if operateValue and " - " in operateValue:
                        # 如果 operateValue 变量不为空,说明有操作值
```

```
                    # 从数据源表中根据坐标获取对应单元格的数据
                    colsNo = int(operateValue.split("-")[1])
                    operateValue = excelObj.getCellOfValue(
                        dataSourceSheetObj,
                        rowNo = idx + 2, colsNo = colsNo)
                # 构造需要执行的 Python 表达式,此表达式对应的
                # 是 PageAction.py 文件中的页面动作函数调用的字符串表示
                tmpStr = "'%s', '%s'" % (locationType.lower(),
                    locatorExpression.replace('"', "'")
                    ) if locationType and locatorExpression else ""
                if tmpStr:
                    tmpStr += \
                      ", '" + operateValue + "'" if operateValue else ""
                else:
                    tmpStr += \
                      "'" + operateValue + "'" if operateValue else ""
                runStr = keyWord + "(" + tmpStr + ")"
                # print runStr
                try:
                    # 通过 eval 函数,将拼接的页面动作函数调用的字符串表示
                    # 当成有效的 Python 表达式执行,从而执行测试步骤的 sheet
                    # 中关键字在 ageAction.py 文件中对应的映射方法,
                    # 来完成对页面元素的操作
                    if operateValue != "否":
                        # 当 operateValue 值为"否"时,
                        # 表示不单击星标联系人复选框
                        eval(runStr)
                except Exception as err:
                    print("执行步骤'%s'发生异常"
                        % rowObj[testStep_testStepDescribe - 1].value)
                    print(traceback.print_exc())
                else:
                    successStep += 1
                    print("执行步骤'%s'成功"
                        % rowObj[testStep_testStepDescribe - 1].value)
            if stepRowNums == successStep + 1:
                successDatas += 1
                # 如果成功执行的步骤数等于步骤表中给出的步骤数
                # 说明第 idx+2 行的数据执行通过,写入通过信息
                writeTestResult(sheetObj = dataSourceSheetObj,
                    rowNo = idx + 2, colsNo = "dataSheet",
                    testResult = "pass")
            else:
                # 写入失败信息
                writeTestResult(sheetObj = dataSourceSheetObj,
                    rowNo = idx + 2, colsNo = "dataSheet",
                    testResult = "faild")
        else:
            # 将不需要执行的数据行的执行时间和执行结果单元格清空
            writeTestResult(sheetObj = dataSourceSheetObj,
                rowNo = idx + 2, colsNo = "dataSheet",
```

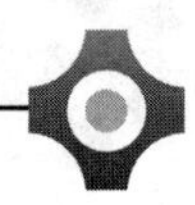

```
                    testResult = "")
        if requiredDatas == successDatas:
            # 只要当成功执行的数据条数等于被设置为需要执行的数
            # 据条数,才表示调用数据驱动的测试用例执行通过
            return 1
        # 表示调用数据驱动的测试用例执行失败
        return 0
    except Exception as err:
        raise err
```

(18) 在 testScripts 包中新建一名为 TestSendMailAndCreateContacts. py 的 Python 文件,用于实现关键字与数据驱动逻辑部分,登录 126 邮箱,然后向邮箱中添加联系人并发送一封带附件的邮件,具体内容如下。

```
#encoding = utf - 8
from . import *
from . import CreateContacts
from .WriteTestResult import writeTestResult

def TestSendMailAndCreateContacts():
    try:
        # 根据 Excel 文件中的 sheet 名获取 sheet 对象
        caseSheet = excelObj.getSheetByName("测试用例")
        # 获取测试用例 sheet 中是否执行列对象
        isExecuteColumn = excelObj.getColumn(caseSheet, testCase_isExecute)
        # 记录执行成功的测试用例个数
        successfulCase = 0
        # 记录需要执行的用例个数
        requiredCase = 0
        for idx, i in enumerate(isExecuteColumn[1:]):
            # 因为用例 sheet 中第一行为标题行,无须执行
            caseName = excelObj.getCellOfValue(caseSheet,
                rowNo = idx + 2, colsNo = testCase_testCaseName)
            # 循环遍历"测试用例"表中的测试用例,执行被设置为执行的用例
            if i.value.lower() == 'y':
                requiredCase += 1
                # 获取测试用例表中,第 idx + 1 行中
                # 用例执行时所使用的框架类型
                useFrameWorkName = excelObj.getCellOfValue(
                    caseSheet, rowNo = idx + 2,
                    colsNo = testCase_frameWorkName)
                # 获取测试用例表中,第 idx + 1 行中执行用例的步骤 sheet 名
                stepSheetName = excelObj.getCellOfValue(
                    caseSheet, rowNo = idx + 2,
                    colsNo = testCase_testStepSheetName)
                print(" ---- ", stepSheetName)
                if useFrameWorkName == "数据":
                    print(" ****** 调用数据驱动 ****** ")
                    # 获取测试用例表中,第 idx+1 行,执行框架为
                    # 数据驱动的用例所使用的数据 sheet 名
                    dataSheetName = excelObj.getCellOfValue(
```

```
            caseSheet, rowNo = idx + 2,
            colsNo = testCase_dataSourceSheetName)
        # 获取第 idx + 1 行测试用例的步骤 sheet 对象
        stepSheetObj = excelObj.getSheetByName(stepSheetName)
        # 获取第 idx + 1 行测试用例使用的数据 sheet 对象
        dataSheetObj = excelObj.getSheetByName(dataSheetName)
        # 通过数据驱动框架执行添加联系人
        result = CreateContacts.dataDriverFun(dataSheetObj, stepSheetObj)
        if result:
            print("用例 %s 执行成功" % caseName)
            successfulCase += 1
            writeTestResult(caseSheet, rowNo = idx + 2,
                colsNo = "testCase", testResult = "pass")
        else:
            print("用例 %s 执行失败" % caseName)
            writeTestResult(caseSheet, rowNo = idx + 2,
                colsNo = "testCase", testResult = "faild")
    elif useFrameWorkName == "关键字":
        print(" ****** 调用关键字驱动 ******* ")
        caseStepObj = excelObj.getSheetByName(stepSheetName)
        stepNums = excelObj.getRowsNumber(caseStepObj)
        successfulSteps = 0
        print("测试用例共 %s 步" % stepNums)
        for index in range(2, stepNums + 1):
            # 因为第一行是标题行,无须执行
            # 获取步骤 sheet 中第 index 行对象
            stepRow = excelObj.getRow(caseStepObj, index)
            # 获取关键字作为调用的函数名
            keyWord = stepRow[testStep_keyWords - 1].value
            # 获取操作元素定位方式作为调用函数的参数
            locationType = stepRow[testStep_locationType - 1].value
            # 获取操作元素的定位表达式作为调用函数的参数
            locatorExpression = stepRow[
                testStep_locatorExpression - 1].value
            # 获取操作值作为调用函数的参数
            operateValue = stepRow[testStep_operateValue - 1].value
            if isinstance(operateValue, int):
                # 如果 operateValue 值为数字型,
                # 将其转换为字符串,方便字符串拼接
                operateValue = str(operateValue)
            # 构造需要执行的 Python 表达式,此表达式对应的
            # 是 PageAction.py 文件中的页面动作函数调用的字符串表示
            tmpStr = "'%s', '%s'" % (locationType.lower(),
                 locatorExpression.replace("'", '"')
                 ) if locationType and locatorExpression else ""
            if tmpStr:
                tmpStr += \
                    ", '" + operateValue + "'" if operateValue else ""
            else:
                tmpStr += \
                    "'" + operateValue + "'" if operateValue else ""
```

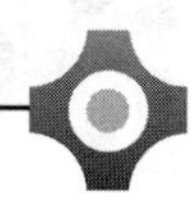

```
                        runStr = keyWord + "(" + tmpStr + ")"
                        # print runStr
                        try:
                            # 通过 eval 函数,将拼接的页面动作函数调用的字符串表示
                            # 当成有效的 Python 表达式执行,从而执行测试步骤的 sheet
                            # 中关键字在 ageAction.py 文件中对应的映射方法,
                            # 来完成对页面元素的操作
                            eval(runStr)
                        except Exception as err:
                            print("执行步骤'%s'发生异常"
                                  % stepRow[testStep_testStepDescribe - 1].value)
                            # 截取异常屏幕图片
                            capturePic = capture_screen()
                            # 获取详细的异常堆栈信息
                            errorInfo = traceback.format_exc()
                            writeTestResult(caseStepObj, rowNo = index,
                                colsNo = "caseStep", testResult = "faild",
                                errorInfo = str(errorInfo), picPath = capturePic)
                        else:
                            successfulSteps += 1
                            print("执行步骤'%s'成功"
                                  % stepRow[testStep_testStepDescribe - 1].value)
                            writeTestResult(caseStepObj, rowNo = index,
                                colsNo = "caseStep", testResult = "pass")
                    if successfulSteps == stepNums - 1:
                        successfulCase += 1
                        print("用例'%s'执行通过" % caseName)
                        writeTestResult(caseSheet, rowNo = idx + 2,
                            colsNo = "testCase", testResult = "pass")
                    else:
                        print("用例%s执行失败" % caseName)
                        writeTestResult(caseSheet, rowNo = idx + 2,
                            colsNo = "testCase", testResult = "faild")
            else:
                # 清空不需要执行用例的执行时间和执行结果,
                # 异常信息,异常图片单元格
                writeTestResult(caseSheet, rowNo = idx + 2,
                    colsNo = "testCase", testResult = "")
        print("共%d条用例,%d条需要被执行,成功执行%d条"
            % (len(isExecuteColumn) - 1, requiredCase, successfulCase))
    except Exception as err:
        print(traceback.print_exc())
```

(19) 在工程 KeyWordAndDataDrivenFrameWork 根目录下新建一个名为 RunTest.py 的 Python 文件,用于编写框架入口代码,具体内容如下。

```
#encoding = utf-8
from testScripts.TestSendMailAndCreateContacts import \
    TestSendMailAndCreateContacts

if __name__ == "__main__":
    TestSendMailAndCreateContacts()
```

执行 RunTest.py 文件，可以看到程序自动打开浏览器，登录 126 邮箱，添加联系人并发送一封带附件的邮件。部分执行结果输出信息如图 15-17 所示。

```
C:\Python27\python.exe D:/PythonProject/KeyWordAndDataDrivenFrameWork/RunTest.py
---- 登录
****** 调用关键字驱动 *******
测试用例共13步
执行步骤‘打开浏览器’成功
执行步骤‘访问被测试网址http://www.126.com’成功
执行步骤‘最大化窗口’成功
执行步骤‘等待126邮箱登录主页加载完成’成功
执行步骤‘断言当前活动页面源码中是否包含“126网易免费邮--你的专业电子邮局”’成功
执行步骤‘显示等待id属性值为x-URS-iframe的frame框的出现，然后切换进入该frame框中’成功
执行步骤‘输入登录用户名’成功
执行步骤‘输入登录密码’成功
执行步骤‘单击登录按钮’成功
执行步骤‘等待’成功
执行步骤‘切回默认会话窗体’成功
执行步骤‘断言登录成功后的页面标题是否包含“网易邮箱6.0版”关键内容’成功
用例‘登录126邮箱’执行通过
---- 创建联系人
****** 调用数据驱动 ******
开始添加联系人“sr@qq.com”
执行步骤‘单击“通讯录”链接，进入通讯录页面’成功
执行步骤‘显示等待“新建联系人”按钮在页面上可见’成功
执行步骤‘单击“新建联系人”按钮’成功
执行步骤‘显示等待输入联系人姓名框是否在页面上可见’成功
执行步骤‘输入联系人姓名’成功
执行步骤‘输入电子邮箱’成功
执行步骤‘设为星标联系人’成功
执行步骤‘输入手机号’成功
执行步骤‘输入联系人备注信息’成功
执行步骤‘单击“确认”按钮，保存新联系人’成功
执行步骤‘等待2秒’成功
```

图 15-17

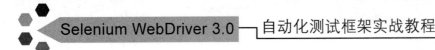

上面我们只是把测试过程及结果总结信息输出到控制台，但不方便错误排查以及后期结果信息查看，为此我们加入打印日志功能。

(20) 通过 logging 模块，为关键字 & 数据混合驱动框架加入打印日志功能。在 config 包中新建一个名叫 logger.conf 的文件，用于配置日志基本信息，具体内容如下。

```
#logger.conf
###############################################
[loggers]
keys = root,example01,example02
[logger_root]
level = DEBUG
handlers = hand01,hand02

[logger_example01]
handlers = hand01,hand02
qualname = example01
propagate = 0

[logger_example02]
handlers = hand01,hand03
```

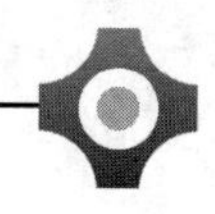

```
qualname = example02
propagate = 0

###############################################
[handlers]
keys = hand01,hand02,hand03

[handler_hand01]
class = StreamHandler
level = INFO
formatter = form01
args = (sys.stderr,)

[handler_hand02]
class = FileHandler
level = DEBUG
formatter = form01
args = ('log\\Mail126TestLogfile.log', 'a')

[handler_hand03]
class = handlers.RotatingFileHandler
level = INFO
formatter = form01
args = ('log\\Mail126TestLogfile.log', 'a', 10 * 1024 * 1024, 5)

###############################################
[formatters]
keys = form01,form02

[formatter_form01]
format = %(asctime)s %(filename)s[line:%(lineno)d] %(levelname)s %(message)s
datefmt = %Y-%m-%d %H:%M:%S

[formatter_form02]
format = %(name)-12s: %(levelname)-8s %(message)s
datefmt = %Y-%m-%d %H:%M:%S
```

(21) 在 util 包中新建一个名叫 Log.py 的 Python 文件，用于初始化日志对象，具体内容如下。

```
# encoding = utf-8
import logging
import logging.config
from config.VarConfig import parentDirPath

# 读取日志配置文件
logging.config.fileConfig(parentDirPath + "\config\Logger.conf")
# 选择一个日志格式
logger = logging.getLogger("example02")                    # 或者 example01

def debug(message):
```

```
        # 定义 dubug 级别日志打印方法
        logger.debug(message)

    def info(message):
        # 定义 info 级别日志打印方法
        logger.info(message)

    def warning(message):
        # 定义 warning 级别日志打印方法
        logger.warning(message)
```

(22) 在 DataDrivenFrameWork 工程根目录下创建一名为 log 的目录，然后修改 testScripts 包中的 CreateContacts. py 和 TestSendMailAndCreateContacts. py 文件的内容如下。

CreateContacts. py 修改后的内容如下。

```
#encoding = utf - 8
from . import *
from .WriteTestResult import writeTestResult
from util.Log import *

def dataDriverFun(dataSourceSheetObj, stepSheetObj):
    try:
        # 获取数据源表中是否执行列对象
        dataIsExecuteColumn = excelObj.getColumn(
            dataSourceSheetObj,
            dataSource_isExecute)
        # 获取数据源表中"电子邮箱"列对象
        emailColumn = excelObj.getColumn(
            dataSourceSheetObj, dataSource_email)
        # 获取测试步骤表中存在数据区域的行数
        stepRowNums = excelObj.getRowsNumber(stepSheetObj)
        # 记录成功执行的数据条数
        successDatas = 0
        # 记录被设置为执行的数据条数
        requiredDatas = 0
        for idx, data in enumerate(dataIsExecuteColumn[1:]):
            # 遍历数据源表,准备进行数据驱动测试
            # 因为第一行是标题行,所以从第二行开始遍历
            if data.value == "y":
                logging.info("开始添加联系人" %s"" %emailColumn[idx + 1].value)
                requiredDatas += 1
                # 定义记录执行成功步骤数变量
                successStep = 0
                for index in range(2, stepRowNums + 1):
                    # 获取数据驱动测试步骤表中
                    # 第 index 行对象
                    rowObj = excelObj.getRow(stepSheetObj, index)
                    # 获取关键字作为调用的函数名
                    keyWord = rowObj[testStep_keyWords - 1].value
```

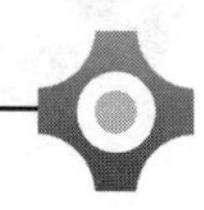

```
            # 获取操作元素定位方式作为调用函数的参数
            locationType = rowObj[testStep_locationType - 1].value
            # 获取操作元素的定位表达式作为调用函数的参数
            locatorExpression = rowObj[
                testStep_locatorExpression - 1].value
            # 获取操作值作为调用函数的参数
            operateValue = rowObj[testStep_operateValue - 1].value
            if isinstance(operateValue, int):
                operateValue = str(operateValue)
            if operateValue and " - " in operateValue:
                # 如果 operateValue 变量不为空,说明有操作值
                # 从数据源表中根据坐标获取对应单元格的数据
                colsNo = int(operateValue.split(" - ")[1])
                operateValue = excelObj.getCellOfValue(
                    dataSourceSheetObj,
                    rowNo = idx + 2, colsNo = colsNo)
            # 构造需要执行的 python 表达式,此表达式对应的
            # 是 PageAction.py 文件中的页面动作函数调用的字符串表示
            tmpStr = "'%s', '%s'" %(locationType.lower(),
                locatorExpression.replace("'", '"')
                ) if locationType and locatorExpression else ""
            if tmpStr:
                tmpStr += \
                  ", '" + operateValue + "'" if operateValue else ""
            else:
                tmpStr += \
                  "'" + operateValue + "'" if operateValue else ""
            runStr = keyWord + "(" + tmpStr + ")"
            # print runStr
            try:
                # 通过 eval 函数,将拼接的页面动作函数调用的字符串表示
                # 当成有效的 Python 表达式执行,从而执行测试步骤的 sheet
                # 中关键字在 ageAction.py 文件中对应的映射方法,
                # 来完成对页面元素的操作
                if operateValue != "否":
                    # 当 operateValue 值为"否"时,
                    # 表示不单击星标联系人复选框
                    eval(runStr)
            except Exception as err:
                logging.debug("执行步骤'%s'发生异常"
                    % rowObj[testStep_testStepDescribe - 1].value)
                print(traceback.print_exc())
            else:
                successStep += 1
                logging.info("执行步骤'%s'成功"
                    % rowObj[testStep_testStepDescribe - 1].value)
        if stepRowNums == successStep + 1:
            successDatas += 1
            # 如果成功执行的步骤数等于步骤表中给出的步骤数
            # 说明第 idx + 2 行的数据执行通过,写入通过信息
            writeTestResult(sheetObj = dataSourceSheetObj,
```

```
                    rowNo = idx + 2, colsNo = "dataSheet",
                    testResult = "pass")
            else:
                # 写入失败信息
                writeTestResult(sheetObj = dataSourceSheetObj,
                    rowNo = idx + 2, colsNo = "dataSheet",
                    testResult = "faild")
        else:
            # 将不需要执行的数据行的执行时间和执行结果单元格清空
            writeTestResult(sheetObj = dataSourceSheetObj,
                rowNo = idx + 2, colsNo = "dataSheet",
                testResult = "")
    if requiredDatas == successDatas:
        # 只要当成功执行的数据条数等于被设置为需要执行的数
        # 据条数,才表示调用数据驱动的测试用例执行通过
        return 1
    # 表示调用数据驱动的测试用例执行失败
    return 0
except Exception as err:
    raise err
```

TestSendMailAndCreateContacts.py 文件修改内容如下。

```
#encoding = utf - 8
from . import *
from . import CreateContacts
from .WriteTestResult import writeTestResult
from util.Log import *

def TestSendMailAndCreateContacts():
    try:
        # 根据 Excel 文件中的 sheet 名获取 sheet 对象
        caseSheet = excelObj.getSheetByName("测试用例")
        # 获取测试用例 sheet 中是否执行列对象
        isExecuteColumn = excelObj.getColumn(caseSheet, testCase_isExecute)
        # 记录执行成功的测试用例个数
        successfulCase = 0
        # 记录需要执行的用例个数
        requiredCase = 0
        for idx, i in enumerate(isExecuteColumn[1:]):
            # 因为用例 sheet 中第一行为标题行,无须执行
            caseName = excelObj.getCellOfValue(caseSheet,
                rowNo = idx + 2, colsNo = testCase_testCaseName)
            # 循环遍历"测试用例"表中的测试用例,执行被设置为执行的用例
            if i.value.lower() == 'y':
                requiredCase += 1
                # 获取测试用例表中,第 idx + 1 行中
                # 用例执行时所使用的框架类型
                useFrameWorkName = excelObj.getCellOfValue(
                    caseSheet, rowNo = idx + 2,
                    colsNo = testCase_frameWorkName)
```

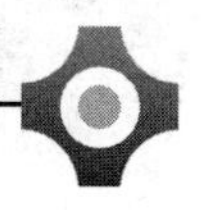

```
# 获取测试用例表中,第 idx + 1 行中执行用例的步骤 sheet 名
stepSheetName = excelObj.getCellOfValue(
    caseSheet, rowNo = idx + 2,
    colsNo = testCase_testStepSheetName)
logging.info(" ---- %s" % stepSheetName)
if useFrameWorkName == "数据":
    logging.info(" ****** 调用数据驱动 ******")
    # 获取测试用例表中,第 idx+1 行,执行框架为
    # 数据驱动的用例所使用的数据 sheet 名
    dataSheetName = excelObj.getCellOfValue(
        caseSheet, rowNo = idx + 2,
        colsNo = testCase_dataSourceSheetName)
    # 获取第 idx+1 行测试用例的步骤 sheet 对象
    stepSheetObj = excelObj.getSheetByName(stepSheetName)
    # 获取第 idx+1 行测试用例使用的数据 sheet 对象
    dataSheetObj = excelObj.getSheetByName(dataSheetName)
    # 通过数据驱动框架执行添加联系人
    result = CreateContacts.dataDriverFun(dataSheetObj,stepSheetObj)
    if result:
        logging.info("用例 %s 执行成功" % caseName)
        successfulCase += 1
        writeTestResult(caseSheet, rowNo = idx + 2,
            colsNo = "testCase", testResult = "pass")
    else:
        logging.info("用例 %s 执行失败" % caseName)
        writeTestResult(caseSheet, rowNo = idx + 2,
            colsNo = "testCase", testResult = "faild")
elif useFrameWorkName == "关键字":
    logging.info(" ****** 调用关键字驱动 *******")
    caseStepObj = excelObj.getSheetByName(stepSheetName)
    stepNums = excelObj.getRowsNumber(caseStepObj)
    successfulSteps = 0
    logging.info("测试用例共 %s 步" % stepNums)
    for index in range(2, stepNums + 1):
        # 因为第一行是标题行,无须执行
        # 获取步骤 sheet 中第 index 行对象
        stepRow = excelObj.getRow(caseStepObj, index)
        # 获取关键字作为调用的函数名
        keyWord = stepRow[testStep_keyWords - 1].value
        # 获取操作元素定位方式作为调用函数的参数
        locationType = stepRow[testStep_locationType - 1].value
        # 获取操作元素的定位表达式作为调用函数的参数
        locatorExpression = stepRow[
            testStep_locatorExpression - 1].value
        # 获取操作值作为调用函数的参数
        operateValue = stepRow[testStep_operateValue - 1].value
        if isinstance(operateValue, int):
            # 如果 operateValue 值为数字型,
            # 将其转换为字符串,方便字符串拼接
            operateValue = str(operateValue)
        # 构造需要执行的 Python 表达式,此表达式对应的
```

```
            # 是 PageAction.py 文件中的页面动作函数调用的字符串表示
            tmpStr = "'%s', '%s'" % (locationType.lower(),
                locatorExpression.replace("'", '"')
                ) if locationType and locatorExpression else ""
            if tmpStr:
                tmpStr += \
                    ", '" + operateValue + "'" if operateValue else ""
            else:
                tmpStr += \
                    "'" + operateValue + "'" if operateValue else ""
            runStr = keyWord + "(" + tmpStr + ")"
            # print runStr
            try:
                # 通过 eval 函数,将拼接的页面动作函数调用的字符串表示
                # 当成有效的 Python 表达式执行,从而执行测试步骤的 sheet
                # 中关键字在 ageAction.py 文件中对应的映射方法,
                # 来完成对页面元素的操作
                eval(runStr)
            except Exception as err:
                logging.debug("执行步骤'%s'发生异常"
                    % stepRow[testStep_testStepDescribe - 1].value)
                # 截取异常屏幕图片
                capturePic = capture_screen()
                # 获取详细的异常堆栈信息
                errorInfo = traceback.format_exc()
                writeTestResult(caseStepObj, rowNo = index,
                    colsNo = "caseStep", testResult = "faild",
                    errorInfo = str(errorInfo), picPath = capturePic)
            else:
                successfulSteps += 1
                logging.info("执行步骤'%s'成功"
                    % stepRow[testStep_testStepDescribe - 1].value)
                writeTestResult(caseStepObj, rowNo = index,
                    colsNo = "caseStep", testResult = "pass")
        if successfulSteps == stepNums - 1:
            successfulCase += 1
            logging.info("用例'%s'执行通过" % caseName)
            writeTestResult(caseSheet, rowNo = idx + 2,
                colsNo = "testCase", testResult = "pass")
        else:
            logging.info("用例%s执行失败" % caseName)
            writeTestResult(caseSheet, rowNo = idx + 2,
                colsNo = "testCase", testResult = "faild")
    else:
        # 清空不需要执行用例的执行时间和执行结果,
        # 异常信息,异常图片单元格
        writeTestResult(caseSheet, rowNo = idx + 2,
            colsNo = "testCase", testResult = "")
logging.info("共%d条用例,%d条需要被执行,成功执行%d条"
    % (len(isExecuteColumn) - 1, requiredCase, successfulCase))
except Exception as err:
    logging.debug(traceback.print_exc())
```

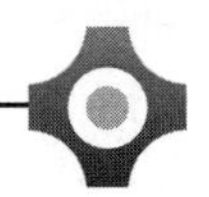

(23) 执行 RunTest. py 文件,执行结束后可以看到在混合框架的工程目录 log 目录中,自动创建一个名叫 Mail126TestLogfile. log 的日志文件,里面记录的都是整个测试过程中的一些测试信息及异常信息,方便后期进行测试结果分析及错误排查。

至此,关键字 & 数据混合驱动框架已经搭建完成,在 Pycharm 工具中,整个工程的结构如图 15-18 所示。

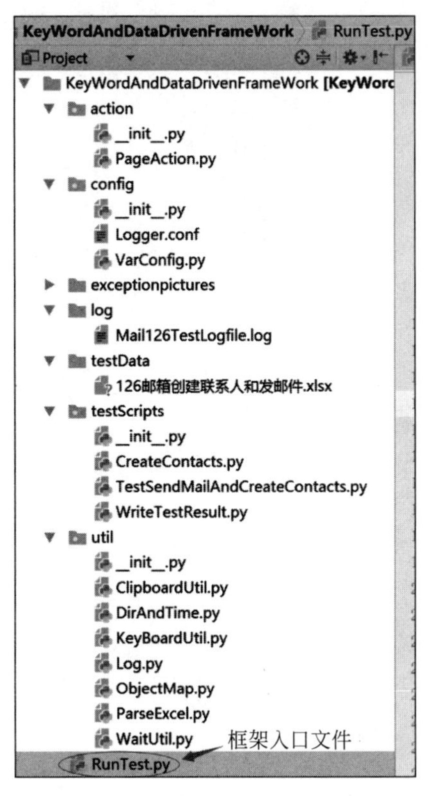

图　15-18

第四篇 常见问题和解决方法

自动化测试常见问题和解决方法

本章主要总结了在自动化测试实践过程中的常见问题、异常及解决方法，请读者在遇到脚本执行异常时查阅本章内容获取相关解决信息或思路。

16.1 如何让 WebDriver 支持 IE 11

使用 IE 浏览器实施自动化测试的过程中，可能会遇到驱动 IE 时报 WebDriverException: Message: Unexpected error launching Internet Explorer. Protected Mode settings are not the same for all zones. Enable Protected Mode must be set to the same value (enabled or disabled) for all zones. 的错误，如果遇到上述问题，请尝试按照下述方法解决。

实现步骤如下：

(1) 在网址 http://www.microsoft.com/en-us/download/details.aspx? id=44069 下载 Windows 的更新包，并在本机安装。

(2) 对于 32 位的 Windows 操作系统，需要检查注册表的信息是否是如下信息：

```
HKEY_LOCAL_MACHINE\SOFTWARE\Microsoft\Internet Explorer\Main\FeatureControl
\FEATURE_BFCACHE
```

对于 64 位的 Windows 操作系统，需要检查注册表的信息是否是如下信息：

```
HKEY_LOCAL_MACHINE\SOFTWARE\Wow6432Node\Microsoft\Internet Explorer\Main\
FeatureControl\FEATURE_BFCACHE
```

FEATURE_BFCACHE 可能存在也可能不存在，如果不存在则需要创建，选择 DWORD 类型，值设定为 0。

(3) 在 IE 浏览器的"工具"菜单下选择"Internet 选项"命令，在弹出的对话框中选择"安全"选项卡，并勾选"Internet""本地 Intranet""受信任的站点"和"受限制的站点"中的"启动保护模式"复选框，如图 16-1 所示。

选择"高级"选项卡，取消勾选"启用增强保护模式"复选框，如图 16-2 所示。

(4) 执行上述操作后，如果脚本在 Firefox 浏览器下可以正常执行，但是在 IE 11 下执行还是报页面元素无法找到的错误，请将 Windows 中在 2014 年 11 月以后的所有 Windows 系统补丁包删除(刚才安装的补丁包除外)，应该可以解决上述问题。

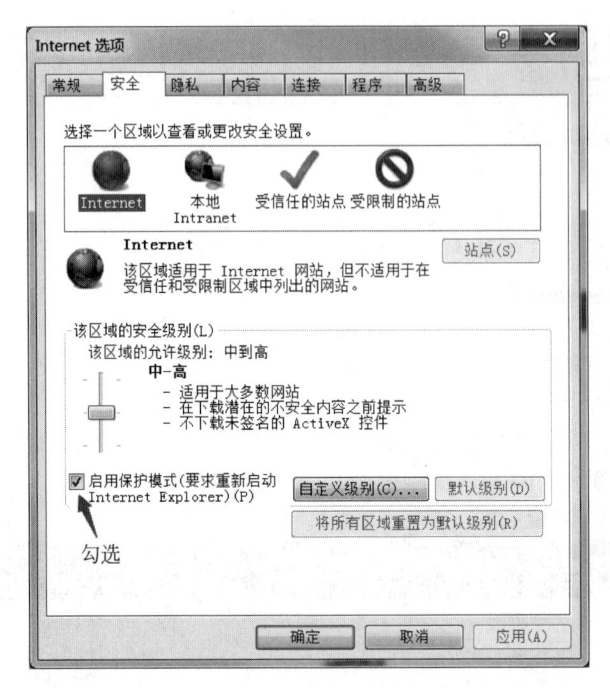

图 16-1

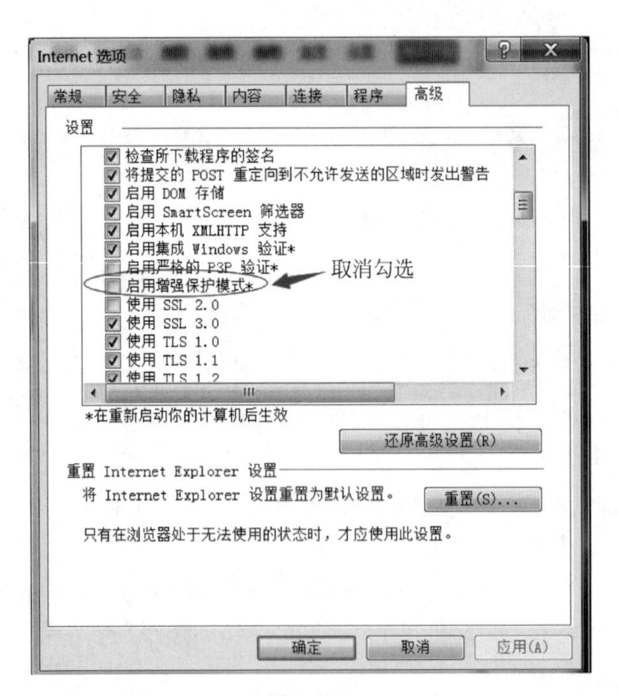

图 16-2

有些版本的 Windows 操作系统默认使用 IE 11，且无须上述配置和安装补丁包。建议读者先尝试使用 IE 11 执行简单的自动化测试脚本，验证 WebDriver 是否支持 IE 11。

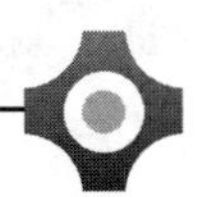

16.2 解决 Unexpected error launching Internet Explorer. Browser zoom level was set to 75%（或其他百分比）的错误

出现此问题的原因是因为浏览器设定了显示区域的缩放百分比，只需要将缩放比例重新设定为 100% 即可解决此类错误。具体的设定方法如图 16-3 所示。

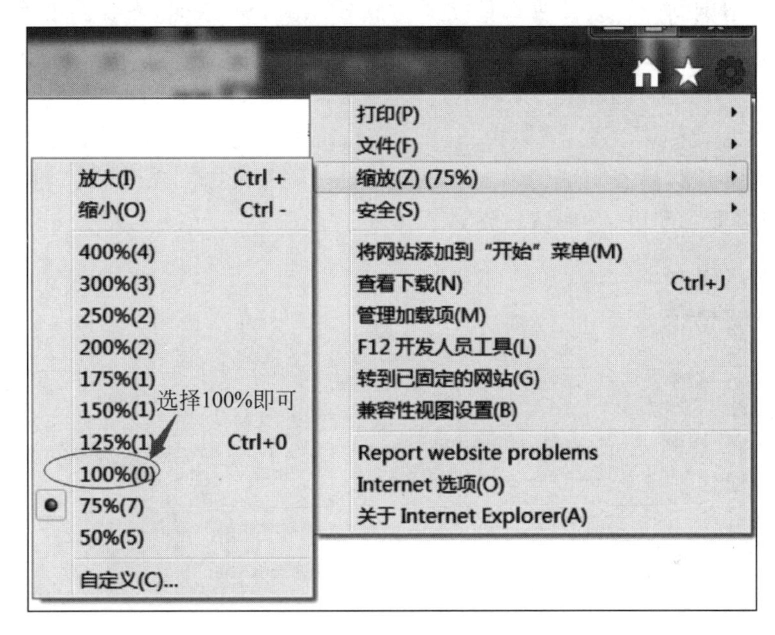

图　16-3

16.3 解决某些 IE 浏览器中输入数字和英文特别慢的问题

在某些 IE 浏览器中，使用 WebDriver 在输入框中输入数字和英文会特别慢，大概 2 秒才能输入一个英文或者数字，大大降低了自动化测试的执行速度。此问题一般出在 64 位的 Windows 操作系统中。

解决方法：将 IEDriverServer.exe 的版本从 64 位换为 32 位即可解决数字或英文或中文字符输入慢的问题。32 位版本的 IEDriverServer.exe 文件下载地址为 http://www.seleniumhq.org/download/，下载链接如图 16-4 所示。

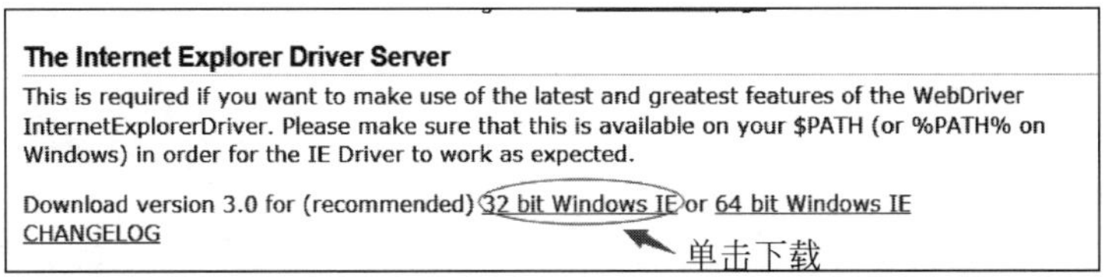

The Internet Explorer Driver Server

This is required if you want to make use of the latest and greatest features of the WebDriver InternetExplorerDriver. Please make sure that this is available on your $PATH (or %PATH% on Windows) in order for the IE Driver to work as expected.

Download version 3.0 for (recommended) 32 bit Windows IE or 64 bit Windows IE
CHANGELOG

单击下载

图　16-4

16.4 解决 Firefox 浏览器的 can't access dead object 异常

使用 Selenium 3 启动 Firefox 浏览器实施自动化测试时，可能会抛出 WebDriverException：Message：can't access dead object 异常信息，并且经常发生在登录跳转到新页面后操作新页面元素时发生，这是因为没有从 iframe 框架中切换到默认会话窗体导致的，因为一般网站的登录模块都是放到一个新的 iframe 里面，当我们填写登录信息前，切换进入该 iframe，点完登录页面后并未从该 iframe 中切回到默认会话窗体，由此导致了该异常的发生。一般从 Selenium 2 版本过渡到 Selenium 3 版本，并且使用 Firefox 浏览器的话，通常会遇到此问题，因为 Selenium 2 及 Chrome 浏览器和 IE 浏览器均没有此问题。如果遇到这样的问题，只需调用代码 driver.switch_to.default_content()，将当前会话窗体切换到默认窗体即可解决，详细请参阅如下实例。

```
# encoding = utf - 8
from selenium import webdriver
import time
from selenium.webdriver.support.ui import WebDriverWait
from selenium.webdriver.support import expected_conditions as EC
from selenium.webdriver.common.by import By
from selenium.webdriver.common.keys import Keys

# 创建 Firefox 浏览器的实例
driver = webdriver.Firefox(executable_path = "c:\geckodriver")
# 最大化浏览器窗口
driver.maximize_window()
driver.get("http://mail.126.com")
# 暂停 5 秒,以便邮箱登录页面加载完成
time.sleep(5)
assert u"126 网易免费邮 -- 你的专业电子邮局" in driver.title
# 创建显示等待
wait = WebDriverWait(driver, 30)
# 检查 id 为 x - URS - iframe 的 frame 是否存在,存在则切换进 frame 控件
wait.until(EC.frame_to_be_available_and_switch_to_it((By.ID, "x - URS - iframe")))
# 获取用户名输入框
userName = driver.find_element_by_xpath('//input[@name = "email"]')
# 输入用户名
userName.send_keys("AgilityToSR")
# 获取密码输入框
pwd = driver.find_element_by_xpath("//input[@name = 'password']")
# 输入密码
pwd.send_keys("AutoTest123")
# 发送一个回车键
pwd.send_keys(Keys.RETURN)
# 等待 5 秒,以便登录成功后的页面加载完成
time.sleep(5)
# 切换至默认句柄,以规避 can't access dead object 异常
driver.switch_to.default_content()
```

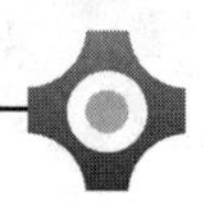

```
assert u"网易邮箱" in driver.title
driver.find_element_by_link_text("退出").click()
time.sleep(3)
driver.find_element_by_link_text("重新登录").click()
time.sleep(2)

# 检查 id 为 x-URS-iframe 的 frame 是否存在,存在则切换进 frame 控件
wait.until(EC.frame_to_be_available_and_switch_to_it((By.ID, "x-URS-iframe")))
# 获取用户名输入框
userName = driver.find_element_by_xpath('//input[@name="email"]')
# 输入用户名
userName.clear()
userName.send_keys("AgilityToSR")
# 获取密码输入框
pwd = driver.find_element_by_xpath("//input[@name='password']")
# 输入密码
pwd.send_keys("AutoTest123")
# 发送一个回车键
pwd.send_keys(Keys.RETURN)
# 等待 5 秒,以便登录成功后的页面加载完成
time.sleep(5)
# 切换至默认句柄,以规避 can't access dead object 异常
driver.switch_to.default_content()
assert u"网易邮箱" in driver.title
driver.quit()
print u"测试通过"
```

执行上面实例代码,即可看到 Selenium 3 驱动 Firefox 浏览器,切换新页面时不再抛出 WebDriverException：Message：can't access dead object 异常。

16.5 常见异常和解决方法

1. NoSuchElementException

解决方法：

(1) 检查页面元素的定位表达式是否编写正确。

(2) 如果等待很长时间依旧没有找到页面元素,建议尝试使用其他定位方式。

2. NoSuchWindowException

解决方法：

(1) 检查浏览器页面元素的定位方式是否正确。

(2) 在检查浏览器页面元素定位方式前,等待一段时间让页面加载完成。

3. NoAlertPresentException

解决方法：

(1) 确认 JavaScript 的 Alert 框是否显示在界面上。

(2) 在处理 Alert 前,先等待几秒。

4. NoSuchFrameException

解决方法：

(1) 检查 Frame 的定位表达式是否编写正确。

(2) 检查 Frame 是否有父 Frame。如果有，则需要先转换到父 Frame 中再进行此 Frame 的操作。

(3) 在转换到此 Frame 前，确保 WebDriver 已经转换到 default content。

(4) 在转换到 Frame 前，先等待几秒。

5. UnhandledAlertException

解决方法：

(1) 检查界面上是否还显示 JavaScript 的提示框。如果还有提示框显示，需单击“确定”或者“取消”按钮。

(2) 如果没有显示 JavaScript 的提示框，则可能是打开某些开发工具造成的，关闭浏览器中打开的开发工具插件即可。

6. UnexpectedTagNameException

解决方法：

(1) 检查目标元素的标签名称是否编写正确。

(2) 等待几秒后，再进行标签名称的相关操作。

7. StaleElementReferenceException

解决方法：

再次重新查找页面元素(因为页面已经刷新，导致页面元素不复存在)。

8. TimeoutException

解决方法：

(1) 检查等待条件的定位表达式是否编写正确。

(2) 增加等待时间。